9th Russian National Conference on Irreversible Processes in Nature and Technics 2017 (9RNC-IPNT)

Journal of Physics: Conference Series Volume 918

Moscow, Russia
25 – 28 January 2017

ISBN: 978-1-5108-5361-4
ISSN: 1742-6588

Printed from e-media with permission by:

Curran Associates, Inc.
57 Morehouse Lane
Red Hook, NY 12571

Some format issues inherent in the e-media version may also appear in this print version.

Printed by Curran Associates, Inc. (2018)

For permission requests, please contact the Institute of Physics
at the address below.

Institute of Physics
Dirac House, Temple Back
Bristol BS1 6BE UK

Phone: 44 1 17 929 7481
Fax: 44 1 17 920 0979

techtracking@iop.org

Additional copies of this publication are available from:

Curran Associates, Inc.
57 Morehouse Lane
Red Hook, NY 12571 USA
Phone: 845-758-0400
Fax: 845-758-2633
Email: curran@proceedings.com
Web: www.proceedings.com

Table of contents

Volume 918

9th Russian National Conference on Irreversible Processes in Nature and Technics (9RNC-IPNT)

25–27 January 2017, Bauman MSTU, Moscow, Russian Federation

Accepted papers received: 27 October 2017
Published online: 28 November 2017

Preface

Papers

IOP Conf. Series: Journal of Physics: Conf. Series **918** (2017) 011001 doi:10.1088/1742-6596/918/1/011001

Preface for "Irreversible processes in nature and technology"

At the Bauman Moscow State Technical University, since 2001, All-Russian conferences under the general title "Irreversible processes in nature and technology" have been held regularly. In 2017, the ninth conference was held with the purpose to analyze and discuss new theoretical and experimental research results of irreversible processes in physical systems and technical devices.

Participation in the conference was traditionally attended by young, promising scientists interested in fundamental and applied problems of describing and applying irreversible processes.

At the first section of the conference were discussed results obtained in the research of irreversible processes in the condensed matters, liquids, under various electromagnetic influences. The fundamental processes occurring in the nuclei of atoms, under astrophysical phenomena and under the mutual influence of no equilibrium physical systems were investigated. In description, a synergetic approach and fractal theory were used.

The second section was devoted to the problems of developing mathematical methods and mathematical models used in the study of various irreversible processes. The authors of the reports proposed methods for modeling irreversible processes in gas states, viscous liquids and solids. The authors described processes in various technical devices, in particular, in optical interferometers.

In the materials of the third section irreversible processes in technical devices were analyzed. A description was given of the processes in pulse launch of nanosatellites, in high-voltage power plants and in the implementation of various technological processes. The idea of utilizing irreversible processes for increasing the efficiency of technical devices was discussed.

The fourth section was devoted to the study of irreversible optical processes. Great attention was paid to the Raman scattering of light, the description of its passage through the photonic crystals, amorphous and crystalline medium. The research results of the physical properties of water by optical methods have been presented, optical effects in nanomaterials and viruses have also been described, as well as the fractional high-order Talbot effect.

In the fifth section the scientific-methodological problems of teaching of natural science disciplines were discussed. The issues of improving the quality of education in the framework of its modernization, application of information technologies and methodological aspects of teaching specific sections of physics and chemistry were considered. Much attention is paid for the preparation of

IOP Conf. Series: Journal of Physics: Conf. Series **918** (2017) 011001 doi:10.1088/1742-6596/918/1/011001

qualified personnel with multidimensional knowledge, necessary for work, both in the field of fundamental physics as well as in solving important practical problems.

We hope that the work of the conference demonstrated the growing interest of scientists and teachers in the study of irreversible processes in nature and their use in technology.

Chairman of the organizing Committee A. N. Morozov

Chairman of the program Committee V. S. Gorelik

IOP Conf. Series: Journal of Physics: Conf. Series **918** (2017) 011002 doi:10.1088/1742-6596/918/1/011002

Peer review statement

All papers published in this volume of *Journal of Physics: Conference Series* have been peer reviewed through processes administered by the proceedings Editors. Reviews were conducted by expert referees to the professional and scientific standards expected of a proceedings journal published by IOP Publishing.

A criterion for estimating the probability of microporosity formation in castings

S N Poliakov[1], **A Yu Korotchenko**[2], **S L Timchenko**[3]

[1] Freiberg, Germany

[2] Professor, Bauman Moscow State Technical University, Russian Federation

[3] Assist. Professor, Bauman Moscow State Technical University, Russian Federation

E-mail: svtimchenko@yandex.ru

Abstract. The study tested a new dimensionless criterion for microporosity formation, which takes a number of significant technological factors into account, including atmospheric and metallostatic pressure, and alloy gas saturation. We recommend using the criterion in systems that model casting solidification to predict microporosity formation more accurately

Keywords: microporosity criterion, porosity estimation, Darcy law, mushy zone, melt filtration, Niyama criterion

1. Introduction

The most effective computational method of microporosity prediction during solidification is a direct simulation of pore formation based on hydrodynamic equations and physical mechanisms of pore formation. However, existing mathematical models that this simulation requires are not yet accurate enough for practical application and require further improvement.

Various forms of pore formation criteria are used in practice. The most common of these is the Niyama criterion [1, 2, 4, 5, 7]. However, all of the criteria developed do have disadvantages. One of the main problems of the existing criteria, including the Niyama criterion, is that they do not take into account a number of significant technological factors that can have a substantial effect on microporosity formation. Paper [2] lists some practical results of employing a refined Niyama criterion that takes the technological factors into account.

When developing a new criterion, we took technological factors into consideration and we did it by introducing the following into Darcy's filtering equation: metallostatic pressure, partial pressure of gases dissolved in the liquid, and capillary pressure in a dendritic lattice. To calculate the capillary pressure, we used the geometric mean distance between secondary dendritic arms as a pore radius.

2. Model

Let us assume that dendrites form during alloy solidification. Then a Piwonka - Flemings inequality [3] can represent the condition of porosity formation at any point x of a one-dimensional mushy zone $0 \le x \le L$ containing dendrites (Fig. 1):

$$\Delta P_{cr}(x) < \Delta P(x), \qquad (1)$$

where $\Delta P(x) = P_L - P(x)$ is the pressure drop in the melt at a given point x, $\Delta P_{cr}(x) = P_L - P_{cr}(x)$ is the critical pressure drop that creates the conditions for pore formation, P_L is the melt pressure at the inlet of the mushy zone for the crystallization onset temperature (temperature T_L), P_S is the fluid pressure for the approximate crystallization end temperature (temperature T_S).

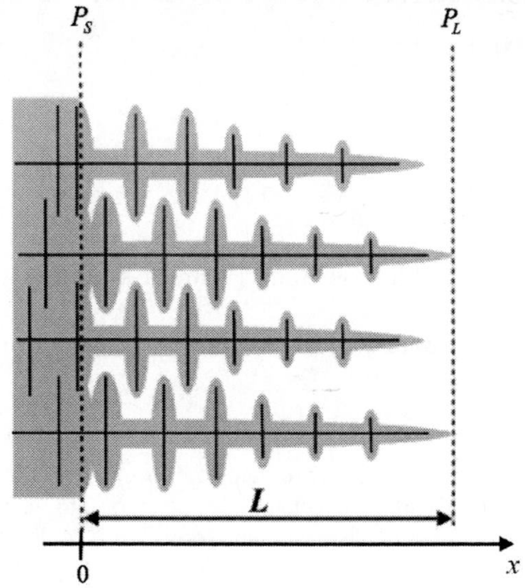

Figure 1. One-dimensional mushy zone model.

The inequality (1) enables us to construct a pore formation criterion for an arbitrary point x of the mushy zone:

$$Por(x) = \frac{\Delta P_{cr}(x)}{\Delta P(x)} < 1 \tag{2}$$

If the inequality (2) is true, then we say that porosity exists, if it is not true, then we say that there is no porosity.

Starting with the paper [1], works [2-7] and others reflect the assumption that the following equation determines the critical pressure $P_{cr}(x)$ at a certain point of the mushy zone:

$$P_{cr}(x) = P_g - P_\sigma \tag{3}$$

Here P_g is the gas pressure in cavitation bubbles, and P_σ is the capillary pressure, which is equal to:

$$P_\sigma = \frac{2\sigma}{r_0}, \tag{4}$$

where σ is the surface tension, r_0 is the radius of a "viable" pore nucleus.

If we take the expression (4) into consideration, the equation (3) for calculating the critical pressure takes the form:

$$P_{cr}(x) = P_g - \frac{2\sigma}{r_0}, \tag{5}$$

and the critical pressure differential is given by:

$$\Delta P_{cr}(x) = P_L - P_g + \frac{2\sigma}{r_0} \tag{6}$$

The pressure in a gas bubble located in liquid metal depends on the gas diffusion rate from the liquid phase into the gas bubble. At a high diffusion rate and with a known equation of state for the gas in the bubble, the gas pressure in the bubble can be expressed in terms of its radius: $P_g = P_g(r_0)$.We should note the fact that it is very difficult to determine in practice the value of the pore nucleus radius for a liquid metal. As a rule, the value of the pore nucleus radius remains unknown, so this fact is a constraint for employing the new criterion to accurately evaluate porosity.

In this paper we propose to estimate the value r_0 according to the distance between secondary dendritic arms λ_2, which represents a space in which a micropore can be possibly located in terms of geometry. This space can be characterized by the size $r_{max} = 0.5 f_l \lambda_2$, where f_l is the proportion of the liquid phase. In fact r_{max} is the radius of the sphere inscribed in the liquid portion of the interdendritic space between secondary dendritic arms (Fig. 2).

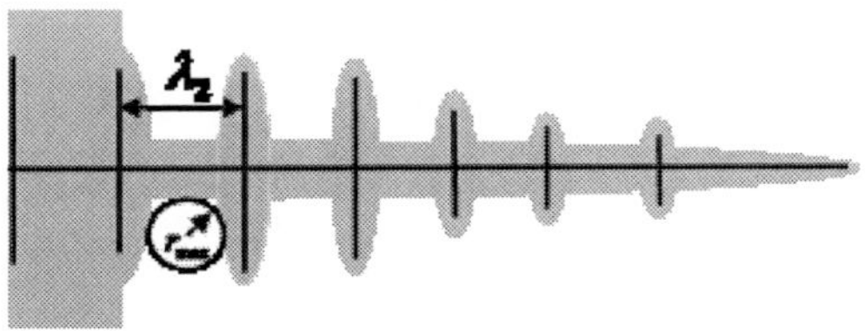

Figure 2. A geometric model of placing micropores in a dendritic lattice.

So, if there is a "viable" critical pore nucleus in the mushy zone, the condition $r_0 \le r_{max}$ is true. Taking into account the ratio for the micropore radius, the critical pressure differential is equal to:

$$\Delta P_{cr}(x) = P_L - P_g + \frac{4\sigma}{f_l(x)\lambda_2} \tag{7}$$

To describe the filtration flow of the melt in a one-dimensional mushy zone, we use the following form of Darcy's equation:

$$u f_l(x) = -\frac{K}{\mu}\frac{dP}{dx}, \tag{8}$$

where u is the filtering rate, μ is the dynamic viscosity, $K = \frac{\lambda_2^2}{180}\frac{f_l^3}{(1-f_l)^2}$ is the Carman - Kozeny filtration coefficient [8,9], λ_2 is the distance between secondary dendritic arms (fig. 2).

From the mass conservation equation for a stationary one-dimensional mushy zone the following is derived:

$$u(x) = -\beta V = const, \tag{9}$$

where V is the crystallization front advancement velocity (fig. 1); $\beta = (\rho_s - \rho_l)/\rho_l$ is the volumetric shrinkage coefficient, ρ_s and ρ_l are the densities of the solid and liquid phases respectively.

If we insert (9) into (8), we get:

$$\beta V f_l(x) = \frac{K}{\mu}\frac{dP}{dx} \tag{10}$$

If the crystallization front velocity is constant, separating the variables and integrating the expression (10) over x, we obtain the required pressure differential at an arbitrary point x of the one-dimensional mushy zone $0 \le x \le L$ (fig. 1):

$$\Delta P(x) = \mu\beta V \int_x^L \frac{f_l(x)}{K(f_l)} dx \tag{11}$$

Substituting a variable f_l for the variable x as well as assuming that the following coefficients μ, β, λ_2 are constant within the mushy zone, and that a value $G = (T_L - T_S)/L$ is the average temperature gradient, taking into account the filtration coefficient $K(f_l)$, we obtain the expression for the pressure differential in the mushy zone:

$$\Delta P(x) = \mu\beta\Delta T \frac{V}{\lambda_2^2 G} I(f_l(x)), \tag{12}$$

where $I(f_l(x)) = 180 \int_{f_l(x)}^{1} \left(\frac{1-f_l}{f_l}\right)^2 df_l$.

Then, taking (12) and (7) into consideration, the criterion of porosity formation (2) at any point in the stationary mushy zone takes the form:

$$Por(x) = \frac{\Delta P_{cr}(x)}{\Delta P(x)} = \frac{P_L - P_g + \dfrac{4\sigma}{f_l(x)\lambda_2}}{\mu\beta\Delta T \dfrac{V}{\lambda_2^2 G} I(f_l(x))} < 1 \tag{13}$$

If the inequality (13) is true for at least one point of the stationary mushy zone, porosity will appear.

3. Conclusion

1. We obtained a new criterion for microporosity formation in castings. The criterion has a dimensionless form and takes into account not only thermal conditions of the solidification process, but also significant technological parameters, such as metallostatic pressure and alloy gas saturation.
2. In formulating the conditions of microporosity formation, instead of the critical pore nucleus radius we suggest using its estimation determined by the distance between secondary dendritic arms and the proportion of the liquid phase. This eliminates technically complex experimental determination of the critical micropore nucleus radius in metal alloys.
3. The resulting criterion does not contain difficult to measure physical parameters and does not require "experimental" tuning, which is of practical use in systems simulating casting processes when evaluating the conditions of microporosity formation in the castings.

References

[1] 1982 *AFS International Cast Metals Journal* **7**(3) 52-63
[2] 2012 *Liteyschik Rossii - The Russian Caster* **4** 23-30
[3] Piwonka T S, Flemings M C 1966 *TMS-AIME* **236** 1157
[4] Carlson K D, Beckermann C 2009 *Metallurgical and Materials Transactions* A **40A** 163-175
[5] Pequet C, Gremaud M, Rappaz M 2002 *Metallurgical and Materials Transactions* A **33A** 2095-06
[6] Balandin G F 1979 *Osnovy teorii formirovaniya otlivki* [Fundamentals of the castings formation theory] vol 2 (Moscow; Mashinostroenie) 335 p
[7] Campbell J 2011 *Complete Casting* (Oxford) 1130 p
[8] Kozeny J 1927 *Sitzungsber Akad. Wiss. Wien* **136** (2a) 271-306
[9] Carman P C 1937 *Transactions, Institution of Chemical Engineers* **15** (London) 150-166

About magnetism (review)

I N Aliev[1] , M Yu Dokukin[1], Z A Samedova[1], D G Melikyants[1] and I S Kopylov[1]

[1] Physics Department, Bauman Moscow Technical University, 5, 2nd Baumanskaya street, Moscow 107005 Russia

E-mail: alievprof@yandex.ru

Abstact: In this paper, the problem was considered contradictions magnetic Coulomb's law and the Bio-Savar-Laplass, and the question was raised about the physical sense of the magnetic vectors. The problem of finding the equilibrium described by the distribution of the surface and volume of stationary currents in a solid body. We analyzed the equilibrium state of a two-phase thermodynamic equilibrium system in the case of the classical theory of the two-fluid Gorter-Casimir, taking into account the magnetic field. It is expected that the results obtained in the course of research, can be used for further development of a modernized classical theory of superconductivity

1. Introduction

Despite huge progress in practical use of magnetism (from compasses to computer memory), the nature of this complex structure remains not clear till present. It is enough to mention a problem, the bound that still it is not possible to reduce to uniform physical and consistent model magnetic Coulomb's laws (permanent magnets) and Bio-Savar-Laplass (the field of direct currents) which are the cornerstone of phenomenological approach to magnetism. And though they describe in a resultant way the same material physical structure - the magnetic field, and the nature of this phenomenon common, however are not connected among themselves at all that very surprisingly (in any case this communication till today did not manage to be found). In work of authors [1] the modified set of equations of Maxwell is presented in the complex form. In it an attempt to three pairs of the free electric, polarized and magnetic density of charges and currents is made: $(\overrightarrow{j_e}, \rho_e)$, $(\overrightarrow{j_p}, \rho_p)$, $(\overrightarrow{j_m}, \rho_m)$ to add the formal molecular current density which does not result in clear physical results $\vec{j}_{mol}$ in a look that.

$$\frac{\partial \rho_{mol}}{\partial t} = -div\,\vec{j}_{mol} = -divrot\,\vec{M} = 0$$

From our point of view one of the reason is the following circumstance. Law B.-S.-L. differentially in essence it cannot be checked. It is experimentally possible to check only an integral form. From mathematics it is well known that on value of a particular integral it is impossible to remove a type of an integrand. Therefore, most likely, it is impossible to be differential ratios the basis for the theory of magnetism. This circumstance indicates that, probably, the padding research with audit of basic concepts is required here.

Constitutive equations of a magnetistatics as a result of initial empirical laws and defining characteristics of permanent magnetic fields look as follows: $\begin{cases} rot\,\vec{H} = 0 \\ div\,\vec{B} = 0 \end{cases}$. Taking into account $\frac{\vec{B}}{\mu_0} = \vec{H} + \vec{M}$ and $\vec{j}_{mol} = rot\vec{M}$ easily the equations to which a magnetic induction submits turn out: $\begin{cases} rot\,\vec{B} = \mu_0\,\vec{j}_{total} \\ div\,\vec{B} = 0 \end{cases}$. Here $\vec{j}_{total} = \vec{j} + \vec{j}_{mol}$, from where follows that the single source of a magnetic field is the total electric current which is the sum of the routine current of conduction flowing in the environment estimated conducting, and molecular current. Let's notice that the similar form of record is convenient that in this case it is possible to consider current of superconducting electrons. This circumstance was taken into consideration at calculation of the surface currents of a superconducting sphere [2-4], a current distribution in the superconducting site of a cylindrical wire [5] and rigorous conclusion of a depth of penetration of a magnetic field in a superconductor (a constant London's) [6].

For a magnetic intensity taking into account $\rho_m = -div\vec{M}$ the system $\begin{cases} rot\,\vec{H} = \vec{j} \\ div\vec{H} = \rho_m \end{cases}$ from which follows that sources of a magnetic field are the latent magnetic electricities and currents of conduction turns out.

Analyzing the received systems it is possible to try to formulate physical sense of magnetic vectors and as follows. What of this magnetic fields should be considered the true magnetic field of the magnetized exemplar? Often in the modern physical literature answer this question as follows. Magnetization of an exemplar, on Ampere's hypothesis, is manifestation of the available intramolecular and intratomic electric currents, that is, eventually, is bound to magnetism of currents. Therefore the magnetic field of $\vec{B}$ is the true magnetic field, it makes actual physical sense while the magnetic field of $\vec{H}$ in an exemplar does not make physical sense. This formal mathematical field. Such categorical answer, apparently, is not right as from the atomic and molecular point of view, except magnetism of intramolecular and intratomic electric currents, there is a magnetism of spin magnetic electrical dipole moments of electrons and atomic kernels. The last, as we know, to magnetism of electric currents cannot be reduced, apparently. To answer a question of that, properly to define the true magnetic field in the magnetized material, it is necessary to leave, apparently, phenomenological approach and to be engaged in examination in each concrete situation of the microscopic atomic and molecular physical nature of a vector of magnetization of $\vec{M}$. As it is represented, only thus it is possible to hope to understand the true physics of vectors $\vec{H}$and $\vec{B}$.

2. Theoretical analysis

The thermodynamic hypothesis of Gibbs allowing to solve a problem by means of the magnetic principle of the virtual works was applied to finding of an equilibrium distribution of the surface and volume stationary currents in a continuous body in work [7]. The variation of a magnetic field energy was considered with the side conditions defining constancy of currents, two of which having differential character are necessary and sufficient for the solution of a task in case of the singly connected body. If the considered body doubly-connected (a torus, a thick ring) appears it is necessary still **one-integral** condition. In work it was shown what this condition at the same time providing a condition of uniqueness of the decision can be or constancy of current of the torus proceeding through a transverse section or a task of constant rate of a magnetic induction through a torus opening. At problem definition the first option as more visual was chosen. The problem was solved with the help of a method of Lagrangian multiplicities.

The basic and a little unexpected result was that circumstance that an induction of a magnetic field and volume current in volume address in zero. I.e. the magnetic field together with currents is squeezed out on a surface. The free current has the surface character and for searching of its density the boundary value problem on finding of a certain multifunction is solved. The issue of a uniqueness is resolved as follows. It is shown that in case of the singly connected bodies except two mentioned above it is not necessary to impose any side conditions – the decision and so it appears the unique. The doubly-connected area requires a side condition in a look: or the task of integral current which uniquely determinates circulation of a magnetic displacement vector or is set a stream of a magnetic induction through a torus opening.

In work [8] it is shown that as such side condition it is also possible to record value of magnetic electrical dipole moment of an equilibrium distribution of currents. Considering that realization of the main result requires and enough only constancy of currents which can be reached, apparently, or an adjustable feeding device or considering Wednesday with a small ohmic resistance (in particular, superconducting structures), it was interesting to consider communication of the received results with Meyssner-Oksenfeld's effect and the equation London's applied in the classical theory of superconductivity. The further research of the received results in of connection with the appeared analogy to made mention effect is executed in work [9]. The detailed analysis of the classical phenomenological theory of superconductivity London's was for this purpose carried out and the incorrectness of a conclusion of a basic set of equations and unsubstantiality of the assumption of an irrotational flow of liquid of superconducting electrons was shown. On the basis of what the conclusion was drawn that for superconductors it is impossible to uniquely determinate a vector of a magnetic intensity as it is not clear how to divide superconducting current into current of conduction and current of magnetization.

Taking into account the results received for constants in time of currents in materials with big conduction the conclusion is drawn that currents and a magnetic field are squeezed out on a surface, i.e. there are only surface currents. The actually surface current just needs to be considered as the current flowing in thin, but terminating thickness a layer which and can be considered as a field depth of penetration. Thickness of a made mention layer was received proceeding from dimensional reasons. The result in received at the same time a case of different force systems coincided with thickness London's that indicates consistency of an initial hypothesis.

Since, a phenomenological electrodynamics of superconductors is constructed on the basis of classical representations, and, despite outstanding successes in an explanation of this phenomenon, the partial classical theory demands essential specifications and improvements. Taking into account the received results it was important to analyse basic laws of an electrodynamics of liquid of superconducting electrons again. In work [10] are removed in a Si-system of a Poynting theorem (law of conservation of energy) and Abragam (law of conservation of momentum) for a true liquid of superconducting electrons. It was shown that an apparent density of a mechanical impulse in an immediate look is not included into the end

IOP Conf. Series: Journal of Physics: Conf. Series **918** (2017) 012002 doi:10.1088/1742-6596/918/1/012002

result, it is considered through a stress tensor. Let's notice that purely convective stream of an impulse which flows in volume together with electronic liquid is also considered.

Using concepts of a fluksoid and superpotential of work [11] it was established that the ratios connecting density of superconducting current in each point and vector potential of a constant magnetic field in case of the singly connected superconductors are prime record of the constitutive equations London's. As fluksoid, being function, the bound to superconducting carriers, it is unambiguous along any selfcontained trajectory describing an opening there was opportunity for analogy to a case of wave functions of atomic electrons to accept a quantization rule of the Bohr – Sommerfeld and to demand that for each superconducting electron the condition was satisfied: $\oint_L \vec{p}_s d\vec{l} = n\hbar$ where $\vec{p}_s = e\Lambda\vec{j} + e\vec{A} = m\vec{v} + +e\vec{A}$ a so-called "initial" impulse of a superconducting electron in an electromagnetic field.

I.e.: $\oint_L e\Lambda\vec{j}d\vec{l} + \oint_L e\vec{A}d\vec{l} = \oint_L e\Lambda\vec{j}d\vec{l} + e\Psi = n\hbar$, where$\Psi$ – the complete magnetic flux. At the long distances from an opening exceeding a field depth of penetration the first item it is close to zero and we receive a quantization rule of a magnetic flux in superconductors:$\Psi = \frac{n\hbar}{e}$. Thus, the well-known condition of quantization for superconductors, as expected, can take place at a small, but terminating thickness of the surface layer.

Further the analysis of equilibrium standing of equilibrium system two-phase thermodynamic in case of the classical double-fluid theory of Gorter-Kazimir taking into account a magnetic field was carried out [12].

Rigorous mathematical conclusion of conditions of thermodynamic magnetic equilibrium is considered for the isolated system for that prime reason that for it the known thermodynamic principle of Gibbs allowing to find in general any equilibrium conditions in what – it is necessary physical and physical and chemical situations, in particular conditions of interphase equilibrium is fair. At the solution of an objective the method of Lagrangian multiplicities was used by analogy with the magnetic principle of the virtual works for the free currents [7].

Due to the aforesaid there was a need of the detailed analysis of quantum mechanical behavior of ensemble of electrons for wide temperature range, in particular, the further analysis of thermodynamic characteristics [13]. In this work an attempt of the detailed analysis of the specified properties was made, having taken as a basis the quantum mechanical description on the basis of operators of secondary quantization, how to do it for electric properties of a solid body at low temperatures.

Taking into account the previous researches conditions of a thermodynamic equilibrium of expression for thermodynamic potentials, entropies and thermal capacities in wide temperature range were removed. As it seems to us, it makes sense to apply the received results to further attempts of development of the upgraded classical theory of superconductivity.

The method of secondary (double) quantization was also applied in work [14] at a conclusion taking into account a perturbation theory in the second order of smallness of a diamagnetic constant of Landau.

Let's note that authors in the quoted works (as far as possible) applied an international system of units (SI) though as it seems to us, in sections of physics of the directly bound to an electromagnetism, use of an absolute (Gaussian) force system as the last meets the modern expectations of the nature of the electromagnetic phenomena much more would be most expedient. Electric and magnetic fields are bound among themselves, moreover, they can be considered two symmetric characteristics of the same physical object. But at the same time dimensions of all sizes (strength and an induction of this field) different that in itself is already unnatural. For overcoming this discrepancy it is also necessary to enter simulated coefficients which discussion of physical sense continues and now. Especially, as it is visible, is shown by consideration of classical questions of magnetism, such as monopoles of Dirac or classical phenomenological theory of superconductivity. Probably, that these concepts are entered and discussed in books and articles on theoretical physics which qualified authors, as a rule, adhere to absolute system in which the received results look more naturally and therefore work with these sections in a Si-system constantly leads business to selection of necessary coefficients.

References

[1] Aliev I, Kopylov I 2015 Use of a formalism of monopoles of Dirac in some problems of magnetism. J. Herald of BMSTU.Natural sciences №6 pp 25-39

[2] Aliev I and Melikyants D 2016 About magnetism and the surface currents of a superconducting sphere J.Electromagnetic waves T.21. №9 pp 13-19

[3] Aliev I and Melikyants D 2016 About magnetization of a superconducting sphere J.Herald of BMSTU: Natural Sciences series №3(66) pp 82-92

[4] Aliev I and Samedova Z 2017 Minimum of a magnetic field energy of a superconducting sphere J. Electromagnetic waves №4

[5] Aliev I and Samedova Z 2016 Distribution of current in the superconducting site of a cylindrical wire J.Herald of the Moscow regional state university: series physics mathematician No. 1 pp 88-97

[6] Aliev I and Samedova Z 2017 About boundary conditions in an electrodynamics of Maxwell and London's J. Surface №6

[7] Aliev I and Kopylov I 2015 Application of a method of Lagrangian multiplicities to calculation of a magnetic field of a direct current J. Dynamics of the composite systems №4 pp 3-10

[8] Aliev I and Samedova Z 2017 Principle of a minimum of energy of the surface superconducting currents J. Electromagnetic waves №6

[9] Aliev I and Kopylov I 2017 On the electrodynamic London's model and Gortera-Kazimir's theories J. Surface No.2 pp 64-72

[10] Aliev I and Melikyants D 2015 About Poynting theorems and Abragama in an electrodynamics of superconductors London's J. Herald of the Moscow regional state university: series physics mathematician No. 4 pp 83-91

[11] Aliev I and Melikyants D 2016 About potentials in an electrodynamics London's J. Herald of BMSTU: Natural Sciences series №2(65) pp 42-51

[12] Aliev I, Dokukin M and Samedova. Z 2016 About an interphase thermodynamic equilibrium of magnetics J. The dynamics of complex system №4 pp 3-9

[13] Aliev I, Dokukin M and Samedova Z 2016 Thermodynamics of system of noninteracting free electrons J. Herald of the Moscow regional state university: series physics mathematician No. 3 pp 57-71

[14] Aliev I, Dokukin M and Samedova Z 2016 Application of double quantization in Landau's diamagnetism J. Herald of BMSTU: Natural Sciences series №4(67) pp 14-27

IOP Conf. Series: Journal of Physics: Conf. Series **918** (2017) 012003 doi:10.1088/1742-6596/918/1/012003

Advanced response of Baikal macroscopic nonlocal correlation detector to solar activity

S M Korotaev[1,2], N M Budnev[3] and V O Serdyuk[2]

[1] Physics Department, Bauman Moscow State Technical University, 5, 2nd Baumanskaya street, Moscow 107005 Russia

[2] Geoelectromagnetic Research Centre, Schmidt Institute of Physics of the Earth, Russian Academy of Sciences, P.O. Box 30 Troitsk, Moscow 142190 Russia

[3] Institute of Applied Physics, Irkutsk State University, 20 Gagarin boulevard, Irkutsk 664003 Russia

E-mail: korotaev@igemi.troitsk.ru

Abstract. The long-term Baikal Deep-Sea Experiment on study of macroscopic entanglement and advanced response of nonlocal correlation detectors to natural random dissipative processes is conducting. There are both advanced and retarded nonlocal correlations, which correspond to the same direction of causality. However, time reversal causality prevails over usual time respecting one. It is the most prominent property of macroscopic entanglement and manifestation of quantum principle of weak causality. In particular, there is an advanced response to the random component of solar activity, which can be used for the forecast of this component.

1. Background

Macroscopic quantum entanglement is phenomenon studied in some special cases only; its consistent theory is not developed yet. However, the heuristic equation of macroscopic entanglement was suggested [1], according to which the processes prove to be correlated with symmetric retardation and advancement. For diffusion entanglement swapping the retardation and advancement can be large. The symmetry of the retarded and advanced correlation components is broken by a shielding medium, suppressing the retarded component in a greater extent. The advanced correlation meets the quantum principle of weak causality: for the uncontrolled entangled states advanced nonlocal correlations through a timelike interval and hence time reversal causality are possible. As the causes in these states are the random processes, such a time reversal causality does not imply the well-known classical paradoxes. It was experimentally demonstrated in the low-dimensional quantum systems [2-5].

In the previous experiments (reviewed in Ref. [1]) macroscopic nonlocal correlations of the probe processes in the detectors (providing their isolation from any classical local impacts) with some natural large-scale dissipative processes with big random component. Solar activity proved to be the dominant source-process. The most important results were detection of advanced nonlocal correlations and experimental proof of time reversal causality for the random processes. The mathematical tool for this proof is causal analysis. It is based on definition for any variables X and Y, via Shannon entropies S, the independence functions $i_{Y|X} = S(Y\,|\,X)/S(Y)$, $i_{X|Y} = S(X\,|\,Y)/S(X)$, $0 \le i \le 1$ and causality function $\gamma = i_{Y|X}/i_{X|Y}$, $0 \le \gamma < \infty$. By definition X is the cause and Y is the effect if $\gamma < 1$. And inversely, Y is the cause and X is the effect if $\gamma > 1$. If τ is time shift of Y relative to X, then the principle of classical causality is formulated as follows: $\gamma < 1 \Rightarrow \tau > 0$, $\gamma > 1 \Rightarrow \tau < 0$, $\gamma \to 1 \Rightarrow \tau \to 0$. Only in case of nonlocal correlation, one can observe violation of these inequalities. Such a violation allowed to forecast the random components of solar and geomagnetic activity [6].

Since macroscopic nonlocal correlations manifest themselves at extremely low frequencies, for their detection the long-terms experiments with the maintenance of highly stable conditions in detectors are necessary. This is implemented under the conditions of the Baikal deep-sea experiment, which has already brought a number of important results on correlations with large-scale geophysical processes [7]. The experimental setup includes three electrode detector of macroscopic nonlocal correlations: *Ub* and *Ut* at depths of 50 m and 1300 m respectively, and remote at 4200 km land detector *Ul*. The detector *Ut* has the maximal signal-to-noise ratio with respect to external heliogeophysical processes.

2. Response of a macroscopic correlation detector to solar activity

Below we analyze the response of *Ut* detector signal to heliogeophysical processes by data of the first continuous two-year measurement series (2013/2015). The detector *Ut* turned out optimal for the signal-to-noise ratio in the study of these processes (the greatest noise is contained in the detector on the Earth surface *Ul*, the smallest – in the bottom one *Ub*, but in the latter the signal from the external processes have significantly shielded the water column).

In Figure 1 the normalized amplitude spectra of the detector *Ut,* solar radio wave flux *R* (at 2800 MHz) and *X*-ray flux (at 0.5-4 Å) at the period *T* range is from 10 to 461 days are shown. As was to be expected *Ut* practically does not respond to deterministic component of solar activity variations, which is represented in *R* and *X* by split 27-day variation and its harmonics. Over longer periods, where the random component of solar activity is dominates, the response in *Ut* is clearly visible. Detailed analysis showed the greatest similarity of the spectra of *Ut* with *R* in the band of periods 365>*T*>59 days, and with *X* in the band 365>*T*>77 days. Next to causal and correlation analysis data from such band-pass filtration. were used.

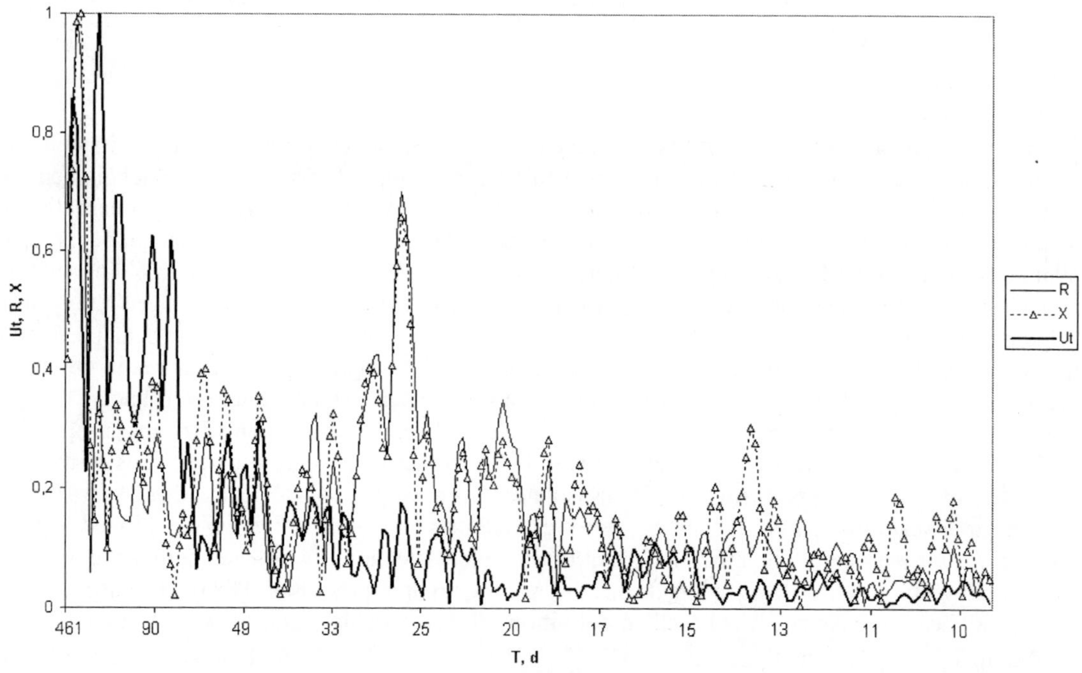

Figure 1. Normalized amplitude spectra of detector signal *Ut*, solar radio wave flux *R* and *X*-ray flux.

The causal and correlation computation results are presented in Figures 2 and 3. Their stability was tested by alternate noising of the time series by the flicker noise (21% power). As one can see $\gamma > 1$, i.e. *Ut* is an effect of *R, X* at any τ. In a couple of *Ut-R* the highest maximum of causality $\gamma = 1.5^{+0.0}_{-0.1}$ is observed at advancement of *Ut* with respect to *R* for 250 days. At the same advancement the deepest minimum of independence $i_{Ut|R} = 0.35^{+0.05}_{-0.00}$ and maximal correlation $r = -0.79^{+0.02}_{-0.01}$ are observed. In a couple of *Ut-X* the results are close: the extreme $\gamma = 1.5^{+0.1}_{-0.0}$, $i_{Ut|R} = 0.28^{+0.02}_{-0.00}$, $r = -0.79^{+0.02}_{-0.01}$ are at advancement 230 days.

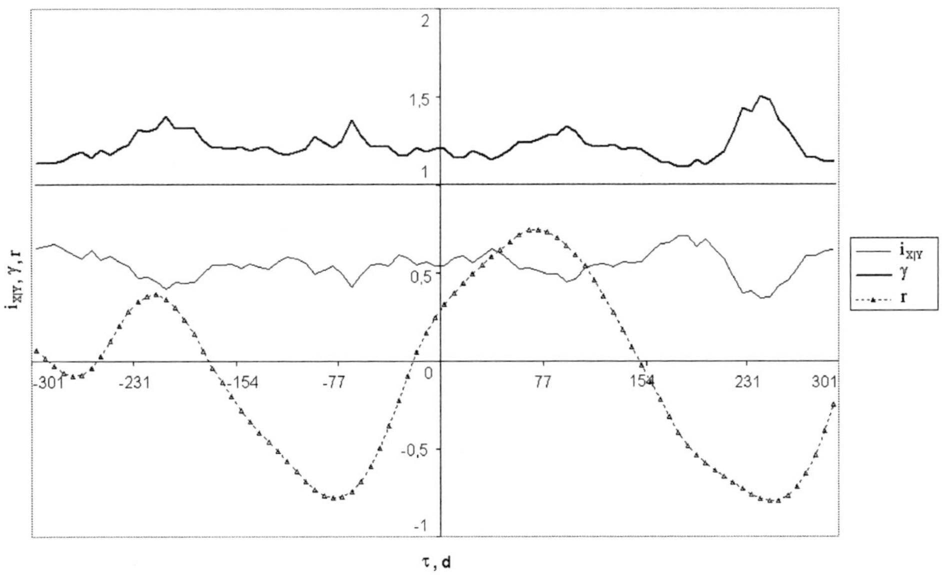

Figure 2. Causal and correlation analysis of *Ut* (X) and *R* (Y). τ<0 corresponds to retardation of *Ut* relative to *R*, τ>0 – to advancement.

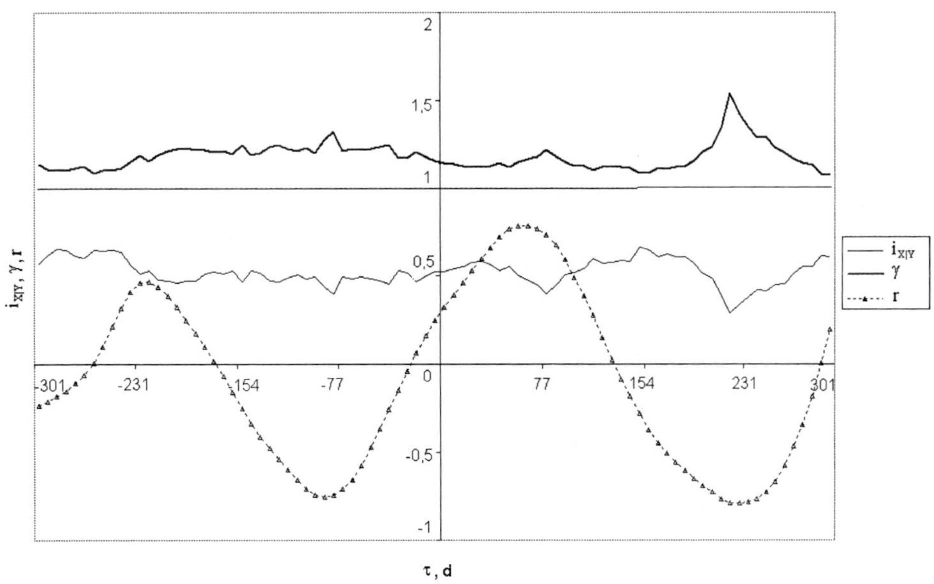

Figure 3. Causal and correlation analysis of *Ut* (X) and *X* (Y). τ<0 corresponds to retardation of *Ut* relative to *X*, τ>0 – to advancement.

The level of advanced correlation is sufficient to forecast solar activity with advancement about 250 days. However, the length of the series *Ut* so far is insufficient to apply the effective forecasting algorithm [6]. But the very possibility of solar activity forecast is easily demonstrated by the series shift. It is shown in Figure 4 by the example of *R*.

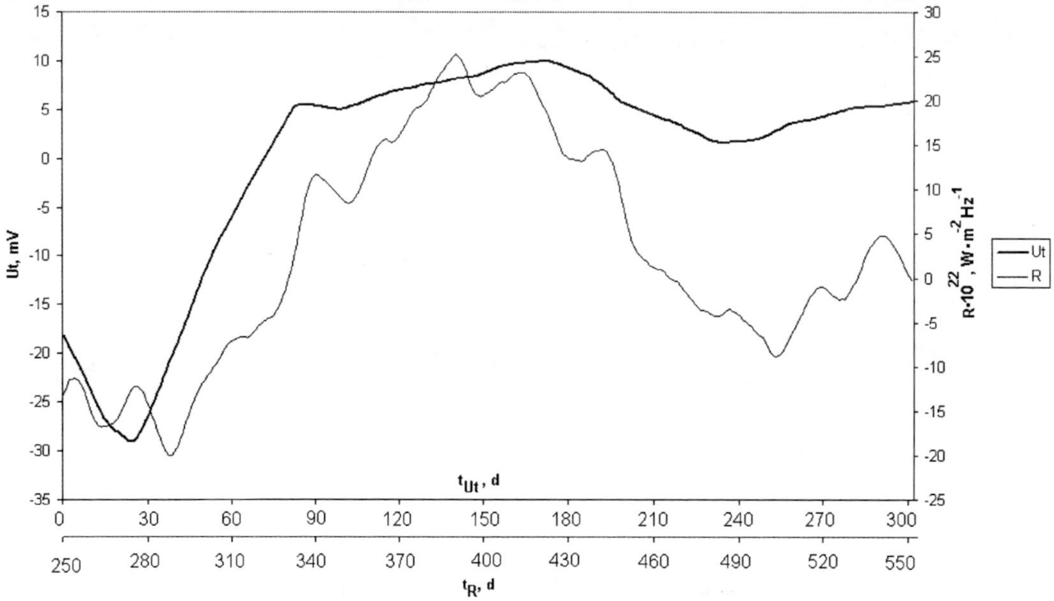

Figure 4. *Ut* approximately forecasts *R* with advancement 250 days.

In agreement with the results of all previous experiments of such a kind [1], there is also a weaker advanced connection of the detector signal with the global geomagnetic activity characterized by the *Dst*-index (this index stresses a random component of the process). With filtration 365>*T*>59 days in the couple *Ut*- *Dst* at advancement of *Ut* with respect to *Dst* for 110 days the extreme values $\gamma = 1.3^{+0.0}_{-0.1}$, $i_{Ut|Dst} = 0.51^{+0.08}_{-0.00}$, $r = -0.65^{+0.01}_{-0.00}$ are observed; with filtration 365>*T*>77 days extreme values are observed at the same advancement and amount to $\gamma = 1.2^{+0.1}_{-0.0}$, $i_{Ut|Dst} = 0.47^{+0.00}_{-0.01}$, $r = -0.68^{+0.03}_{-0.00}$.

3. Verification of nonlocal nature of correlations

The observed violation of the principle of strong (classical) causality described above is a sufficient condition of nonlocality. However, taking into consideration the singularity of this entanglement evidence, let us now verify a necessary condition of nonlocality and besides do it in more traditional manner not appealing to time shifts (as this is done through violation of Bell-like inequalities and steering inequalities). We are interested in a steering inequality, because it is just appropriate necessary condition of locality. The combination of connections of *Ut* with geomagnetic and solar activities provides an opportunity for this.

Suppose some process C acts upon a distant process A by means of any local interaction along the causal chain $C \to B \to A$. The intermediate process B is situated so that local carriers of interaction cannot come A avoiding B (e.g. B occupies a spherical layer around A). Then the claim of locality implies the steering inequality (derived e.g. in Ref. [1]):

$$i_{A|C} \geq \max\left(i_{A|B}, i_{B|C}\right) \qquad (1)$$

In our case $A = Ut$, $B = Dst$, $C = R$ or X. Note, classical local influence of solar activity, indexed by R or X on geomagnetic activity, indexed by *Dst*, is well known. Let us verify Ineq. (1) by substitution of experimental values of independence functions under the same filtration of the series $A = Ut$, $B = Dst$, $C = R$ or X.

With filtration 365>*T*>59 days: for $C = R$: $i_{Ut|R} = 0.35^{+0.05}_{-0.00}$, $i_{Ut|Dst} = 0.51^{+0.08}_{-0.00}$, $i_{Dst|R} = 0.57^{+0.00}_{-0.01}$; for $C = X$: $i_{Ut|X} = 0.36^{+0.14}_{-0.00}$, $i_{Ut|Dst} = 0.51^{+0.08}_{-0.00}$, $i_{Dst|X} = 0.55^{+0.00}_{-0.00}$. With filtration 365>*T*>77 days: for $C = R$: $i_{Ut|R} = 0.34^{+0.00}_{-0.05}$, $i_{Ut|Dst} = 0.47^{+0.00}_{-0.01}$, $i_{Dst|R} = 0.54^{+0.00}_{-0.00}$; for $C = X$: $i_{Ut|X} = 0.28^{+0.02}_{-0.00}$, $i_{Ut|Dst} = 0.47^{+0.00}_{-0.01}$, $i_{Dst|X} = 0.55^{+0.01}_{-0.01}$. In all four variants Ineq. (1) is reliably violated.

Thus, the nonlocal nature of the observed in the Baikal experiment correlations between the detector signal and solar activity is verified through the violation of the steering inequality without resorting to time relations.

4. Conclusion

Advanced response of Baikal macroscopic nonlocal correlation detector to solar activity has been revealed. This remarkable manifestation of quantum principle of weak causality can be used for the forecast of random component of solar activity. Nonlocal nature of the observed correlations has also been confirmed by violation of the steering inequality.

Acknowledgment

This work was supported by RFBR (grant 15-05-00609).

References

[1] Korotaev S M 2011 *Causality and Reversibility in Irreversible Time* (Irvine, CA: Scientific Research Publishing)
[2] Laforest M, Baugh J and Laflamme R 2006 *Phys. Rev. A* **73** 032323
[3] Ma X-S, Zotter S, Kofler J, Ursin R, Jennewien T, Brukner Č, and Zeilinger A 2012 *Nature Physics* **8** 479
[4] Megidish E, Halevy A, Shacham T, Dvir T, Dovrat L, and Eisenberg H S 2013 *Phys. Rev. Lett.* **110** 210403
[5] Danan A, Farfurnik D, Bar-Ad S and Vaidman L 2013 *Phys. Rev. Lett.* **111** 240402
[6] Korotaev S M, Serdyuk V O and Gorohov J V 2007 *Doklady Earth Sciences* **415A** 975
[7] Korotaev S M, Serdyuk V O, Kiktenko E O, Budnev N M and Gorohov J V 2015 in *Unified Field Mechanics* ed R L Amoroso, L H Kauffman and P Rowlands (Singapore: World Scientific) pp 366–373

IOP Conf. Series: Journal of Physics: Conf. Series **918** (2017) 012004 doi:10.1088/1742-6596/918/1/012004

Analysis of pulse-launched nanosatellite stability in the atmosphere

Yu V Gerasimov[1], E A Ivanov[2], G K Karetnikov[1], I A Konstantinova[2] and A B Selivanov[1]

[1] Associated professor, Bauman Moscow State Technical University, Moscow, RF
[2] Master Degree, Bauman Moscow State Technical University, Moscow, RF

E-mail: drgerasimov.y@gmail.com

Abstract. The article shows the effects of various aerodynamic parameters, the possibilities of their change are determined.The increase in the length of the lightweight stern, the diameter of the skirt and the thinning of the tube and skirt material improve the stability, but adversely affect the strength or aerodynamic properties of the product in question. The uniform distribution of mass over the nanosatellite volume is preferable in the process of acceleration with high-speed throw and exit from the railgun, but it negatively affects the stability in flight.

1. Introduction

The creation of nanosatellite cluster space systems (satellite weight below 10 kg) is a perspective trend in space technology development. The transition to small vehicles requires changes in the technology of satellite delivery into orbit.

One of the most effective ways to launch a nanosatellite is a pulse launch. Advantages of throwing satellites into orbit are widely discussed in the contemporary literature, for example, in [1] - [3]. In the long term this type of transport benefits economically, technologically and environmentally.

The nanosatellite is assumed to acquire an initial starting velocity in a pulse accelerator, and then a passive suborbital section of the path follows. In the process of a nanosatellite movement, its trajectory can be corrected according to the specified program by means of a correcting impulse device and an ion engine. The launch pad can be on the surface of the Earth, on a hill, or at heights of up to 40 km in an air start system.

In the paper, the main parameters of the nanosatellite structure providing stabilization in the atmosphere in the altitude range of 30-80 km above the Earth are investigated. Models with a continuous mass distribution over the volume of a nanosatellite and with a lightweight stern are considered.

2. Analysis of aerodynamic moments and forces

A nanosatellite is an axisymmetric high-fineness-ratio configuration body, stabilized in flight by an aerodynamic skirt. The aerodynamic scheme is shown in Figure 1. Nanosatellite dynamics is determined by the principal vector of aerodynamic forces and the moment with respect to a certain pole. In the body coordinates, the principal vector of aerodynamic forces can be decomposed into axial, normal and side components (in the wind coordinates - into the drug, lift and lateral forces, respectively) [4].

The calculation of flow parameters is defined by Reynolds and Knudsen numbers. The parameters of the atmosphere are defined according to the tables of the standard atmosphere [5]. Density ρ, temperature T, the air dynamic viscosity μ and the mean free path λ are specified in the Table 1 for several points in altitude range. Relevant Reynolds $\mathrm{Re} = \frac{\rho \cdot V \cdot l}{\mu}$ and Knudsen $Kn = \frac{\lambda}{\delta_{погр}}$ numbers for a body of the length $l = 1.2$ m moving with the velocity $V = 8\ 000$ m/s are given. According to [4], the thickness of the boundary layer reaches $\delta_{погр} = 4.64\ {}^{l}\!/\!\sqrt{\mathrm{Re}}$.

Table 1.

Altitude(km)	Density(kg/m^3)	Temperature(K)	Viscosity(mkPa $\cdot$ c)	Re	Kn
30	$1.84 \cdot 10^{-2}$	226.5	14.753	$1.50 \cdot 10^{7}$	0.00307
40	$4.00 \cdot 10^{-3}$	250.4	16.009	$3.00 \cdot 10^{6}$	0.00632
50	$1.03 \cdot 10^{-3}$	270.7	17.037	$7.23 \cdot 10^{5}$	0.0121
60	$3.10 \cdot 10^{-4}$	247.0	15.837	$2.35 \cdot 10^{5}$	0.0228
70	$8.28 \cdot 10^{-5}$	219.6	14.377	$6.91 \cdot 10^{4}$	0.0463
80	$1.85 \cdot 10^{-5}$	198.7	14.426	$1.53 \cdot 10^{4}$	0.0980

The table shows that at altitudes up to 80 km the air molecule mean free path is much smaller than the thickness of the boundary layer and the calculation is made using the continuum model. At the preliminary stage, viscous friction and bottom resistance are not considered. These parameters are supposed to be considered in a selective numerical experiment.

Aerodynamic coefficients calculation is carried out according to the theory of Newtonian flow. Such a model leads to an estimate of the aerodynamic forces, which is in satisfactory agreement with the experiment in the hypersonic stream [4]. The aerodynamic coefficients of the nanosatellite are estimated in body coordinates, under hypersonic flow conditions.

The following parameters of the nanosatellite are used:

- The midsection diameter d_{mid};
- Fore body length X_l;
- Cylindrical part length X_{cyl};
- Bottom diameter Ds and skirt length H_s.

The coefficient of pressure at the surface is calculated by the condition that the molecules of the incident flow lose momentum component normal to the surface:

$$\bar{p} = 2(\sin\beta \cdot \cos\alpha - \sin\alpha \cdot \cos\beta \cdot \cos\gamma)^2, \quad (1)$$

where: β is a local slope angle of streamlined surface to the nanosatellite axis;

α isthe angle of attack;

γ is the integration parameter equal to the angle between the plane of the angle of attack and the meridian plane. The range of $\gamma > \gamma_{Shadow}$ corresponds to the wind shadow area, it is marked by shading in Figure 1.

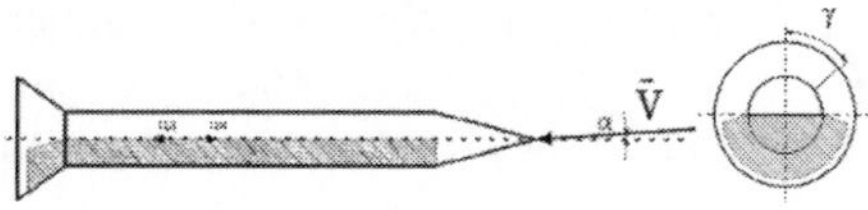

(a) Nanosatellite aerodynamic scheme.

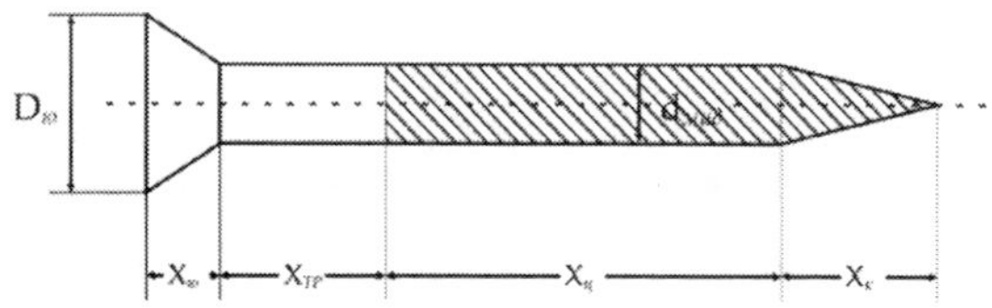

(b) Layout of the nanosatellite with a lightweight stern.

Figure 1.

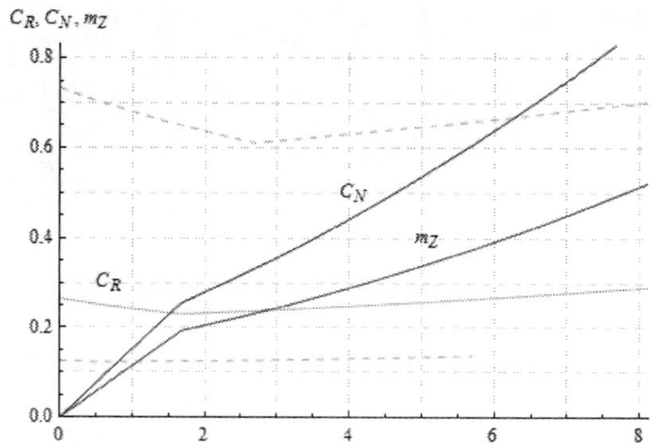

Figure 2. The dependence of aerodynamic coefficients on the angle of attack.

The formulae for the coefficients of axial and normal forces and pitching moment are [4]:

$$C_R = \frac{2}{S_{mid}} \int_0^l dx \cdot r \cdot tg\beta \int_0^{\gamma_{Shadow}} d\gamma \cdot \bar{p}$$

$$C_N = \frac{2}{S_{mid}} \int_0^l dx \cdot r \int_0^{\gamma_{Shadow}} d\gamma \cdot \bar{p} \cdot \cos\gamma$$

$$m_Z = \frac{2}{l \cdot S_{mid}} \int_0^l dx \cdot r \cdot x \int_0^{\gamma_{Shadow}} d\gamma \cdot \bar{p} \cdot \cos\gamma \quad (2)$$

where: r is the local radius of the axially symmetrical body;
l is the length of nanosatellite.

The coefficients of axial C_R, normal C_N forces and pitching moment m_z are calculated depending on the angle of attack α in the body coordinates. The corresponding graphs are shown in Figure 2. The following parameters are chosen:

$$d_{mid} = 60\ mm;\ X_1 = 2d_{mid};\ X_{cyl} = 10d_{mid};\ X_s = 1,1d_{mid};\ D_s = 1,5d_{mid}. \quad (3)$$

The graph $C_R(\alpha)$ shows that growth of the wind shadow of the cylindrical skirt part slightly reduces the axial resistance of the incident stream. The knee on the graphs corresponds to the critical angle of attack α_*, where the wind shadow from the cylindrical part reaches the edge of the skirt. At the chosen fineness-ratio configuration $\alpha_* \approx 1.7^o$. The graphs of the coefficient of axis force C_R at $D_s = 1,3d_{mid}$ and $D_s = 1,8d_{mid}$ are shown in the dashed lines. With increasing diameter of the skirt there is a significant increase in air resistance.

IOP Conf. Series: Journal of Physics: Conf. Series **918** (2017) 012004 doi:10.1088/1742-6596/918/1/012004

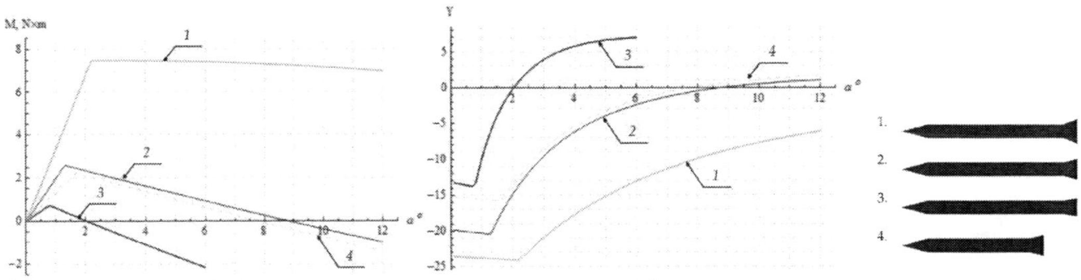

(a) (b) (c)

Figure 3.

The nature of the wind shadow influence on the nanosatellite dynamics in flight is clearly visible in the graphs of Figure 3. Figure 3a shows the dependencies of the pitching moment relative to the center of mass on the angle of attack for different geometric configurations (the proportions are shown in Figure 3c) of a 10-kg nanosatellite in the hypersonic flow with dynamic pressure 33 kPa. According to [5], the selected dynamic pressure corresponds to a flight at the altitude of 50 km at a speed of 8000 m/s. In the set of geometrical parameters (3) the diameter of aerodynamic skirt was varied. On Figure 3b the dependence of the stability coefficient on the angle of attack is shown at different diameters of the aerodynamic skirt: $D_s = 1.3d_{mid}$; $1.5d_{mid}$; $1.85d_{mid}$.

The moment of external aerodynamic forces relative to the center of mass is calculated by the formula:

$$M_z = \left(m_z - C_N \cdot \frac{X_{CM}}{l}\right) \cdot l \cdot S_{mid} \cdot q, \qquad (4)$$

where:

X_{CM} is a position of the center of mass relative to nanosatellite fore body;

$q = \frac{\rho \cdot V^2}{2}$ is the dynamic pressure.

Factor of static stability is calculated by the formula [4]:

$$S = \frac{X_{CM}}{l} - \frac{m_Z}{C_N}.$$

Graphs in Figure 4 show that the necessary stability of the nanosatellite is provided by a significant diameter of the skirt.

Another parameter that allows varying streamliningof thenanosatellite in flight is the shape of skirt. Figure 4a shows the dependence of the axial force on the angle of attack. The length of the skirt is increased from sample 1 to 3 by means of decreasing the cylindrical part while retaining all other dimensions, as it is shown in Figure 4c. Graphs in Figure 4b show the dependence of the corresponding coefficients of stability. A comparison of the graphs suggests that improving the flow conditions of the skirt leads to regimes with loss of stability.

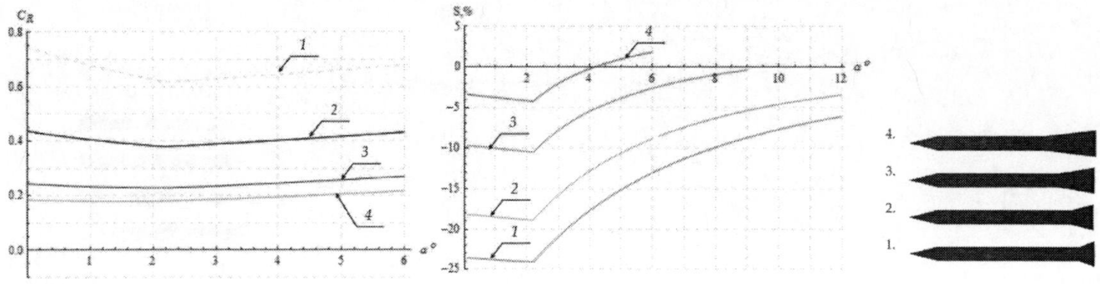

(a)(b)(c)

Figure 4.

Thus, with a continuous mass distribution, an acceptable stability of a nanosatellite is ensured by the deterioration of the flow patterns.

3. Dependence of the stability coefficient on the nanosatellite layout

An alternative way to ensure the nanosatellite stability is the displacement of the center of mass relative to the pressure center by redistribution of the mass and selection of the lightweight stern of the length X_{light}, as shown in Figure 5. The average material density of the working compartment and the walls of the lightweight part is assumed to be equal to 6000 kg / m3. The average weight of the lightweight part varies due to the average density ρ ; the wall thickness is assumed to be 2 mm. Fuel and oxidizer is located in the stern for changing the orbit and reducing the bottom resistance.

Graphs in Figure 5 show the static stability coefficient dependence on the angle of attack for different ratios between the geometric characteristics of a nanosatellite with a lightweight stern.

Basic set of parameters (d_{mid} = 60 mm; $X_1 = 2\ d_{mid}$; $X_{cyl} = 10\ d_{mid}$; $X_s = 1{,}1 d_{mid}$; $D_s = 1{,}5\ d_{mid}$) was varied while mass (10 kg) and density (6 000 kg/m3) were constant.

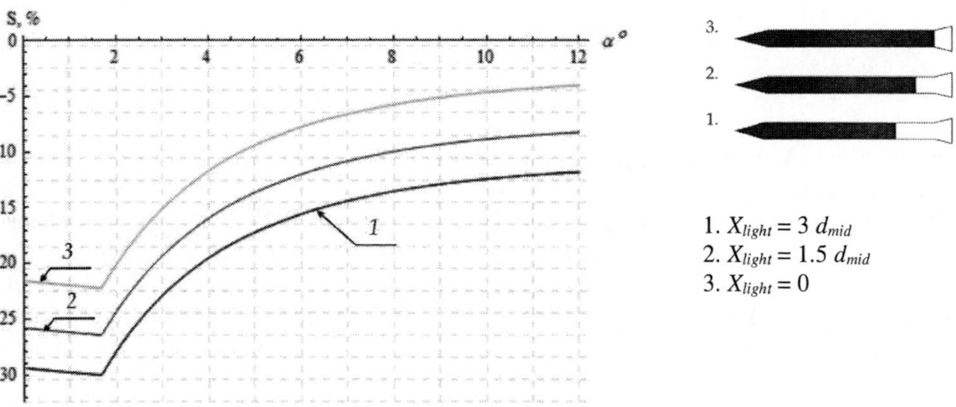

1. $X_{light} = 3\ d_{mid}$
2. $X_{light} = 1.5\ d_{mid}$
3. $X_{light} = 0$

(a)

IOP Conf. Series: Journal of Physics: Conf. Series **918** (2017) 012004 doi:10.1088/1742-6596/918/1/012004

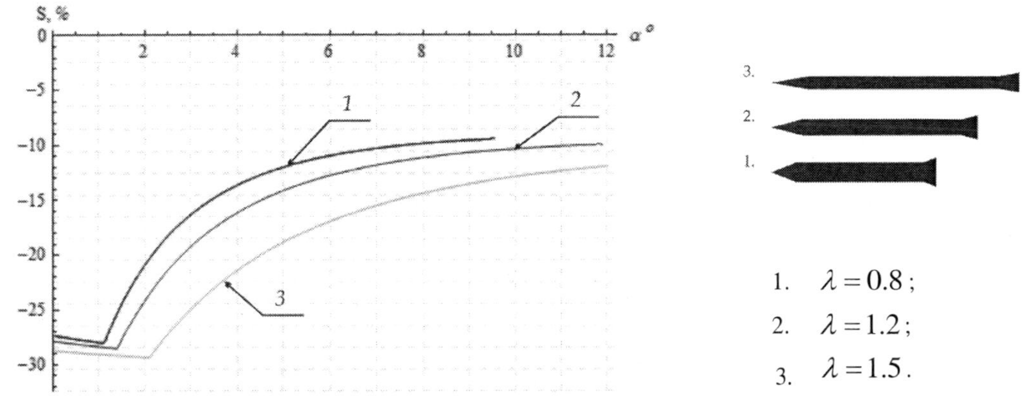

(b)

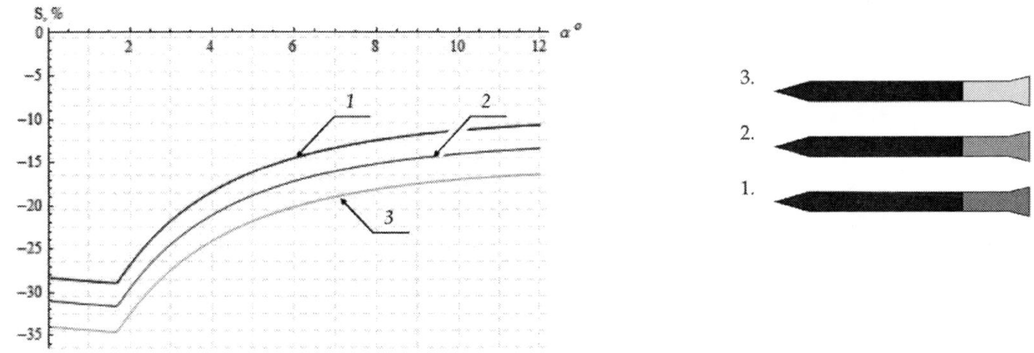

(c)

Figure 5.

The Figure 5a shows graphs of the stability coefficient dependence on the angle of attack for different lengths of the lightweight stern X_{light} =0, 1.5 d_{mid}, 3 d_{mid}. From the graphs it is seen that without a lightweight cylindrical part, stability is acceptable in the range of angles of attack up to $\alpha \approx 5^o$; at X_{light} = 1.5 d_{mid} it is up to 8°, at X_{light} = 3 d_{mid} the acceptable range is more than 12°.

Each curve in Figure 5b corresponds to the particular fineness-ratio configuration. The structure length is increased by λ = 0.8, 1.2, 1.5 due to the reduction of the midsection, while the total mass of the nanosatellite and the density of the material remain constant.

Figure 5c shows the effect of mass redistribution in a nanosatellite while reducing thickness of the lightweight central shell.

4. Conclusion

The article shows the effects of various aerodynamic parameters, the possibilities of their change are determined.The increase in the length of the lightweight stern, the diameter of the skirt and the thinning of the tube and skirt material improve the stability, but adversely affect the strength or aerodynamic properties of the product in question [1], [6]. The uniform distribution of mass over the nanosatellite volume is preferable in the process of acceleration with high-speed throw and exit from the railgun, but it negatively affects the stability in flight.

The tasks to be solved in the following works: development of a rational aerodynamic scheme and layout; the problem of the nanosatellite stability in the orbit.

IOP Conf. Series: Journal of Physics: Conf. Series **918** (2017) 012004 doi:10.1088/1742-6596/918/1/012004

References

[1] Gerasimov Yu V, Karetnikov G K, Selivanov A B and Fionov A S 2013 Assessment of the Relative Payload Mass of a Nanosatellite Launched into near-Earth Space with a Gun Launch and Corrective Engines *Herald of the Bauman Moscow State Technical University. Series Mechanical Engineering***3**

[2] McNabJ 2003 Launch to space with an electromagnetic railgun *IEEE transactions on magnetics***39** no.1

[3] BozicO, EggersT and Wiggen S2011 Aerothermal and flight mechanic considerations by development of small launchers for low orbit payloads started from lorentz rail accelerator*Progress in Propulsion Physics***2** 765-784

[4] Krasnov NF 1970*Aerodynamics of bodies of revolution*(New York)

[5] StandardISO 2533:1975 1975Standard Atmosphere

[6] Gerasimov YuV, Gerasimov MYu, Karetnikov GK, Selivanov AB and Fionov AS 2014A comparative assessment of the parameters of a nano-satellite launched into near-Earth Space with a pulse launch and corrective engines *Polet***3**

IOP Conf. Series: Journal of Physics: Conf. Series **918** (2017) 012005 doi:10.1088/1742-6596/918/1/012005

Approximate solution to a singular perturbed boundary value problem of thermal shielding

I I Latypov
Bashkir State University, 32, Validy str., 450076, Ufa, Russia
E-mail: LatypovII@rambler.ru

Abstract. The paper aims to investigate the problem of distribution of a non-regular, non-steady-state thermal field in the porous thermal shield material irradiated by a high flow of energy. A mathematical model of the original problem is stated in the form of a singular perturbed boundary value problem of a thermal conductivity equation with the nonlinear boundary conditions on moving boundaries. Its solution is obtained as asymptotic Poincare-type expansions in powers of small parameters.

1. Introduction

The current development of power engineering is characterized by the increase in the maximum temperatures and pressures of the heat conducting and the working media. In particular, the generating units are subjected to the temperatures far exceeding the levels specified for the materials of these structures. Under the circumstances, a cooling system is very important for a reliable operation of the heat-stressed parts.

One of the most promising cooling systems is porous cooling. It combines the best features of evaporative thermal shielding with advantages of heat accumulation systems. Evaporative thermal shielding comprises two processes: the process of internal heat exchange and the process of external heat exchange. During the internal heat exchange, the gas draws the heat away from the porous layer, filtering it to the external surface. During the external heat exchange, the cooled gas, flowing away from the wall, diffuses through the boundary layer diluting and pushing the high-temperature gas flow away from the surface. It is the second process that enables a higher productivity of the porous cooling in comparison with the heat accumulation systems [1].

This paper states and solves the problem of determining the non-regular non-steady-state thermal field of the porous thermal shield material irradiated by a laser with I intensity with the account for the radiation-convective heat losses and the temperature function of the material absorptivity [2].

2. The problem statement

According to [1], the thermal shielding composite material is assumed intact, so an energy conservation equation is written as

$$(\rho c)_\Sigma \frac{\partial U}{\partial \tau} + C_g G_g \frac{\partial U}{\partial \xi} + A(U) = \frac{\partial}{\partial \xi}\left(\lambda_\Sigma \frac{\partial U}{\partial \xi} \right) \qquad (1)$$

where Σ is an index denoting thermal physical properties of the following system: porous medium + gaseous component; C_g, G_g are thermal capacity and gaseous component mass flow; λ_Σ is the system's thermal conductivity; $A(U)$ is the system's absorptivity; U is the point $(\xi;\tau)$ temperature.

Let us assume that the thermal physical properties of the system are independent from both the temperature and the space-time coordinates. Then [3, 4], the original problem may be formulated as a singular perturbed boundary value problem of the thermal conductivity equation with the nonlinear boundary conditions:

$$\frac{\partial U(x,t)}{\partial t} = Fo \cdot \frac{\partial^2 U(x,t)}{\partial x^2} + k(t) \cdot \frac{\partial U(x,t)}{\partial x} + A\big(U(x,t)\big),$$

$$U(x,t) = u_0(x), \quad t \to 0,$$
$$\frac{\partial U(x,t)}{\partial x} = \gamma_1 \left[U(x,t) - u_{1c}(t)\right] - \sigma_1 \left[U^4(x,t) - u_{1c}^4(t)\right] + A_\upsilon \cdot I(t), \quad x = 0,$$
$$\frac{\partial U(x,t)}{\partial x} = \gamma_2 \cdot \left[U(x,t) - u_{2c}(t)\right] - \sigma_2 \left[U^4(x,t) - u_{2c}^4(t)\right], \qquad x = L,$$
$$(x,t) \in \Omega' = \{(x,t): 0 < x < L, 0 < t < t_0 < +\infty\},$$
$$Fo = a^2 \overline{t} / \overline{x}^2, \quad a = \lambda_\Sigma / (\rho c)_\Sigma, \quad k = C_g G_g / (\rho c)_\Sigma$$

where x, t are dimensionless coordinates; γ_1, γ_2, σ_1, σ_2 are constant coordinates; a is the system's thermal conductivity; $\overline{t}$, $\overline{x}$ are, correspondingly, time and space scales; $u_0(x)$ is the initial temperature distribution in the porous material; u_{1c}, u_{1c} are the temperatures of the working medium and the gas coolant; A_v is the amount of radiation energy emitted per unit time, per unit volume; F_0 is Fourier series expansion parameter: $Fo \ll 1$. The smallness of Fo is ensured by the low thermal conductivity of the structural material, a massive body or a short time interval.

In this paper the laser pulse intensity is approximated using Gaussian function

$$I(t) = I_0 \cdot \exp\left\{ -\frac{(t-t_1)^2}{t_{10}^2} \right\}, \quad I_0,\ t_1,\ t_{10} = const,$$

and the absorptivity of the material is approximated using the following type function:

$$A(U) = A_0 + A_1 \cdot U(x,t), \quad A_i = const, \ i = 0,1.$$

Let us convert the original boundary value problem to an equivalent problem with moving boundaries [5-10]. To do this, we transpose the variables $x = \xi - \int_0^\tau k(s)ds$, $t = \tau$. Let us introduce the notation: $U(x,t) = T(\xi,\tau)$, $u_0(x) = T_0(\xi)$, $\varphi_1(\tau) = \int_0^\tau k(s)ds$, $\varphi_2(\tau) = L + \varphi_1(\tau)$. For the purposes of uniformity instead of ξ, τ let us write x and t, respectively. Then we obtain

$$\frac{\partial T(x,t)}{\partial t} = Fo \cdot \frac{\partial^2 T(x,t)}{\partial x^2} + A\left[T(x,t)\right], \tag{2}$$
$$T(x,t) = T_0(x), \qquad t \to 0, \tag{3}$$
$$\frac{\partial T(x,t)}{\partial x} = \gamma_1 \cdot \left[T(x,t) - u_{1c}(t)\right] - \sigma_1 \cdot \left[T^4(x,t) - u_{1c}^4(t)\right] + A_\upsilon \cdot I(t), \quad x = \varphi_1(\tau), \tag{4}$$
$$\frac{\partial T(x,t)}{\partial x} = \gamma_2 \cdot \left[T(x,t) - u_{2c}(t)\right] - \sigma_2 \left[T^4(x,t) - u_{2c}^4(t)\right], \quad x = \varphi_2(\tau), \tag{5}$$
$$(x,t) \in \Omega = \{ (x,t): \varphi_1(\tau) < x < \varphi_2(\tau), 0 < t < t_0 < +\infty\}.$$

A physical model is a plate with L thickness and T_0 initial temperature. The material thermal physical parameters are assumed constant. The plate surface ($x = \varphi_1(\tau)$) is exposed to a variable heat flux. As the surface ($x = \varphi_1(\tau)$) reaches the melting point (T_m), melting begins. At that, the liquid phase is assumed to disappear directly. On the surfaces ($x = \varphi_1(\tau)$) and ($x = \varphi_2(\tau)$) the radiative-convective heat transfer is maintained [2,3]. In practice, the temperature dependence of materials' absorbing capacity is approximated using the following type function: $A(T(x,t)) = A_0 + A_1 \cdot T(x,t)$.

As a result, the original boundary value problem is reduced to a singular perturbed boundary value problem of the thermal conductivity equation with nonlinear boundary conditions on moving boundaries.

Let us write down the integral expression of solution to the boundary value problem

$$T(x,t)=\int_0^L T(y,0)\cdot\Gamma\left(x,t;y,0\right)dy+\gamma_1 Fo\int_0^t u_{1c}(s)\cdot\Gamma\left(x,t;\ \varphi_1(s),s\right)ds+$$

$$+A_\upsilon Fo\int_0^t I(s)\cdot\Gamma\left(x,t;\ \varphi_1(s),s\right)ds+Fo\int_0^t \nu_1(s)\cdot\Gamma\left(x,t;\ \varphi_1(s),s\right)ds-$$

$$-\gamma_2 Fo\int_0^t u_{2c}(s)\Gamma\left(x,t;\ \varphi_2(\tau),s\right)ds-Fo\int_0^t \nu_2(s)\Gamma\left(x,t;\ \varphi_2(s),s\right)ds+\int_0^t\int_{\varphi_1(s)}^{\varphi_2(s)} A(T)\Gamma\left(x,t;y,s\right)dyds=$$

$$=T_{oL}(x,t)+T_{1L}(x,t)+T_I(x,t)+T_{1nL}(x,t)+T_{2L}(x,t)+T_{2nL}(x,t)+T_A(x,t)=$$

$$=T_L(x,t)+T_I(x,t)+T_{1nL}(x,t)+T_{2nL}(x,t)+T_A(x,t)=T^*(x,t)+T_A(x,t) \qquad (6)$$

where Γ $(x,\ t;\ y,\ s)$ is a Green's function of the corresponding homogeneous boundary value problem for Ω region; $T_L(x,\ t)$ is the solution to the linear boundary value problem.

We will obtain the Green's function Γ $(x,\ t;\ y,\ s)$ and the approximate solution of the boundary value problem (6) using a "geometrical optics" asymptotic technique proposed by Nenesenko [5, 11] as asymptotic Poincare-type expansions in powers of small parameters [4-12].

The solution algorithm of the boundary value problem (2) - (5) is as follows:

- find the Green's function corresponding to the homogeneous boundary value problem in the region with the moving boundaries [5, 7, 11];
- transpose the obtained expression of the Green's function (or its asymptotic expansion) into the integral expression, and find the approximate solution $\overline{T}_L(x,t)=T_L(x,t)+T_I(x,t)$ of the linear boundary value problem [5-7] by;
- find the unknown functions $\nu_i(s),\ i=1,2$ using the boundary conditions (4) - (5), whereupon the problem is reduced to solving a system of nonlinear integral equations [5, 7, 11];
- transpose the obtained asymptotic expansions of functions $\nu_i(s),\ i=1,2$ into the integrals $T_{1nL}(x,t),\ T_{2nL}(x,t)$, analyze these integrals asymptotically [7, 12], and find the approximate solution $T^*(x,t)$ of nonlinear boundary value problem;
- use $T^*(x,t)$ solution to obtain the approximate solution $T(x,t)$ of the original nonlinear boundary value problem.

3. The solution of the boundary value problem

To simplify the calculations let us assume

$$T_0(x)=T_0=const,\ u_{ic}=const,\ i=1,2;\ k(t)=k=const>0. \qquad (7)$$

Then, the Green's function of the homogeneous boundary value problem can be obtained in the following form [7, 9, 11]

$$\Gamma\left(x,t;y,s\right)=G_0\left(x,t;y,s\right)+\sum_{m=0}^{\infty}\sum_{i=1}^{4}\exp\left\{-\frac{\left[h_{i0}^{(m)}\right]^2}{4Fo\left(t-s\right)}+\frac{R_i^{(m)}}{Fo}\right\}\left[\frac{1}{2\sqrt{\pi Fo\left(t-s\right)}}+\right.$$

$$\left.+\sum_{j=1}^{2}\exp\left\{\frac{\left[h_{ij}^{(m)}\right]^2}{4Fo\left(t-s\right)}\right\}\left\{p_{i,2j-1}^{(m)}erfc\left[\frac{h_{ij}^{(m)}}{2\sqrt{Fo\left(t-s\right)}}\right]+\overline{p}_{i,2j}^{(m)}ierfc\left[\frac{h_{ij}^{(m)}}{2\sqrt{Fo\left(t-s\right)}}\right]\right\}\right],$$

where $G_0\left(x,t;y,s\right)=\dfrac{1}{2\sqrt{\pi Fo\left(t-s\right)}}\exp\left\{-\dfrac{\left[x-y\right]^2}{4Fo\left(t-s\right)}\right\}$, $h_{ij}^{(m)}=h_{i0}^{(m)}+\left[\left(-1\right)^i k-\left(-1\right)^j a_j\right]\left(t-s\right)$,

$$\left[\begin{array}{l} h_{10}^{(m)}=C+D+2mL, \\ h_{20}^{(m)}=B+F+2mL, \end{array}\right. \left[\begin{array}{l} h_{30}^{(m)}=B+D+\left(2m+1\right)L, \\ h_{40}^{(m)}=C+F+\left(2m+1\right)L, \end{array}\right. \left[\begin{array}{l} F=kt+L-x,\ D=x-kt, \\ B=ks+L-y,\ C=y-ks, \end{array}\right.$$

$$\left[\begin{array}{l} R_1^{(m)}=k\left[C+mL\right],\ R_2^{(m)}=-k\left[B+mL\right], \\ R_3^{(m)}=k\left(m+1\right)L,\ R_4^{(m)}=-k\left(m+1\right)L, \end{array}\right. \left[\begin{array}{l} a_j=k+2Fo\cdot\gamma_j,\ a=a_2-a_1=-2Fo\cdot b, \\ b=\gamma_1-\gamma_2,\ \overline{p}_{i,2j}^{(m)}=p_{i,2j}^{(m)}\sqrt{\dfrac{t-s}{Fo}},\ j=1,2, \end{array}\right.$$

$$m=0:\left[\begin{array}{l} p_{11}^{(0)}=-\dfrac{a_1}{2Fo}, p_{23}^{(0)}=\dfrac{a_2}{2Fo},\ p_{12}^{(0)}=p_{14}^{(0)}=p_{21}^{(0)}=p_{22}^{(0)}=p_{24}^{(0)}=0, \\ p_{41}^{(0)}=p_{31}^{(0)}=-\dfrac{a_1 b}{a_1+a_2},\ p_{43}^{(0)}=p_{33}^{(0)}=-\dfrac{a_2 b}{a_1+a_2},\ p_{32}^{(0)}=p_{34}^{(0)}=p_{42}^{(0)}=p_{44}^{(0)}=0, \end{array}\right.$$

$$m=1:\left[\begin{array}{l} p_{11}^{(1)}=\dfrac{aba_2}{\left(a_1+a_2\right)^2}-\dfrac{a_1-a}{2Fo},\ p_{12}^{(1)}=\dfrac{2ba_1^2}{a_1+a_2},\ p_{13}^{(1)}=-\dfrac{aba_2}{\left(a_1+a_2\right)^2}, \\ p_{21}^{(1)}=\dfrac{aba_1}{\left(a_1+a_2\right)^2},\ p_{22}^{(1)}=-\dfrac{2ba_2^2}{a_1+a_2},\ p_{23}^{(1)}=-\dfrac{aba_1}{\left(a_1+a_2\right)^2}+\dfrac{a+a_2}{2Fo}, \end{array}\right.$$

$$p_{31}^{(1)}=2b\left(-1+\dfrac{3a_2^2\left(a_1+a_2\right)+2a_1a_2\left(a+a_2\right)}{\left(a_1+a_2\right)^3}\right),\ p_{32}^{(1)}=p_{44}^{(1)}=\dfrac{-2aba_1^2}{\left(a_1+a_2\right)^2},$$

$$p_{34}^{(1)}=p_{42}^{(1)}=\dfrac{-2aba_2^2}{\left(a_1+a_2\right)^2},$$

$$p_{33}^{(1)}=\dfrac{ba_2}{\left(a_1+a_2\right)}\left(-3a_2+\dfrac{\left(2a^2-a_1a_2\right)\left(a_1+a_2\right)-2aa_2}{\left(a_1+a_2\right)^2}\right),\ p_{41}^{(1)}=\dfrac{2ba_1}{\left(a_1+a_2\right)^2}\left(a-\dfrac{4a_1a_2}{\left(a_1+a_2\right)}\right),$$

$$p_{43}^{(1)}=2b\left(-1+\dfrac{3a_1^2\left(a_1+a_2\right)-2a_1a_2\left(a-a_2\right)}{\left(a_1+a_2\right)^3}\right).$$

Transposing the Green's function into expression (6), we obtain the solution of the boundary value linear problem $T_l(x,t)$ in the form of

$$\overline{T}_L(x,t)=T_L(x,t)+T_l(x,t),$$

$$T_L(x,t)\,\square\,T_0+\sum_{i=1}^{2}\frac{\left[u_{ic}-T_0\right]}{2}\cdot B_{i0}(x,t);\quad T_{iL}(x,t)\,\square\,\frac{u_{ic}}{2}\cdot B_{i0}(x,t),\ Fo\to 0, \tag{8}$$

$$B_{i0}(x,t)\,\square\,erfc\left[\frac{(i-1)L+(-1)^{i+1}\cdot x}{2\sqrt{Fo\cdot t}}\right]+\exp\left\{-\frac{k[x-kt]}{Fo}+(i-1)\frac{kL}{Fo}\right\}\times$$

$$\times\left[\frac{a_i-k}{a_i+k}erfc\left[\frac{(i-1)L-(-1)^i(x-2kt)}{2\sqrt{Fo\cdot t}}\right]-\frac{2a_i}{a_i+k}\exp\left\{\frac{(a_i+k)(x-kt)}{2Fo}\right\}\times\right.$$

$$\times\exp\left\{\frac{(a_i+k)\left[Fo\cdot t\gamma_i-(i-1)L\right]}{2Fo}\right\}erfc\left[\frac{(i-1)L+(-1)^{i+1}\cdot(x+2Fo\cdot t\gamma_i)}{2\sqrt{Fo\cdot t}}\right]\Bigg]+$$

$$+\frac{4a_i^2(a_i-k)}{(a_1+a_2)(a_{3-i}^2-k^2)}\cdot\exp\left\{\frac{a_{3-i}+k}{2Fo}\cdot\left[x-kt+Fo\gamma_{3-i}+(-1)^i(3-i)L\right]\right\}\times$$

$$\times\exp\left\{-\frac{k[x-kt]}{Fo}-\frac{kL}{Fo}(i-2)\right\}erfc\left[\frac{(3-i)L+(-1)^i(x+2Fo\cdot t\gamma_{3-i})}{2\sqrt{Fo\cdot t}}\right],\quad Fo\to 0;$$

$$T_I(x,t)=A_\upsilon Fo\int_0^t I(s)\cdot\Gamma\left(x,t;\ \varphi_1(s),s\right)ds \,\square\, A_\upsilon FoI_0t_{10}\frac{\sqrt{\pi}}{2}\left[erfc\left[\frac{t_1-t}{t_{10}}\right]-erfc\left[\frac{t_1}{t_{10}}\right]\right]+$$

$$+A_\upsilon FoI_0\exp\left\{\frac{t^2-t_1^2}{t_{10}^2}\right\}\left\{\frac{k-a_1}{k+a_1}\left[\exp\left\{-\frac{k\left[x-kt\right]}{Fo}\right\}erfc\left[\frac{x-2kt}{2\sqrt{Fot}}\right]-erfc\left[\frac{L+x}{2\sqrt{Fot}}\right]\right]-\right. \tag{9}$$

$$\left.-erfc\left[\frac{L-x}{2\sqrt{Fot}}\right]-erfc\left[\frac{x}{2\sqrt{Fot}}\right]\right\}\times F(t),\quad Fo\to 0,$$

$$F(t)=\begin{cases}\dfrac{t}{2}\left[-1+\dfrac{1}{3}\left(\dfrac{t}{t_{10}}\right)^2-\dfrac{1}{5}\left(\dfrac{t}{t_{10}}\right)^4+\dfrac{1}{7}\left(\dfrac{t}{t_{10}}\right)^6-\ldots\right], & t=t_1\\[2ex] \dfrac{t_{10}^2}{8(t-t_1)}\left[\dfrac{2k}{k-a_1}\left(1-\dfrac{1}{\sqrt{1-g}}\right)-2\right],\quad g=\dfrac{8Fo(t-t_{10})}{k^2t_{10}^2}, & t\neq t_1\end{cases}$$

We obtain

$$T^*(x,t)=\overline{T}_L(x,t)+Fo\int_0^t \nu_1(s)\cdot\Gamma\left(x,t;k\cdot s,s\right)ds+Fo\int_0^t \nu_2(s)\cdot\Gamma\left(x,t;k\cdot s+L,s\right)ds, \tag{10}$$

where $\nu_i(t)$ is the solution of the system of nonlinear integral equations

$$\nu_i(t)=(-1)^{i+1}\cdot\sigma_i\cdot\left[\left[\overline{T}_L(x,t)+Fo\int_0^t \nu_i(s)\cdot\Gamma\left(x,t;\varphi_i(s),s\right)ds\right]^4-u_{ic}^4\right],\quad x=\varphi_i(t),\ i=1,2. \tag{11}$$

Let us suppose that $k\sqrt{t}=O\left(Fo^{q_1}\right),\ 0<q_1<\frac{1}{2},\ Fo\to 0,$ then for $x=\varphi_i(t),\ i=1,2$ we have

$$T_L(x,t)\Big|_{x=\varphi_i(t)}\,\square\,\frac{T_0k+u_{ic}\gamma_iFo}{k+Fo\gamma_i}=b_{i0}\,\square\,T_0,\quad \int_0^t\Gamma\left(x,t;\varphi_i(s),s\right)ds\,\square\,\frac{1}{k+Fo\gamma_i}\,\square\,\frac{1}{k},\quad Fo\to 0,$$

$$T_I\left(x,t\right)\Big|_{x=\varphi_1(t)}\,\square\,\frac{Fo}{k}A_9I_0\exp\left\{-\frac{\left[t-t_1\right]^2}{t_{10}^2}\right\},\quad b_1=\frac{A_9I_0}{k},\quad Fo\to 0,$$

and the solution to the system of integral equations can be written as asymptotic series in powers of F_0 small parameter:

$$\nu_i(t)=\sum_{j=0}^{n}p_{ij}\cdot Fo^j+O\left(Fo^{n+1}\right),\quad Fo\to 0,\ i=1,2, \tag{12}$$

where $p_{i0}=\beta_i$, $\beta_i=\sigma_i\cdot\left[b_{i0}^4-u_{ic}^4\right]$, $p_{i1}=\frac{2\sigma_i b_{i0}^3}{k+Fo\gamma_i}\beta_i+(2-i)\frac{4b_{i0}^3}{k}\sigma_i A_g I_0\exp\left\{-\frac{\left[t-t_1\right]^2}{t_{10}^2}\right\}$,

$$p_{i2}=\frac{14\sigma_i b_{i0}^6}{(k+Fo\gamma_i)^2}\beta_i^2+(2-i)\left(\frac{16b_{i0}^6}{k}\sigma_i^2 b_1\exp\left\{-\frac{\left[t-t_1\right]^2}{t_{10}^2}\right\}+b_{i0}^2\sigma_i^2 b_1^2\exp\left\{-\frac{2\left[t-t_1\right]^2}{t_{10}^2}\right\}\right).$$

Using the obtained forms, the solution $T^*(x,t)$ will be written as

$$T^*(x,t)\approx\overline{T}_L(x,t)+\sum_{i=1}^{2}\frac{T_{iL}(x,t)}{\gamma_i u_{ic}}\left[\sum_{j=0}^{n}p_{ij}\cdot Fo^j\right],\ Fo\to 0 \tag{13}$$

The solution to the sought-for boundary value problem is determined by the convergence method as follows

$$T^{(m+1)}(x,t)=T^*(x,t)+\int_0^t\int_{\varphi_1(s)}^{\varphi_2(s)}A\left(T^{(m)}(y,s)\right)\Gamma\left(x,t;y,s\right)dyds,\ T^{(0)}=T^*(x,t),m=0,1,2\cdots. \tag{14}$$

$$T^{(1)}(x,t)=T^{(0)}(x,t)+\int_0^t\int_{\varphi_1(s)}^{\varphi_2(s)}\left(A_0+A_1T^{(0)}(y,s)\right)\Gamma\left(x,t;y,s\right)dyds=T^{(0)}(x,t)+T_{A_0}(x,t)+T_{A_1}(x,t)$$

,

$$T^{(0)}=T^*(x,t),$$

$$T_{A_0}(x,t)\approx A_0t+A_0\sum_{i=1}^{2}C_i,$$

$$C_i\approx\left(\frac{[i-1]L-x}{2k}+\frac{Fo}{k[k-a_i]}\right)erfc\left[\frac{[i-1]L+(-1)^{i+1}x}{2\sqrt{Fot}}\right]+$$

$$+\left(\frac{[1-i]L+x-2kt}{2k}\frac{a_i-k}{a_i+k}+\frac{Fo}{k}\frac{a_i-k}{[k+a_i]^2}\right)\exp\left\{-\frac{k\left[x-kt-(i-1)L\right]}{Fo}\right\}\times$$

$$\times erfc\left[(-1)^{i+1}\frac{x-2kt-(i-1)L}{2\sqrt{Fot}}\right]+\frac{4Fo}{a_i+k}\frac{a_i}{\left[a_i^2-k^2\right]}\exp\left\{\frac{\left[x-kt+a_it+(1-i)L\right]^2}{4Fot}\right\}\times$$

$$\times\exp\left\{-\frac{\left[x+(1-i)L\right]^2}{4Fot}\right\}erfc\left[(-1)^{i+1}\frac{x-kt+a_it+(1-i)L}{2\sqrt{Fot}}\right]+$$

$$+\left((-1)^{i+1}\frac{2}{a_i+k}\sqrt{\frac{Fot}{\pi}}\right)\exp\left\{-\frac{\left[x+(1-i)L\right]^2}{4Fot}\right\};$$

$$T_{A_0}(x,t)\approx A_0t+\sum_{i=0}^{1}\sum_{p=0}^{n}\sum_{j=0}^{m}c_{pj}^{(i)}(Fo)^p\left(\frac{iL+(-1)^i[x-kt]}{Fo}\right)^j+O\left((Fo)^{n+1}\left(\frac{iL+(-1)^i[x-kt]}{Fo}\right)^{m+1}\right)$$

;

$$T_{A_1}(x,t) \square \sum_{i=0}^{1}\sum_{p=0}^{n}\sum_{j=0}^{m} d_{pj}^{(i)}(Fo)^p\left(\frac{iL+(-1)^i[x-kt]}{Fo}\right)^j + O\left((Fo)^{n+1}\left(\frac{iL+(-1)^i[x-kt]}{Fo}\right)^{m+1}\right);$$

$$T^{(1)}(x,t) = T^*(x,t) + A_0 t +$$

$$+\sum_{i=0}^{1}\sum_{p=0}^{n}\sum_{j=0}^{m} \bar{d}_{pj}^{(i)}(t)(Fo)^p\left(\frac{iL+(-1)^i[x-kt]}{Fo}\right)^j + O\left((Fo)^{n+1}\left(\frac{iL+(-1)^i[x-kt]}{Fo}\right)^{m+1}\right),$$

$$\bar{d}_{pj}^{(i)}(t) = c_{pj}^{(i)}(t) + d_{pj}^{(i)}(t).$$

Thus, for asymptotic form of the temperature field of the original boundary value problem, the following statement is true.

Theorem.
Under the assumptions (7), (10), depending on "proximity" of the point (x, t) to the boundaries $\varphi_1(t)$, $\varphi_2(t)$, the asymptotic expansion of the solution to the non-steady singular perturbed boundary value problem (2)-(5) takes the following form:

1. In the "*boundary layer*" of the border $x = \varphi_1(t)$, if the following conditions are fulfilled: $x-\varphi_1(t) = O(Fo^q)$, $q>1$, $Fo \to 0$,

$$T(x,t) = \sum_{p=0}^{n}\sum_{j=0}^{m} a_{pj}^{(1)}(x,t) Fo^p \left(\frac{x-\varphi_1(t)}{Fo}\right)^j + O\left(Fo^{n+1}\left(\frac{x-\varphi_1(t)}{Fo}\right)^{m+1}\right),\ Fo \to 0. \quad (15)$$

2. In the "*intermediate layer*" of the border $x = \varphi_1(t)$, if the following conditions are fulfilled: $x-\varphi_1(t) = O(Fo)$ $x-\varphi_2(t) = O(Fo^{q_1})$, $q_1 < 1$, $Fo \to 0$,

$$T(x,t) = \sum_{p=0}^{\bar{n}} b_p^{(1)}(x,t)\cdot Fo^p + O\left(Fo^{\bar{n}+1}\right),\ Fo \to 0. \quad (16)$$

3. In the "*region of the points far away from the boundaries* $x = \varphi_1(t)$, $x = \varphi_2(t)$", where the following condition is fulfilled: $x-\varphi_i(t) = O(Fo^{q_2})$, $q_2 < 1$, $i = 1,2$; $Fo \to 0$,

$$T(x,t) = T_0 + A_0\cdot t,\ Fo \to 0. \quad (17)$$

4. In the "*intermediate layer*" of the border $x = \varphi_2(t)$, if the following conditions are fulfilled: $x-\varphi_2(t) = O(Fo)$, $x-\varphi_1(t) = O(Fo^{q_3})$, $q_3 < 1$, $Fo \to 0$,

$$T(x,t) = \sum_{p=0}^{\bar{n}} b_p^{(2)}(x,t)\cdot Fo^p + O\left(Fo^{\bar{n}+1}\right),\ Fo \to 0. \quad (18)$$

5. In the "*boundary layer*" of the border $x = \varphi_2(t)$, if the following condition is fulfilled: $x-\varphi_2(t) = O(Fo^{q_4})$, $q_4 > 1$, $Fo \to 0$

$$T(x,t) = \sum_{p=0}^{n}\sum_{j=0}^{m} a_{pj}^{(2)}(x,t) Fo^p \left(\frac{x-\varphi_2(t)}{Fo}\right)^j + O\left(Fo^{n+1}\left(\frac{x-\varphi_2(t)}{Fo}\right)^{m+1}\right),\ Fo \to 0. \quad (19)$$

where the coefficients $a_{pj}^{(i)}(x,t)$, $b_p^{(i)}(x,t)$ are defined in an explicit form. They are independent from the small parameters of the expansion. That is, we have the asymptotic Poincare-type expansion.

The resulting expansions (15) - (19) are in accord with the original and boundary conditions of the boundary value problem and correlate with one another.

Conclusion

The proposed approximate solution of the sought-for boundary value problem enables making numerical calculation and estimating the solution error; assessing quantitatively and qualitatively the contribution of key parameters into the approximate solution and obtaining a dynamic pattern of temperature distribution in the material.

References

[1] Polezhaev Y V and Yurevich F B 1976 *Teplovaya zashchita.* (Moscow: Energia) p 392
[2] Prokhorov A M, Konov V I, Ursu I and Miheilesku I N 1988 *Vzaimodeystvie lazernogo izlucheniya s metallami* (Moscow: Nauka)
[3] Uglov A A, Smurov I Y et al. 1991 *Modelirovanie teplofizicheskikh protsessov impul'snogo lazernogo vozdeystviya na metally* (Moscow: Nauka)
[4] Nesenenko G A, Latypov I I and Naselski S P 1993 *Nelineynye kraevye zadachi mat. fiziki i ikh prilozheniya* pp 94-5.
[5] Kravchenko V F, Nesenenko G A and Latypov I I 1999 *Differentsial'nye uravneniya* **35** Iss. 9 pp 1184-92.
[6] Latypov I I 1999 *Priblizhennoe reshenie singulyarno vozmushchennykh kraevykh zadach nestatsionarnoy teploprovodnosti* PhD thesis abstract (Moscow) p 25
[7] Latypov I I 1999 *Priblizhennoe reshenie singulyarno vozmushchennykh kraevykh zadach nestatsionarnoy teploprovodnosti.* PhD thesis (Russia, Birsk) p 178.
[8] Latypov I I and Nesenenko G A *1996 Teplomassoobmen – MMF – 96. Radiatsionnyy i kombinirovannyy teploobmen* vol 2 (Minsk: ANK "ITMO im. A. V. Lykova") pp 167 – 71
[9] Latypov I I 1997 *Proc. Int. Conf on Modelling and Investigation of system stability. (Kiev)* p 68
[10] Latypov I I 2006 *Trudy chetver-toy Rossiyskoy natsional'noy konferentsii po teploobmenu* vol 5 (Moscow: Izdatel'skiy dom MEI) pp 138-42.
[11] Nesenenko G A, Pustovoit V I and Kravchenko V F 2006 *Asimptotika Poincaré resheniy zadach neregulyarnogo teplo- i massoperenosa* (Moscow: Fizmatlit) p 419
[12] Latypov I I, Shakirov R A and Ulitin N V 2014 *Bulletin of Kazan University of Technology* **17**, № 5. pp 80-6.

IOP Conf. Series: Journal of Physics: Conf. Series **918** (2017) 012006 doi:10.1088/1742-6596/918/1/012006

Approximation of mutual information in a bipartite quantum state under single-party decoherence

S V Rozanov[1], **E O Kiktenko**[1,2,3]

[1] Bauman Moscow State Technical University, 5 2nd Baumanskaya St., Moscow 105005, Russia

[2] Steklov Mathematical Institute of Russian Academy of Sciences, 8 Gubkina St., Moscow 119991, Russia

[3] Geoelectromagnetic Research Center of Schmidt Institute of Physics of the Earth, Russian Academy of Sciences, P.O. 30 Troitsk, Moscow Region 142190, Russia

E-mail: evgeniy.kiktenko@gmail.com

Abstract. We consider the problem of approximate calculation of quantum mutual information in a bipartite quantum state where one of subsystems goes through a decoherence process (quantum channel). The approximation employs a reduced density matrix of the subsystem affected by decoherence and the initial value of mutual information before decoherence. We test the approximation on sets of randomly generated qubit-qudit mixed states for three types of qubit decoherence channels: dephasing, depolarizing, and damping. We consider how approximation accuracy is related to decoherence type, decoherence strength, and the dimension of the qudit subsystem.

Decoherence, that is, an uncontrolled interaction of a quantum system with its environment [1, 2], becomes one of the main obstacles to practical realization of quantum information technologies. Study of decoherence includes different problems such as quantitative characterisation [3] and finding efficient ways for protecting quantum states [4, 5].

In this paper we consider decoherence of a subsystem within a bipartite system. In this case, decoherence usually leads to destruction of correlation between subsystems, which is a vital resource for realization of quantum computation and communication. In particular, we consider an approach to a simplified calculation of quantum mutual information, that is, a measure of total correlation (classic and quantum) between subsystems.

Consider an arbitrary finite dimensional quantum state of a bipartite system *AB*, given by its density matrix ρ_{AB} . Let us introduce a decoherence process of the subsystem *A* described by a completely positive trace-preserving map [6] Φ acting in the state space of the subsystem *B*.

Then the state of the whole system after decoherence takes the form

$$\rho_{AB}^{\mathrm{dec}A} = (\Phi \otimes \mathrm{Id})[\rho_{AB}], \quad (1)$$

where Id is an identical channel acting in the state space of the subsystem *B*.

Next, we introduce the definition of quantum mutual information for an arbitrary bipartite state σAB :

$$I[\sigma_{AB}] = S[\sigma_A] + S[\sigma_B] - S[\sigma_{AB}], \quad (2)$$

where $\sigma_A = \mathrm{Tr}_B \sigma_{AB}$ and $\sigma_B = \mathrm{Tr}_A \sigma_{AB}$ are corresponding reduced density matrices of the subsystems *A* and *B* of the state σAB , and

$$S[\sigma] = -\mathrm{Tr}[\sigma \log \sigma] \quad (3)$$

is von Neumann entropy.

To obtain an approximate expression for calculating the quantity $I_{AB}^{\text{dec}A} \equiv I\left[\rho_{AB}^{\text{dec}A}\right]$, that is, mutual information after the decoherence process, we employ a spectral decomposition of the density matrix $\rho_A = \text{Tr}_B \rho_{AB}$ of the initial state ρ_A :

$$\rho_A = \sum_i \lambda_i |\psi_i\rangle_A \langle\psi_i|, \tag{4}$$

where $|\psi_i\rangle_A$ and λ_i denote eigenvectors and eigenvalues of the matrix ρ_A. Next, we employ a purification of the state ρ_A given by

$$|\Psi\rangle_{AR} = \sum_i \sqrt{\lambda_i} |\psi_i\rangle_A \otimes |\psi_i\rangle_R, \tag{5}$$

where R is a reference subsystem with a state space identical to the one of A. It is easy to see that $\text{Tr}_R |\Psi\rangle_{AR}\langle\Psi| = \rho_A$.

Then, we can consider the state realized after the decoherence of the subsystem of the pure state $|\Psi\rangle AR$:

$$\rho_{AR}^{\text{dec}A} = (\Phi \otimes \text{Id})\left[|\Psi\rangle_{AR}\langle\Psi|\right]. \tag{6}$$

We denote its mutual information as $I_{AR}^{\text{dec}A} = I\left[\rho_{AR}^{\text{dec}A}\right]$. Finally, we consider the following hypothesis

$$I_{AB}^{\text{dec}A} / I_{AB} \approx I_{AR}^{\text{dec}A} / I_{AR}, \tag{7}$$

where $I_{AB} \equiv I\left[\rho_{AB}\right]$ and $I_{AR} \equiv I\left[|\Psi\rangle_{AR}\langle\Psi|\right]$ are initial values of the mutual information in the states of AB and AR before decoherence. The intuition behind (7) is that a decoherence process acting on A destroys approximately the same portion of the initial mutual information both in ρ_{AB} and the purified version $|\Psi\rangle_{AR}$.

The purity $|\Psi\rangle_{AR}\langle\Psi|$ of the state implies that its mutual information is given by $I_{AR} = 2S_A$, where $S_A \equiv S\left[\rho_A\right]$. Employing the hypothesis (7), we can introduce an approximation of mutual information after decoherence $I_{AB}^{\text{dec}A}$ in the following way:

$$\tilde{I}_{AB}^{\text{dec}A} = \frac{I_{AB}}{2S_A} I_{AR}^{\text{dec}A}.. \tag{8}$$

We note that the value of $\tilde{I}_{AB}^{\text{dec}A}$ is actually defined by the initial mutual information I_{AB}, the reduced density matrix of the subsystem affected by decoherence, and the particular decoherence channel. As a quantitative measure of the approximation (8) accuracy we introduce the relative deviation

$$\varepsilon = \frac{I_{AB}^{\text{dec}A} - \tilde{I}_{AB}^{\text{dec}A}}{I_{AB}^{\text{dec}A}}. \tag{9}$$

Next, we consider the approximation (8) in the cases where the particle is a qubit and is affected by one of the following channels [7, 8]:

(i) the dephasing channel

$$\Phi^{\text{deph}}(p)[\rho] = (1-p)\rho + p(|0\rangle\langle 0|\rho|0\rangle\langle 0| + |1\rangle\langle 1|\rho|1\rangle\langle 1|), \tag{10}$$

(ii) the depolarizing channel

$$\Phi^{\text{depol}}(p)[\rho] = (1-p)\rho + p/4\sum_{i=0}^{3}\sigma_i\rho\sigma_i^\dagger \tag{11}$$

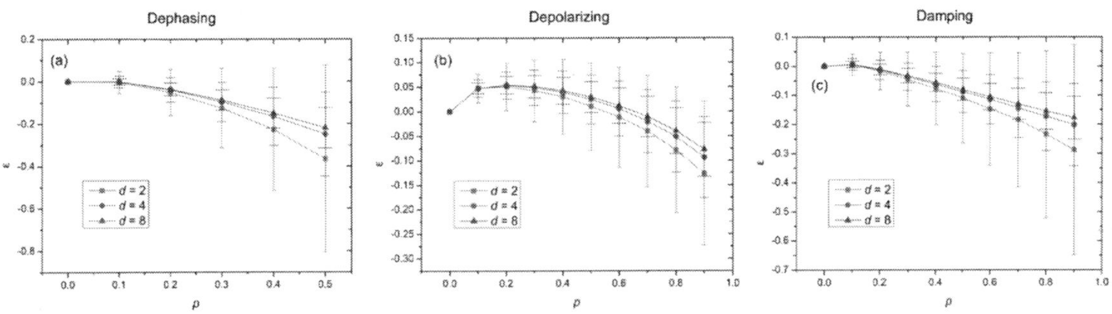

Figure 1. Mean values of relative approximation accuracy ε for randomly generated qubit-qudit mixed states and different types of qubit decoherence: dephasing (a), depolarization (b), and damping (c). "Error bars" stand for standard deviation, d denotes qudit's dimension, and p stands for the strength of decoherence process. Each point is obtained from $N = 10^4$ random states.

(iii) the damping channel

$$\Phi^{\text{damp}}(p)[\rho] = A_1 \rho A_1^{\dagger} + A_2 \rho A_2^{\dagger}, \quad A_1 = |0\rangle\langle 0| + \sqrt{1-p}\,|1\rangle\langle 1|, \quad A_2 = \sqrt{p}\,|1\rangle\langle 0| \qquad (12)$$

Here $\{\sigma_i\}_{i=0}^{3}$ stands for standard Pauli matrices together with the identity matrix, and $p \in [0,1]$ defines the strength of the decoherence process.

To study the accuracy of the approximation (8) we consider randomly and uniformly (according to the Haar measure) generated [9] mixed states ρ_{AB}, where A is a qubit, and B is a qudit of the dimension d. Fig. 1 presents the results of our numerical simulation.

Let us consider the basic patterns that are true for all three decoherence types. Firstly, the standard deviation around the mean value grows with the decoherence strength p. Thereby the approximation proves to be valuable for relatively weak decoherence. Secondly, the accuracy of approximation increases when the dimension of the second subsystem grows. Thirdly, the approximation considered is characterised by a bias which depends on the particular decoherence type: for the dephasing and damping channels usually $\tilde{I}_{AB}^{\text{dec}A} > I_{AB}^{\text{dec}A}$, which yields negative mean values of ε.

Comparing the results of approximation for different decoherence models, we see that the most accurate results are obtained for the depolarizing channel. The worst accuracy appears for the dephasing channel, where the relative deviation could exceed 100% for $p > 0.5$.

To conclude, we point out the main results of our work. We considered an approximation (8) for mutual information in a bipartite state, where one subsystem is affected by a decoherence channel. We obtain an approximate value using the initial mutual information value, the initial density matrix of the subsystem affected by decoherence, and the particular form of the decoherence process. We implemented our approximation for randomly generated sets of states and found out that (i) it has a bias that depends on the decoherence process, (ii) the approximation accuracy decreases with increasing decoherence strength, and (iii) the approximation accuracy increases when the second subsystem (not affected by decoherence) dimension grows. We expect that this method for approximate calculation of mutual information can be useful in assessing the possibilities of using unitary rotation operations over a system for protecting quantum correlations.

Acknowledgments
The work of E.K. was supported by the grant of the President of the Russian Federation (project MK-2815.2017.1).

References

[1] Menskii M B 1998 *Phys.-Usp.* **41** 923.

IOP Conf. Series: Journal of Physics: Conf. Series **918** (2017) 012006 doi:10.1088/1742-6596/918/1/012006

[2] Zurek W H 2003 *Rev. Mod. Phys.* **75** 715.
[3] Fedichkin L, Fedorov A and Privman V 2003 *Proc. SPIE* vol 5105 p 243. [4] Wang S-C et al 2014 *Phys. Rev. A* **89** 022318.
[4] Qin W, Wang C and Zhang X 2015 *Phys. Rev.* A **91** 042303.
[5] Holevo A 2012 *Quantum Systems, Channels, Information. A Mathematical Introduction* (Berlin, Boston: De Gruyter).
[6] Jang S S et al 2006 *Phys. Rev.* A **74** 062112.
[7] Song W 2007 *Phys. Rev.* A **76** 014307.
[8] Zyczkowski K et al 1998 *Phys. Rev.* A **58** 883.

IOP Conf. Series: Journal of Physics: Conf. Series **918** (2017) 012007 doi:10.1088/1742-6596/918/1/012007

Calculating vibration-rotation spectrum intensities of laser-induced fluorescence for the $B^1\Pi - X^1\Sigma^+$ electronic transition in a LiCs molecule

A D Smirnov[1], V S Boldyrev[1]

[1]Bauman Moscow State Technical University, Department of Chemistry, 5 2nd Baumanskaya St., Moscow, 105005 Russia

E-mail: adsmir@mail.ru

Abstract. The purpose of this study was to calculate the intensities of the vibration-rotation spectrum of laser-induced fluorescence for the $B^1\Pi - X^1\Sigma^+$ electronic transition in a LiCs molecule. We find vibration-rotation energies and wave functions as a result of numerically solving the radial wave equation based on potential energy curves plotted in the present work for the ground and excited states. We compare calculated and experimental intensities.

1. Introduction

Over the last decade, the number of experimental and theoretical publications dealing with heteronuclear alkali dimers in ground and excited electronic states has considerably grown. A feature that distinguishes heteronuclear alkali dimers from homonuclear ones is the presence of a permanent electric dipole moment. Work [1] remarks that polar alkali metal dimers are promising candidates for creation of quantum computers. In [2-6] we calculated energetic and radiative parameters for heteronuclear alkali dimers (CsK, CsLi, RbLi, NaK, NaRb, NaCs, KRb, RbCs, KLi). In the present work we calculated vibration-rotation spectrum intensities of laser-induced fluorescence for the $B^1\Pi - X^1\Sigma^+$ electronic transition in a $^7Li^{133}Cs$ molecule. Out of all mixed alkali dimers LiCs has the biggest permanent electric dipole moment [7]. Publications [8, 9] used laser techniques and high resolution equipment to conduct experimental studies of the LiCs molecule spectrum. These studies derived vibration and rotation spectroscopic constants for the ground and excited electronic states of the $B^1\Pi - X^1\Sigma^+$ transition in a LiCs molecule.

2. Calculation

We calculated the intensity for a spontaneous emission on an electric dipole transition $e'v'j' \rightarrow e''v''j''$ using an equation from [10]

$$I(e'v'j' \rightarrow e''v''j'') = \frac{64\pi^4}{3c^3} N_{e'v'j'} \frac{S_{J'J''}}{2J'+1} \nu_{v'j,v''j''}^{4} \left| \int \psi_{v'j'}(r) R_e(r)\, \psi_{v''j''}(r)\, dr \right|^2$$

where $N_{e''v''j''}$ is the energy level population of the excited state; $S_{J'J''}$ is the rotational line strength; $\nu_{v'j',v''j''}$ is the wavenumber of a vibration-rotation line; $\psi_{v'j'}(r)$ and $\psi_{v''j''}(r)$ are rovibrational wave functions for the excited and ground electronic states; $R_e(r)$ is the dependence of the electronic transition dipole moment on internuclear distance. We used the following expression to calculate wavenumbers of a vibration-rotation line:

$$\nu_{v'j',v''j'} = T_e^{'} + E(v',j') - E(v'',j''),$$

where $T_e^{'}$ is electronic energy of the excited state; $E(v', j')$, $E(v'', j'')$ are vibration-rotational energies of the excited and ground electronic states.

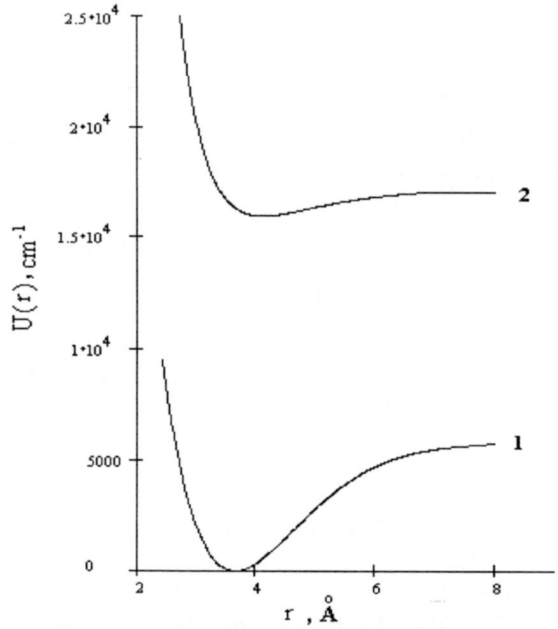

Figure 1. Potential energy curves for the ground (1) and excited (2) electronic states for the $B^1\Pi - X^1\Sigma^+$ transition in a LiCs molecule.

We derived the values of $E(v', j')$, $E(v'', j'')$, $\psi_{v'j'}(r)$ and $\psi_{v''j''}(r)$ as a result of solving the radial wave equation numerically using the Rydberg – Klein – Rees (RKR) potential energy curves, plotted in the present work. Figure 1 shows potential energy curves for the ground and excited electronic states for the $B^1\Pi - X^1\Sigma^+$ transition in a LiCs molecule. We took the function $R_e(r)$ from the publication [11]. We calculated wavenumbers for 77 vibration-rotation lines. Table 1 presents part of the calculated values and compares them with experimental data.

Table 1. Comparison between experimental and calculated wavenumbers of the vibration-rotation lines for the $B^1\Pi - X^1\Sigma^+$ transition in a LiCs molecule.

v''	j''	$\nu^{calc}_{v'j',v''j''}$, cm^{-1}	$\nu^{exp}_{v'j',v''j''}$, cm^{-1}	v''	j''	$\nu^{calc}_{v'j',v''j''}$, cm^{-1}	$\nu^{exp}_{v'j',v''j''}$, cm^{-1}
			$v' = 0$	$j' = 76$			
1	76	15545.091	15545.154	7	76	14539.353	14539.402
2	76	15372.122	15372.169	8	76	14379.407	14379.456
3	76	15201.236	15201.297	9	76	14221.731	14221.785

IOP Conf. Series: Journal of Physics: Conf. Series **918** (2017) 012007 doi:10.1088/1742-6596/918/1/012007

4	76	15032.505	15032.562	10	76	14066.366	14066.418
5	76	14865.931	14865.982	11	76	13913.349	13913.392
6	76	14701.538	14701.582	12	76	13762.722	13762.767
				$v'=2$	$j'=38$		
0	37	16112.333	16112.391	8	37	14722.903	14722.925
0	39	16083.821	16083.889	8	39	14696.039	14696.071
3	37	15575.746	15575.788	9	37	14558.726	14558.747
3	39	15547.833	15547.886	9	39	14532.081	14532.113
4	37	15400.987	15401.025	10	37	14396.741	14396.757
4	39	15373.279	15373.326	10	39	14370.319	14370.345
5	37	15228.300	15228.336	11	37	14236.971	14236.983
5	39	15200.798	15200.844	11	39	14210.777	14210.798
6	37	15057.709	15057.741	12	37	14079.450	14079.448
6	39	15030.417	15030.459	12	39	14053.504	14053.486

Figures 2 and 3 show the results of calculating relative intensities of vibration-rotation lines in the spectrum for the $B^1\Pi - X^1\Sigma^+$ electronic transition in a LiCs molecule and a comparison with an experimental spectrum [9].

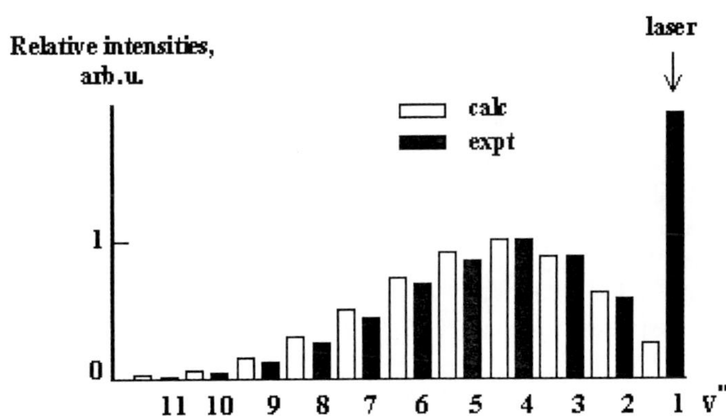

Figure 2. Comparison between experimental and calculated relative intensities of vibration-rotation lines ($v'=0, j'=76$; $v'', j''=76$) for the $B^1\Pi - X^1\Sigma^+$ electronic transition in a LiCs molecule. Calculated and experimental intensities are normalized to be equal for the strongest spectral line, $\nu_{(laser)} = 15545.158$ cm^{-1}.

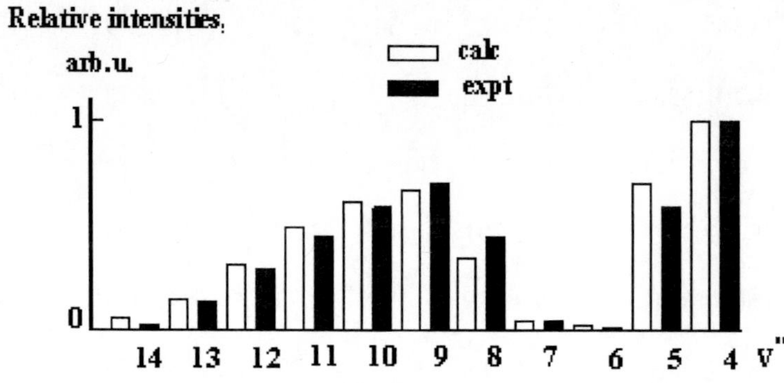

Figure 3. Same as figure 2 for the $v' = 2, j' = 38$; $v'', j'' = 39$ lines.

3. Discussion of results

Good agreement between the calculated and experimental wavenumbers of the vibration – rotation lines confirms the accuracy of the potential curves plotted in this work and the wave functions computed for the ground and excited states of the $B^1\Pi - X^1\Sigma^+$ electronic transition in a LiCs molecule. The calculated wavenumbers deviate from experimental values by less than 0.1 cm^{-1}, which makes for approximately 0.0005 %.

Figures 2 and 3 demonstrate that the intensity distributions of vibration-rotation lines of the calculated spectrum are in good agreement with the experimental laser-induced fluorescence spectrum of the $B^1\Pi - X^1\Sigma^+$ electronic transition in a LiCs molecule.

4. Conclusion

The results show that the wavenumbers and vibration-rotation line intensities calculated using the potential curves plotted in the present paper for the ground and excited electronic states of a LiCs molecule closely agree with the experimental data.

References

[1] De Mille D 2002 *Phys. Rev. Lett.* **88** 067901
[2] Smirnov A D 2007 *J. Struct. Chem.* **48** 15
[3] Smirnov A D 2014 *Opt. Spectrosc.* **117** 373
[4] Smirnov A D 2015 *Herald of the Bauman Moscow State Tech. Univ., Nat. Sci.* **№ 6** 52
[5] Smirnov A D 2016 *Opt. Spectrosc.* **121** 783
[6] Smirnov A D 2017 *Herald of the Bauman Moscow State Tech. Univ., Nat. Sci.* **№ 1** 112
[7] Aymar N and Dileu O 2005 *J. Chem. Phys.* **122** 204302
[8] Staanum P, Pashov A et al 2007 *Phys. Rev. A* **75** 042513
[9] Stein A, Pashov A et al. 2008 *Eur. Phys. Journal D.***48** 177
[10] Zare R N 1964 *J. Chem. Phys.* **40** 1934
[11] Kowalczyk P, Jastrzebski W et al 2015 *J. Chem. Phys.* **142** 234308

IOP Conf. Series: Journal of Physics: Conf. Series **918** (2017) 012008 doi:10.1088/1742-6596/918/1/012008

Calculation of the Intensity of Physical Time Fluctuations Using the Standard Solar Model and its Comparison with the Results of Experimental Measurements

A N Morozov
Head of "Physics" Department, Bauman Moscow State Technical University, Moscow, Russia

E-mail: amor59@mail.ru

Abstract. The article reviews the possibility of describing physical time as a random Poisson process. An equation allowing the intensity of physical time fluctuations to be calculated depending on the entropy production density within irreversible natural processes has been proposed. Based on the standard solar model the work calculates the entropy production density inside the Sun and the dependence of the intensity of physical time fluctuations on the distance to the centre of the Sun. A free model parameter has been established, and the method of its evaluation has been suggested. The calculations of the entropy production density inside the Sun showed that it differs by 2-3 orders of magnitude in different parts of the Sun. The intensity of physical time fluctuations on the Earth's surface depending on the entropy production density during the sunlight-to-Earth's thermal radiation conversion has been theoretically predicted. A method of evaluation of the Kullback's measure of voltage fluctuations in small amounts of electrolyte has been proposed. Using a simple model of the Earth's surface heat transfer to the upper atmosphere, the effective Earth's thermal radiation temperature has been determined. A comparison between the theoretical values of the Kullback's measure derived from the fluctuating physical time model and the experimentally measured values of this measure for two independent electrolytic cells showed a good qualitative and quantitative concurrence of predictions of both theoretical model and experimental data.
Key words: standard solar model, irreversible processes, entropy production, physical time fluctuations, Kullback's measure.

1. Introduction

An actively discussed and developed hypothesis as to the existence of nonlocal communication of quantum macroscopic processes [1-4] and the related experiments [5-7] led to a strong conviction about the existence of mutual influence of dissipative processes in nature. The experimental verification of the hypothesis presented is carried out at the level of establishing mutual influence of electroencephalographic (EEG) brain activity from pairs of people [8], as well as in physical experiments [9, 10].

The theory of non-Markovian processes is developed at the same time, which allows the irreversible processes in a variety of systems to be described, with due regard for fluctuations in dissipation factors [11-13]. The application of a physical time model to describe the impact of dissipative processes on measuring systems was first proposed in the works [14, 15].

The fluctuating physical time model [16], which was used in the work [17] to describe the propagation of light in the Universe to estimate the possibility of recording relic gravitational waves, as well as in the work [18] to explain the causality inversion, may be considered as a hypothesis helping to explain mutual influence of irreversible macroscopic processes.

The work theoretically predicts the intensity of physical time fluctuations using the standard solar model [19] and compares the results of long-term records of the Kullback's measure of voltage fluctuations on electrolytic cells with the predictions of the fluctuating time model [20, 21].

2. Fluctuating physical time model

Let us review the fluctuating physical time model [16], based on the assumption that the physical time under review is a random Poisson process $\tau(t)$ with fluctuations, equal to $\tau_0 = 1/\nu_\tau$, where ν_τ - the intensity of physical time fluctuations. The process is graphically displayed in Fig. 1.

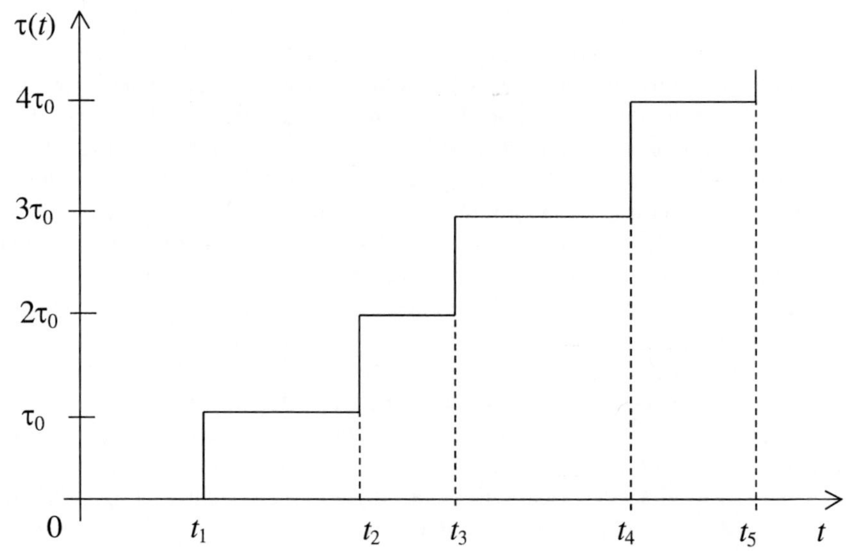

Fig. 1. Dependence of one's own physical time $\tau(t)$ on macroscopic time t

A one-dimensional characteristic function of the Poisson process shown in Fig. 1 is given by [22]

$$g_\tau(\mu_\tau;t) = \exp((\exp(i\tau_0\mu_\tau)-1)\nu_\tau t), \tag{1}$$

and its n-dimensional characteristic function at $t_n \ge ... \ge t_2 \ge t_1$ accordingly takes the form

$$g_{n\tau}(\mu_{1\tau},\mu_{2\tau},...,\mu_{n\tau};t_1,t_2,...,t_n) = \exp\left(\nu_\tau \sum_{j=1}^{n} t_j \left(\exp\left(i\tau_0 \sum_{k=j}^{n} \mu_{k\tau}\right)-1\right)\right). \tag{2}$$

The characteristic functions (1) and (2) make it possible to write the expectation function $\langle\tau(t)\rangle$ and correlation function $\langle\tau(t_1)\tau(t_2)\rangle$ for arbitrary time points t_1 and t_2 as follows:

$$\langle\tau(t)\rangle = t, \tag{3}$$

$$\langle\tau(t_1)\tau(t_2)\rangle = t_1 t_2 + \tau_0 \min(t_1,t_2). \tag{4}$$

For the application of the proposed fluctuating time model, we need to develop a method of calculating the intensity ν_τ of physical time fluctuations.

3. Calculation of the intensity of physical time fluctuations during the production of entropy by irreversible processes

To calculate the intensity ν_τ of physical time fluctuations during the production of entropy by irreversible processes for a quasi-stationary case, we use the following equation

$$\nu_\tau\left(\frac{\partial^2\nu_\tau}{\partial x^2}+\frac{\partial^2\nu_\tau}{\partial y^2}+\frac{\partial^2\nu_\tau}{\partial z^2}\right)=-\frac{c}{k}\rho_S(x,y,z), \tag{5}$$

where: c - speed of light, k - Boltzmann's constant, $\rho_S(x,y,z)$ - volumetric density of the production of entropy by irreversible processes.

For a spherically symmetric case, the equation (5) takes the form

$$\nu_\tau\left(\frac{d^2\nu_\tau}{dr^2}+\frac{2}{r}\frac{d\nu_\tau}{dr}\right)=-\frac{c}{k}\rho_S(r). \tag{6}$$

If we assume that irreversible processes do not occur in space and $\rho_S(r)=0$, the equation (6) takes the form

$$\nu_\tau\left(\frac{d^2\nu_\tau}{dr^2}+\frac{2}{r}\frac{d\nu_\tau}{dr}\right)=0. \tag{7}$$

Its two solutions can be presented as follows

$$\nu_\tau(r)=0 \tag{8}$$

and

$$\nu_\tau(r)=C_1+\frac{C_2}{r}, \tag{9}$$

where C_1 and C_2 - arbitrary constants.

Let us review the case where the production of entropy occurs only at the surface of a sphere of radius R. Such a case in the first approximation corresponds to the production of entropy during solar radiation and the Earth's thermal radiation. At the same time, it is also assumed that all other irreversible processes occurring inside the Sun and Earth make a small contribution to the development of physical time fluctuations.

Let us introduce the equation (6) for this case in the form of

$$\nu_\tau\left(\frac{d^2\nu_\tau}{dr^2}+\frac{2}{r}\frac{d\nu_\tau}{dr}\right)=-\frac{c}{k}\sigma_S\delta(r-R), \tag{10}$$

where σ_S - surface density of the production of entropy, $\delta(r-R)$ - delta function.

Then, within the sphere, we may assume that $\rho_S(r)=0$ and solution (9) is valid (solution (8) is trivial and has no physical meaning). However, to fulfill the condition of finiteness of the intensity ν_τ of physical time fluctuations at any point in space (particularly, in the centre of the sphere), the constant C_2 needs to be taken equal to zero: $C_2=0$. Consequently, at $r<R$ we obtain

$$\nu_\tau(r)=C_1. \tag{11}$$

Outside the sphere, at $r>R$, the solution (9) still remain valid. Based on the vanishing condition of the function $\nu_\tau(r)$ at an infinite distance from the sphere $\nu_\tau(r)\big|_{r\to\infty}=0$, the constant C_1 needs to be taken equal to zero: $C_1=0$. Then the solution outside the sphere at $r>R$ takes the form

$$\nu_\tau(r) = \frac{C_2}{r}. \tag{12}$$

Based on the continuity condition of the function $\nu_\tau(r)$ at $r = R$, we may state the following:

$$C_1 = \frac{C_2}{R}. \tag{13}$$

Substitution of solutions (11) and (12) into equation (10) based on formula (13) allows us to determine the constants C_1 and C_2

$$C_1 = \sqrt{\frac{c}{k} R\sigma_S}, \tag{14}$$

$$C_2 = \sqrt{\frac{c}{k} R^3\sigma_S}. \tag{15}$$

The resulting expressions allow us to estimate in the first approximation the surface density of production of entropy during the thermal radiation using the formula given in the work [23]

$$\sigma_S = \frac{4}{3}\frac{W_C}{T_C}\frac{1}{4\pi R_C^2} = \frac{W_C}{3\pi T_C R_C^2}, \tag{16}$$

where: $W_C = 3{,}85 \cdot 10^{26}$ W – the power of solar radiation, $T_C = 5830$ K – temperature of the Sun, $R_C = 6{,}96 \cdot 10^8$ m – radius of the Sun. Then, the intensity $\nu_\tau(r)$ of physical time fluctuations on the Sun's surface will be equal to: $\nu_{\tau C} = 1{,}48 \cdot 10^{22}$ s^{-1}, and at a distance from the Sun to the Earth - $\nu_{\tau 0} = 6{,}87 \cdot 10^{19}$ s^{-1}, respectively.

4. Calculation of the production of entropy using the standard solar model

The work [19] described the standard model of the internal structure of the Sun, which allowed us to calculate the temperature T_i inside the Sun in relation to the radius r_i and the power of energy W_i production due to the thermonuclear reactions inside the i-th spherical layer, where i is the number of the spherical layer inside the Sun. The data presented here allow us to calculate the density $\rho_S(r_i)$ of production of entropy inside the Sun in relation to the radius r_i. For this purpose, we use the formula similar to the expression (16)

$$\rho_S(r_i) = \frac{4}{3}\left(\frac{\sum_{k=1}^{i+1} W_k}{T_{i+1}} - \frac{\sum_{k=1}^{i} W_k}{T_i}\right)\frac{1}{(V_{i+1} - V_i)}, \tag{17}$$

where the sphere volume V_i of the radius r_i equals to: $V_i = \frac{4}{3}\pi r_i^3$.

Fig. 2 shows the diagram of the dependence of the production of entropy $\rho_S(r)$ calculated by formula (17) with the use of the standard solar model parameters presented in the work [19]. The values of entropy production density $\rho_S(r)$ in Fig. 2 are given in W/K·m^3.

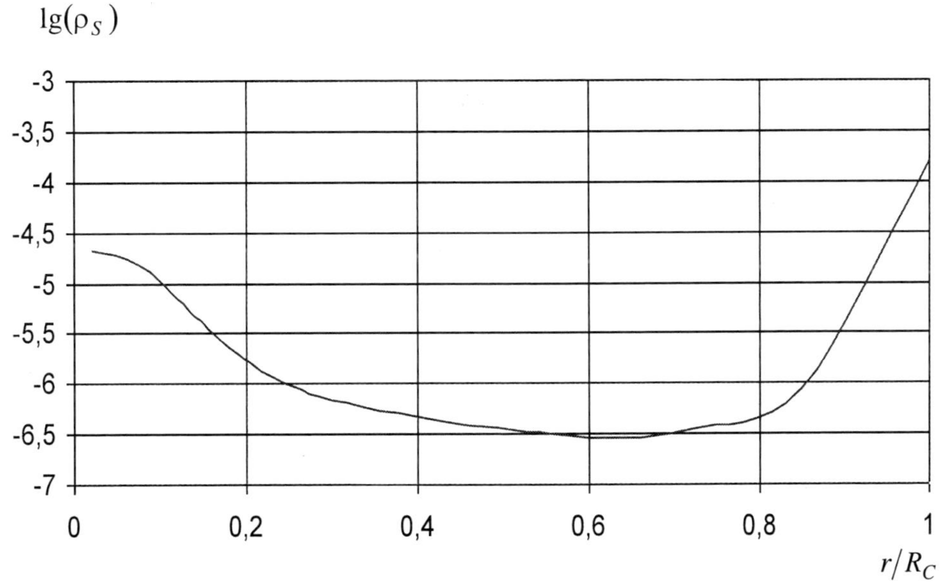

Fig. 2. The Dependence diagram of the entropy production density $\rho_S(r)$

As may be inferred from the diagram, the entropy production density $\rho_S(r)$ inside the Sun first decreases with the increase of the radius r, and then starts to increase sharply within the near-surface layer and reaches its maximum value at the surface of the Sun. The sharp increase in the entropy production density within the near-surface layer of the Sun is associated with a sharp temperature drop within the same layer. However, it should be noted that, according to Fig. 2, the entropy production density differs by 2-3 orders of magnitude in different parts of the Sun.

5. Calculation of the intensity of physical time fluctuations

To calculate the intensity $\nu_\tau(r)$ of physical time fluctuations inside the Sun, we may use the formula (6) presented in the form of the system of equations

$$\nu_\tau\left(\frac{d\kappa_\tau}{dr}+\frac{2}{r}\kappa_\tau\right)=-\frac{c}{k}\rho_S(r), \qquad (18)$$

$$\frac{d\nu_\tau}{dr}=\kappa_\tau. \qquad (19)$$

Outside the Sun, at $r > R_C$, the calculation of the intensity $\nu_\tau(r)$ of physical time fluctuations can be made using the formula similar to (12):

$$\nu_\tau(r)=\frac{\nu_{\tau C}R_C}{r}, \qquad (20)$$

where $\nu_{\tau C}=\nu_\tau(R_C)$ - the intensity of physical time fluctuations on the Sun's surface.

The system of equations (18) and (19) was solved by the Euler numerical method with an allowance for correction [24]. At $r=0$, the value κ_τ was assumed to be equal to zero:

$\kappa_\tau(r)|_{r=0} = 0$, and the value of the intensity $\nu_\tau(r)|_{r=0} = \nu_\tau(0)$ of physical time fluctuations varied.

To evaluate the intensity $\nu_\tau(0)$, we need to use *a priori* information that can not be derived from the model proposed. One of the possible options of evaluating the intensity $\nu_\tau(0)$ is the assumption that the value is determined by the most probable thermonuclear reactions inside the Sun associated with the fusion of helium nuclei from hydrogen nuclei within the hydrogen-helium cycle. The energy of major reactions within the specified cycle has the following range of values: $E = 5{,}5...17{,}3$ MeV. Then, the intensity $\nu_\tau(0)$ in the centre of the Sun may be estimated by the formula:

$$\nu_\tau(0) = \frac{E}{\hbar}, \tag{21}$$

where $\hbar$ - Plank's constant. The calculation gives us the following: $\nu_\tau(0) = (0{,}9...2{,}7)\cdot 10^{22}$ s^{-1}. If we take into consideration the whole cycle of hydrogen-to-helium transformation, the energy of which is: $E = 26{,}7$ MeV, then the value of the intensity $\nu_\tau(0)$ in the centre of the Sun results in: $\nu_\tau(0) = 4{,}1\cdot 10^{22}$ s^{-1}. It is obvious that the values obtained are approximate.

Fig. 3 shows the diagrams for multiple values of the intensity $\nu_\tau(0)$ of physical time fluctuations in the centre of the Sun: $\nu_\tau(0) = (2;3;4)\cdot 10^{22}$ s^{-1}.

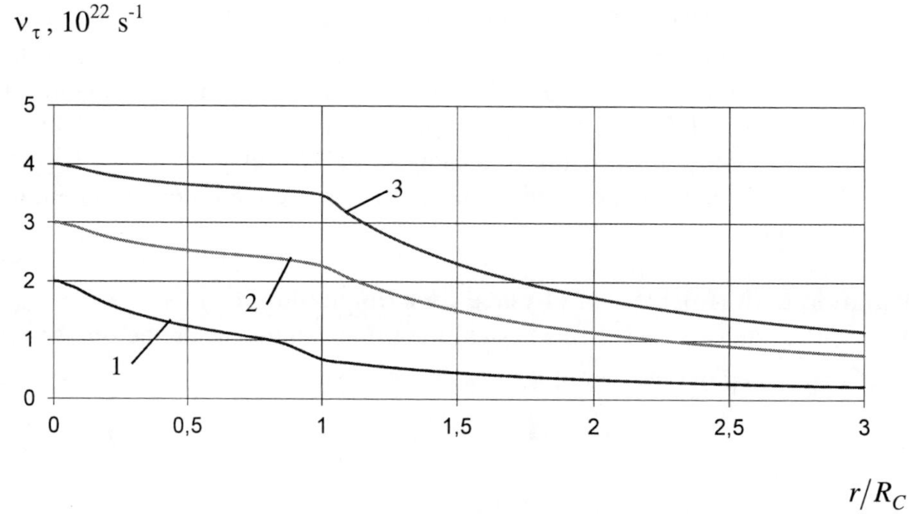

Fig. 3. Dependence of the intensity $\nu_\tau(r)$ of physical time fluctuations at various values of intensity $\nu_\tau(0)$ in the centre of the Sun:
1 - $\nu_\tau(0) = 2\cdot 10^{22}$ s^{-1}; 2 - $\nu_\tau(0) = 3\cdot 10^{22}$ s^{-1}; 3 - $\nu_\tau(0) = 4\cdot 10^{22}$ s^{-1}

As can be seen from the data obtained, the intensity $\nu_{\tau C}$ of physical time fluctuations on the Sun's surface and, consequently, the intensity $\nu_{\tau 0}$ of physical time fluctuations created by the Sun on the Earth's orbit, depend on the adopted value of the intensity $\nu_\tau(0)$ of the physical time fluctuations in the centre of the Sun. The latter may be calculated using formula

(20) with the insertion of distance r equal to the distance from the Sun to the Earth: $r = 1{,}5 \cdot 10^{11}$ m.

Table 1 includes the calculated values of the intensity of physical time fluctuations on the Sun's $\nu_{\tau C}$ and the Earth's $\nu_{\tau 0}$ surfaces in relation to the adopted value of the intensity in the centre of the Sun $\nu_{\tau}(0)$.

Table 1

Intensity of physical time fluctuations on the Sun's $\nu_{\tau C}$ and Earth's $\nu_{\tau 0}$ surfaces

$\nu_{\tau}(0)$, 10^{22} s^{-1}	$\nu_{\tau C}$, 10^{22} s^{-1}	$\nu_{\tau 0}$, 10^{19} s^{-1}
2,0	0,67	3,1
3,0	2,27	10,5
4,0	3,47	16,1

As follows from the calculations performed, the value of the intensity of physical time fluctuations on the Sun's and Earth's surfaces heavily depends on the intensity $\nu_{\tau}(0)$ of physical time fluctuations in the centre of the Sun, the value of which is an arbitrary parameter within the model considered.

6. Calculation of the intensity of physical time fluctuations due to the Earth's thermal radiation

Let us calculate the intensity of physical time fluctuations for the case where the heat radiating sphere is located in a space with a certain value of the intensity $\nu_{\tau 0}$ of physical time fluctuations, generated by external processes. Such a case is observed when calculating the intensity $\nu_{\tau}(r)$ of physical time fluctuations generated by the thermal radiation of the Earth under the Sun's external influence characterized by the intensity $\nu_{\tau 0}$.

Then, the solution of the equation (10) inside the sphere can be presented as follows

$$\nu_{\tau}(r) = C_1 + \nu_{\tau 0}, \tag{22}$$

and outside the sphere

$$\nu_{\tau}(r) = \frac{C_2}{r} + \nu_{\tau 0}, \tag{23}$$

yet the formula between the constants C_1 and C_2 retains its form (13).

Substitution of solutions (22) and (23) in equation (10) allows us to obtain the following expression

$$(C_1 + \nu_{\tau 0})\left(-\frac{C_2}{R^2}\right) = -\frac{c}{k}\sigma_S, \tag{24}$$

which can be represented as an algebraic equation

$$C_1^2 + \nu_{\tau 0} C_1 - \frac{c}{k} R\sigma_S = 0. \tag{25}$$

Solution of equation (25) takes the form

$$C_1 = -\frac{\nu_{\tau 0}}{2} \pm \sqrt{\frac{\nu_{\tau 0}^2}{4} + \frac{c}{k} R\sigma_S}, \tag{26}$$

noting that the physical meaning pertains only to the solution with a plus sign in front of the root in (26).

Then, finally, in accordance with formula (22), the value of the intensity ν_τ of physical time fluctuations on the Earth's surface will look as follows [25]

$$\nu_\tau = \frac{\nu_{\tau 0}}{2} + \sqrt{\frac{\nu_{\tau 0}^2}{4} + \frac{c}{k} R_3 \sigma_S}\ , \tag{27}$$

where $R_3 = 6{,}4 \cdot 10^6$ m – radius of the Earth. The entropy production surface density due to the Earth's re-radiation of sunlight may be calculated by the following formula [23]

$$\sigma_S = \frac{4}{3}\left(\frac{w_3}{T_3} - \frac{w_C}{T_C}\cos\gamma\right), \tag{28}$$

where: $w_3 = \sigma T_3^4$ - power of thermal radiation of the Earth from one square metre, $\sigma = 5{,}67 \cdot 10^{-8}$ W/(K^4·m^2) - Stefan-Boltzmann constant, $T_3 = 254$ K - the average radiation temperature of the Earth, $w_C = 1368$ W/m^2 - the intensity of solar radiation falling on the Earth, $T_C = 5778$ K - solar radiation temperature in the vicinity of the Earth, γ - the angle of sunlight incidence on the Earth's surface, which depends on the time of year and the day.

Substitution of formula (28) in equation (27) allows us to calculate the average intensity ν_τ of physical time fluctuations on the surface of the Earth at $\nu_{\tau 0} = 3{,}1 \cdot 10^{19}$ s^{-1} (see Table 1): $\nu_\tau = 3{,}5 \cdot 10^{19}$ s^{-1}.

7. Calculation of the Kullback's measure of voltage fluctuations on electrolytic cells in relation to the intensity of physical time fluctuations

Let us make a thermodynamic calculation of variations of the Kullback's measure of voltage fluctuations on an electrolytic cell [21]. For this purpose, we use the following equation

$$dH = -\nu_\tau H dt + dW_\tau(t), \tag{29}$$

where the characteristic function of a Poisson process $W_\tau(t)$ may be presented as follows

$$g_\tau(\mu_\tau, t) = \exp[(\exp(iD_\tau \mu_\tau) - 1)\nu_\tau t]. \tag{30}$$

Here we have the symbol introduced

$$D_\tau = \frac{mc^2}{\hbar N}, \tag{31}$$

where: m - ionic mass, N - number of ions in a small amount of the electrolyte.

The characteristic function of the Kullback's measure based on equation (29) is given by [21]

$$g(\lambda_H; t) = \exp\left\{\nu_\tau \int_{-\infty}^{t} [\exp(iD_\tau G(t,\tau)\lambda_H) - 1] d\tau\right\}, \tag{32}$$

where

$$G(t,\tau) = \exp[-\nu_\tau (t - \tau)]. \tag{33}$$

Substitution of formula (33) in equation (32) allows us to determine the first moment of the Kullback's measure:

$$\langle H \rangle = \left.\frac{\partial g(\lambda_H)}{i\partial \lambda_H}\right|_{\lambda_H = 0} = \frac{D_\tau}{\nu_\tau}. \tag{34}$$

Calculation of the coefficient D_τ using formula (31) at a characteristic ionic mass in a distilled water $m = 3{,}2 \cdot 10^{-26}$ kg and the number of ions $N = 10^{10}$ gives the following value of coefficient D_τ:

$$D_\tau = 2{,}7 \cdot 10^{15} \text{ s}^{-1}. \tag{35}$$

Substitution of the value in formula (34) taking into account the evaluation of the intensity of physical time fluctuations on the Earth's surface $\nu_\tau = 3{,}5 \cdot 10^{19}$ s^{-1} allows us to calculate the approximate value of the Kullback's measure: $\langle H \rangle = 7{,}7 \cdot 10^{-5}$.

8. Calculation of the Earth's effective thermal radiation temperature

Later, we will proceed from the model assuming that the solar radiation heats the Earth's surface, and furthermore, due to the heat transfer process, the heat is transferred to the upper atmosphere with subsequent thermal radiation into space. Then the simplest model of the process described above may be represented by the following differential equation

$$\frac{dT_3}{dt} + \alpha T_3 = \alpha(T - \eta w_3), \tag{36}$$

where T_3 - Earth's thermal radiation temperature, α - coefficient characterizing the process of heat transfer from the Earth's surface to the upper atmosphere, T - the Earth's surface temperature, which in the first approximation can be considered equal to the air temperature at the Earth's surface, and the coefficient η can be determined by the heat equilibrium condition: the average density of the radiation falling from the Sun onto the surface of the Earth $\langle w_C \cos\gamma \rangle$ is equal to the average density of the Earth's thermal radiation $\langle w_3 \rangle$:

$$\langle w_C \cos\gamma \rangle \cong \langle w_3 \rangle. \tag{37}$$

The following values of the coefficients were used in the calculations:

$$\alpha = 1{,}17 \cdot 10^{-5} \text{ s}^{-1}, \tag{38}$$

$$\eta = 7{,}9 \cdot 10^{-2} \text{ K/W}. \tag{39}$$

In this case, the entropy production density will be calculated according to formula (28), in which the value of the Earth's thermal radiation temperature T_3 is substituted by the expression obtained in the solution of equation (36).

9. Results of modeling of variations of the Kullback's measure and comparison with experimental diagrams

When calculating the theoretical value of the Kullback's measure $\langle H \rangle$, we shall use the expressions derived above: (27), (28), (34) and (36). For the evaluation the intensity was considered as $\nu_{\tau 0} = 3{,}1 \cdot 10^{19}$ s^{-1}. The atmospheric temperature T data within the near-surface layer of the Earth were taken from the web site "Pogoda i Klimat" (www.pogodaiklimat.ru) for the Moscow meteorological station located at the National Economy Achievement Exhibition (WMO index: 27612).

The works [20, 21] present the results of measurements of the Kullback's measure of voltage fluctuations on two separate electrolytic cells. These papers also provide a detailed description of the electrolytic cells used and their characteristics.

Also presented are the experimental data that have been obtained over the past five years, from March 20, 2011 to June 19, 2016. The overall duration of the experiments was 1,919 days.

The time series obtained experimentally and taken from the web-site "Pogoda i Klimat" were subjected to two types of filtering: band-pass and low-pass [10]. In the case of band-pass filtering, the processes with the periods of less than 24 hours (1 day) and more than 600 hours (25 days) were discarded, and in case of low-pass filtering - the ones with periods of less than 240 hours (10 days) were discarded.

Fig. 4 shows the diagram of the correlation coefficient $K(H_1, H_2)$ of the Kullback's measures H_1 and H_2 for two independent experimental units under band-pass filtering. The maximum correlation coefficient value calculated by the method described in the work [26] is as follows: $K(H_1, H_2) = 0{,}252 \pm 0{,}099$, with a virtually zero shift of the given time series. The probability as to the presence of the specified value of the correlation coefficient is equal to: $p = 0{,}989$ [27].

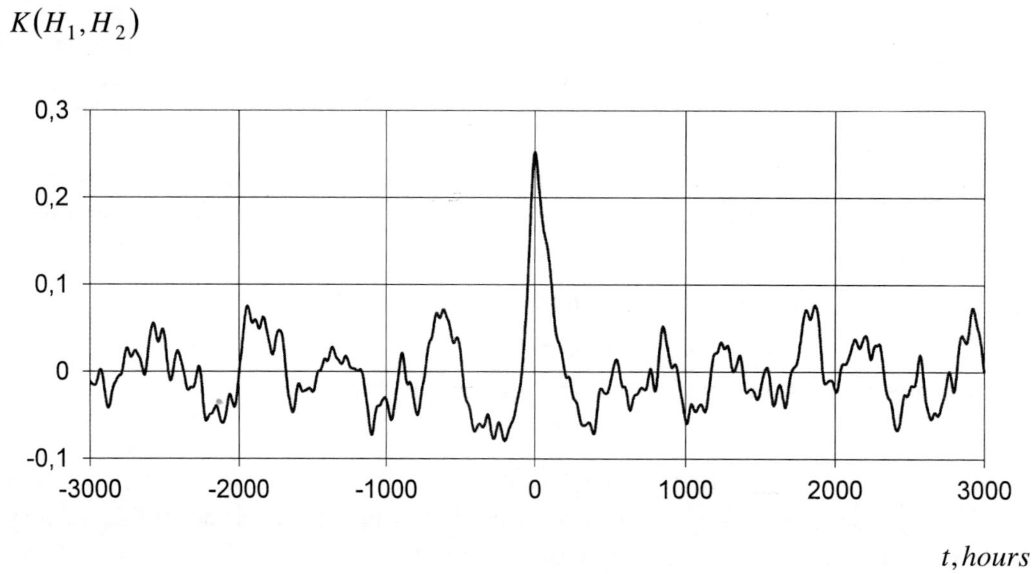

Fig. 4. Correlation coefficient of values of the Kullback's measures H_1 and H_2 for two independent units under band-pass filtering.

In case of low-pass filtering, the correlation coefficient of the Kullback's measures H_1 and H_2 for two independent units takes the following form: $K(H_1, H_2) = 0{,}377 \pm 0{,}094$; probability $p = 0{,}9999$.

Table 2 presents the values of correlation coefficients of experimentally obtained variations of the Kullback's measures H_1 and H_2 for two independent units and the theoretically calculated value of the Kullback's measure $\langle H \rangle$ based on formula (34). Please note that the time shift between the calculated values of variations of the Kullback's measure $\langle H \rangle$ and the experimental values H_1 and H_2 virtually equals to zero, which indicates a direct impact on the experimental units to change the entropy production density during the sunlight-to-Earth's thermal radiation conversion. The table below presents correlation coefficients for two units, the top cells containing the coefficients calculated under band-pass filtering, the bottom cells - the ones calculated under low-pass filtering.

Table 2

Coefficients of correlation of experimental and theoretical values of the Kullback's measures

Correlation coefficients	Unit no. 1	Unit no. 2
$K(H_k, \langle H \rangle)$	0,285±0,098	0,505±0,088
	0,242±0,099	0,838±0,057

The correlation coefficient of the average value of the Kullback's measure $H_{1,2}$ calculated by the formula

$$H_{1,2} = (H_1 + H_2)/2, \tag{40}$$

and its theoretical value $\langle H \rangle$ under band-pass filtering has the value of

$$K(H_{1,2}, \langle H \rangle) = 0{,}485 \pm 0{,}089;\ p = 0{,}999999, \tag{41}$$

and under low-pass filtering equals to

$$K(H_{1,2}, \langle H \rangle) = 0{,}630 \pm 0{,}079;\ p > 0{,}999999. \tag{42}$$

The diagram shown in Fig. 5 illustrates the correlation function of the average value of the Kullback's measure $H_{1,2}$ and the theoretically calculated value of the Kullback's measures $\langle H \rangle$ under band-pass filtering. As may be seen from the figure, the specified values are correlated with a virtually zero time shift.

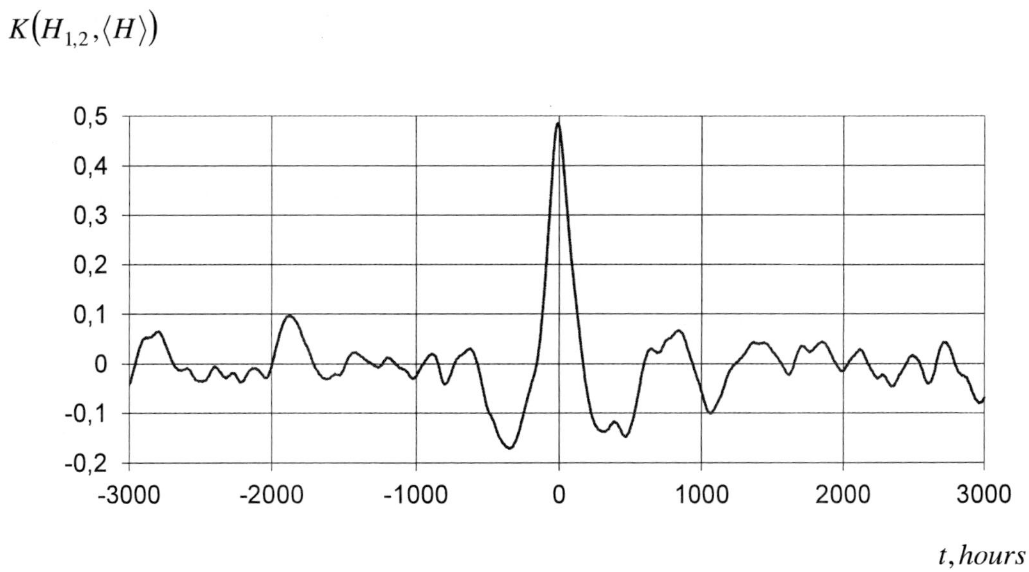

Fig. 5. Correlation coefficient of the average value of the Kullback's measure $H_{1,2}$ and the estimated value of this measure $\langle H \rangle$ under band-pass filtering

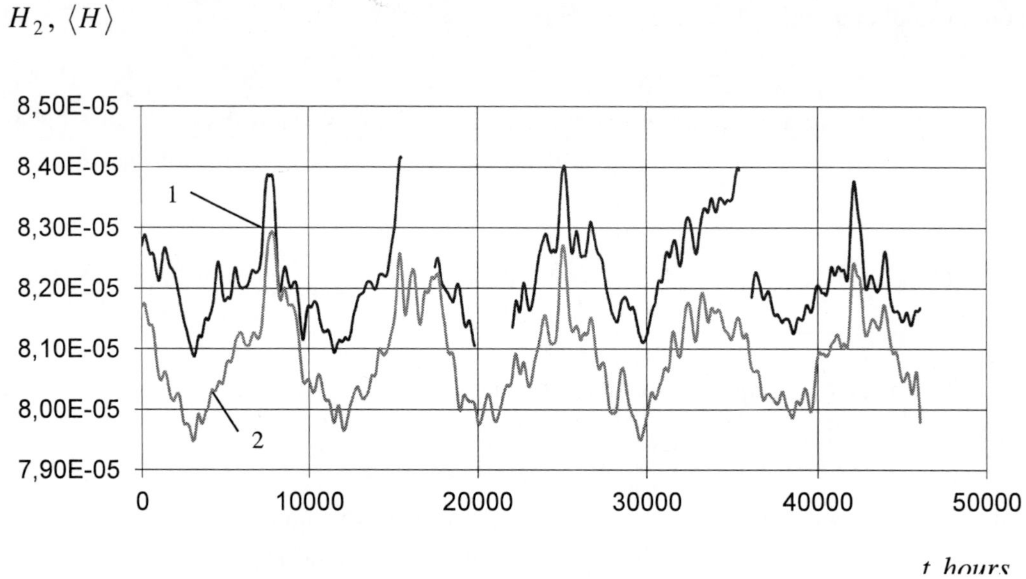

Fig. 6. Diagrams of the dependence of Kullback's measures H_2 (curve 1) and the calculated dependence of this measure $\langle H \rangle$ (curve 2)

Fig. 6 shows the comparison of the experimental time dependence diagram of the Kullback's measure H_2 for the second unit, and the same diagram for the estimated values of this measure $\langle H \rangle$. The diagrams shown are typical for the case of low-pass filtering.

The diagrams shown in Fig. 6 illustrate a clear correlation of the Kullback's measure H_2 and the estimated values of this measure $\langle H \rangle$ within the model of the entropy production during the sunlight-to-Earth's thermal radiation conversion.

10. Conclusion

The application of the standard solar model allowed us to calculate the intensity of physical time fluctuations based on equation (5) describing the dependence of the reported intensity on the production of entropy naturally occurring in irreversible processes. The theoretically calculated values of variations of the Kullback's measure $\langle H \rangle$ agree with the experimentally measured values of the Kullback's measures H_1 and H_2 of voltage fluctuations on two separate electrolytic cells, both qualitatively and quantitatively. To confirm the results obtained, four new installations aimed at exploring the voltage fluctuations in small volumes of electrolyte are being created.

The author gratefully acknowledges the assistance of A.N. Nikitin and S.M. Korotaev in the research and in the preparation of this article for publication.

References

[1] Julsgaard B, Kozhekin A and Polzik E S 2001 *Nature* **413** pp 400-403
[2] Benatti F, Floreanini R and Piani M 2003 *Physical Review Letters* **91** (7) pp 070402-4
[3] Dur W, Briegel H J 2004 *Physical Review Letters* **92** (18) pp 180403-4

[4] Xu H, Strauch F W, Dutta S K, Johnson P R, Ramos R C, Berkley A J, Paik H, Anderson J R, Dragt A J, Lobb C J and Wellstood F C 2005 *Physical Review Letters* **94** (2) pp 027003-1-4

[5] Korotaev S M, Morozov A N, Serdyuk V O and Sorokin M.O 2002 *Russian Physics Journal.* **45** (5) pp 431-444

[6] Korotaev S M, Morozov A N, Serdyuk V O, Gorokhov J V and Machinin V A 2005 *NeuroQuantology* **3** (4) pp 275-294

[7] Korotaev S M, Morozov A N, Serdyuk V O, Gorokhov Yu V, Fillippov B P and Machinin V A 2007 *Russian Physics Journal* **50** (4) pp 333-341

[8] Dotta B T, Mulligan B P, Hunter M D and Persinger M A 2009 *NeuroQuantology* **7** (4) pp 548-551

[9] Morozov A N and Turchaninov S O 1992 *Biophysics* **37** (4) pp 567-568

[10] Morozov A N 2016 *NeuroQuantology* **14** (3) pp 477-483

[11] Morozov A N 1996 *JETP* **82** (4) pp 703-708

[12] Morozov A N and Skripkin A V 2011 *Physics Letters A* **375** (46) pp 4113-4115

[13] Morozov A N and Skripkin A V 2011 *Journal of Engineering Physics and Thermophysics* **84** (6) pp 1201-1208

[14] Kozyrev N A 1958 *Causal or asymmetrical mechanics in linear approximation.* (Pulkovo: Main astronomical observatory Publ.)

[15] Kozyrev N A 1971 *Time is Science and Philosophy* pp 111-132

[16] Morozov A N and Nazolin A L 2001 *Dynamic systems with fluctuating time.* (Moscow: Bauman MSTU)

[17] Morozov A N and Gladyshev V O 2002 *Russian Physics Journal* **45** (2) pp 113-117

[18] Kuraev A A 2011 *Dokladi BGUIR* **2** pp 115-116

[19] Bahcall J N, Huebner W F, Lubow S H, Parker P D and Ulrich R K 1982 *Reviews of Modern Physics* **4** (3) pp 767-799

[20] Morozov A N 2015 *Radiooptics of the Bauman MSTU* **6** pp 62-76

[21] Morozov A N 2016 *Radiooptics of the Bauman MSTU* **1** pp 35-61

[22] Pugachev V S and Sinitsyn I N 1987 *Stochastic differential systems: analysis and filtering.* (Chichester: Wiley Publ.)

[23] Izakov M N 1997 *Physics-Uspekhi* **40** (10) pp 1035-1042

[24] Samarskii A A 1982 *Introduction to numerical methods.* (Moscow: Nauka Publ.)

[25] Morozov A N 2016 *Nonlinear World* **4** pp 59-66

[26] Max J 1989 *Methodes et techniques de traitement du signal et applications aux mesures physiques. Tome 1. Principes generaux et methodes classiques.* (Paris: Masson Publ.)

[27] Jahnke E, Emde F and Losch F 1960 *Tafeln hoherer funktionen. stuttgart.* (Stuttgart: Teubner Publ.)

IOP Conf. Series: Journal of Physics: Conf. Series **918** (2017) 012009 doi:10.1088/1742-6596/918/1/012009

The chiral cosmological models with two components

I V Fomin

Department of Physics, Bauman Moscow State Technical University,
2-nd Baumanskaya street, 5, Moscow, Russia

E-mail: ingvor@inbox.ru

Abstract. The purpose of this paper is to consider cosmological models, containing two scalar fields. The method of exact solutions constructions of the scalar field's dynamical equations on the second stage of the accelerated expansion of the universe with matter on the basis on models including two fields without interactions was considered. A cosmological models are associated with nonlinear sigma models in this work.

1. Introduction

The theory of cosmological inflation successfully describes the state of the accelerated expansion of the universe in the early stages of evolution. Also, inflationary cosmology explains the origin of primary inhomogeneities and predicts their spectrum [1, 2, 3, 4].

The mechanism of the initial inflationary scenario and the second accelerated expansion [5] can be described under the assumption of the existence of a scalar field, or an inflaton ϕ specifying the stage of inflation and the quintessence field, which determines the observed accelerated expansion of the universe [6]. Different cosmological scenarios differ in the choice of effective potential $V(\phi)$ of scalar field.

In four-dimensional space-time with a single scalar field there were obtained the restrictions on state parameter for quintessence $-1 < w < -1/3$ and for phantom scalar field $w < -1$, where w is the state parameter in the equation of state $p = w\rho$, p is the pressure and ρ is the energy density [7].

To describe the stage of cosmological inflation and re-accelerated expansion in the framework of the Einstein gravity theory, models with nontrivial kinetic energy or the k-essence models [8, 9] and models with two scalar fields [10, 11, 12, 13, 14, 15] were investigated. Within the framework of the approach, based on the chiral cosmological models, various types of kinetic and potential interaction between the field of the inflaton and a scalar field simulating dark energy at the inflation stage, as well as interactions of dark matter and dark energy in the subsequent stages of the evolution of the Universe are possible [13, 14]. The cosmological perturbations in the chiral models of inflation were considered in the work [15].

In present paper, the simple way of exact solutions construction for cosmological models with two scalar fields and matter field on the basis of known solutions for models with two noninteracting scalar fields is considered. This approach is associated with nonlinear sigma models that combine kinetic and potential type interactions between scalar fields and lead to Friedmann cosmological models [13, 14, 15].

IOP Conf. Series: Journal of Physics: Conf. Series **918** (2017) 012009 doi:10.1088/1742-6596/918/1/012009

2. Chiral cosmological models

Nonlinear sigma models with a potential of interaction (chiral cosmological models) are widely used to describe various epochs of the evolution of the universe. In such models there is an internal space, the metric of which determines the interaction between scalar fields.

Let us write the action for the nonlinear sigma model with additional material sources [15]

$$S = \int dx^4\sqrt{-g}\left(\frac{R}{2} + \frac{1}{2}h_{ij}\partial_\mu\psi^i\partial_\nu\psi^j g^{\mu\nu} - V(\phi,\chi)\right) + S_m; \tag{1}$$

where R is Ricci scalar, $V(\phi,\chi)$ is the potential of scalar fields, $g^{\mu\nu}$ is the metric tensor of space, h_{ij} is the metric tensor of the fields space or the target (chiral) space, $\psi^1 = \phi$, $\psi^2 = \chi$ and S_m is the action for material fields.

The metric of the field space is defined as follows

$$ds^2 = h_{ij}(\psi^k)d\psi^i d\psi^j \tag{2}$$

A system of Einstein's equations in a flat Friedman-Robertson-Walker (FRW) space for a sigma model with two scalar fields and a material field corresponding to a barotropic perfect fluid in the system of units $8\pi G = c = 1$ can be written as follows

$$3H^2 = \frac{1}{2}h_{11}(\chi)\dot{\phi}^2 + h_{12}(\phi,\chi)\dot{\phi}\dot{\chi} + \frac{1}{2}h_{22}(\phi)\dot{\chi}^2 + V(\phi,\chi) - \frac{\rho_{m0}}{a^n} \tag{3}$$

$$-\dot{H} = \frac{1}{2}h_{11}(\chi)\dot{\phi}^2 + h_{12}(\phi,\chi)\dot{\phi}\dot{\chi} + \frac{1}{2}h_{22}(\phi)\dot{\chi}^2 - \frac{\rho_{m0}}{a^n}, \tag{4}$$

where $\rho_m = \rho_{m0}a^{-n}$ is the energy density of the barotropic perfect fluid, which effectively describes the material components of the universe (dark matter and baryon matter) at the stage of the second accelerated expansion, with following equation of state $p_m = \rho_m(n-3)/3$, where $0 \le n \le 6$ and p_m is the pressure of fluid. Radiation does not have a significant effect on the dynamics of the universe in this era [16]. The components of metric tensor of fields space h_{11} and h_{22} determine the potential interaction between the scalar fields, and $h_{12} = h_{21}$ determines the kinetic interaction. Also, positive components h_{11}, h_{22} correspond to canonical real scalar fields with positive kinetic energy $h_{11}(\chi)\dot{\phi}^2/2 > 0$, for example, and negative components correspond to phantom fields with negative kinetic energy.

Note that in the case of $\rho_{m0} = k$, where k is the coefficient determining the curvature of the FRW space, and $n = 2$, the system (3)–(4) corresponds to the equations of the scalar field's dynamics in the FRW space with nonzero curvature $k = -1, 0, 1$ without matter.

Now, we consider the models with the kinetic interaction $h_{11} = h_{22} = 1$ under the condition

$$h_{12}(\phi,\chi)\dot{\phi}\dot{\chi} = \frac{\rho_{m0}}{a^n} \tag{5}$$

and the system of equations (3)–(4) are

$$3H^2 = \frac{1}{2}\dot{\phi}^2 + \frac{1}{2}\dot{\chi}^2 + V(\phi,\chi) \tag{6}$$

$$-\dot{H} = \frac{1}{2}\dot{\phi}^2 + \frac{1}{2}\dot{\chi}^2 \tag{7}$$

Thus, the system of equations (6)–(7) corresponds to the models with scalar fields without interactions. Further, we will consider exact solutions for polynomial and exponential potentials on the base of known solutions for two scalar field without interactions between them and without material fields.

IOP Conf. Series: Journal of Physics: Conf. Series **918** (2017) 012009 doi:10.1088/1742-6596/918/1/012009

3. Polynomial potential

Following the work [17], we consider the Hubble parameter

$$H = H_0 + H_1\phi + H_2\chi, \tag{8}$$

where

$$H_0^2 = \frac{2}{9}\left(\frac{3}{2}A + E + D\right), H_1^2 = \frac{D}{3}, H_2^2 = \frac{E}{3} \tag{9}$$

As a result, from the equations (6)–(7) one can obtain

$$V(\phi, \chi) = A + B\phi + C\chi + D\phi^2 + E\chi^2 + F\phi\chi, \tag{10}$$

where

$$B^2 = \frac{8D}{3}\left(\frac{3}{2}A + E + D\right), C^2 = \frac{8E}{3}\left(\frac{3}{2}A + E + D\right), F^2 = 4D^2E^2 \tag{11}$$

with scalar fields

$$\phi(t) = \phi_0 - 2H_1 t \tag{12}$$

$$\chi(t) = \chi_0 - 2H_2 t \tag{13}$$

and scale factor

$$a(t) = a_0 \exp\left[(H_0 + H_1\phi_0 + H_2\chi_0)t - (H_1^2 + H_2^2)t^2\right] \tag{14}$$

Using our approach we derive from equation (5) that

$$h_{12}(t) = \frac{\rho_{m0}}{H_1 H_2 a^n} = \frac{\rho_{m0}}{H_1 H_2 a_0^n} \exp\left[-n(H_0 + H_1\phi_0 + H_2\chi_0)t + n(H_1^2 + H_2^2)t^2\right] \tag{15}$$

or

$$h_{12}(\phi, \chi) = \frac{\rho_{m0}}{H_1 H_2 a_0^n} \exp\left[\frac{n(H_0 + H_1\phi_0 + H_2\chi_0)(\phi - \phi_0)}{2H_1} + n(H_1^2 + H_2^2)\left(\frac{\chi - \chi_0}{2H_2}\right)^2\right] \tag{16}$$

$$h_{12}(\phi, \chi) = \frac{\rho_{m0}}{H_1 H_2 a_0^n} \exp\left[\frac{n(H_0 + H_1\phi_0 + H_2\chi_0)(\chi - \chi_0)}{2H_1} + n(H_1^2 + H_2^2)\left(\frac{\phi - \phi_0}{2H_2}\right)^2\right] \tag{17}$$

Thus, from the solutions for models with two noninteracting scalar fields at the stage of inflation [17], new exact solutions were obtained for the stage of second accelerated expansion with a matter on the base of chiral cosmological models.

4. Exponential potential

For an exponential potential of the form [17]

$$V = V_0\left(1 - A_1 e^{-\alpha\phi} + A_2 e^{-2\alpha\phi} - B_1 e^{-\beta\chi} + B_2 e^{-2\beta\chi} + \frac{A_1 B_1}{2} e^{-\alpha\phi - \beta\chi}\right), \tag{18}$$

where

$$\alpha^2 = \frac{3}{2} - \frac{6A_2}{A_1}, \beta^2 = \frac{3}{2} - \frac{6B_2}{B_1} \tag{19}$$

the Hubble parameter [17]

$$H = \sqrt{\frac{V_0}{3}}\left(1 - A_1 e^{-\alpha\phi} - B_1 e^{-\beta\chi}\right), H_0 = \sqrt{\frac{V_0}{3}} \tag{20}$$

Thus, one can obtain

$$\phi(t) = \frac{1}{\alpha}\ln[e^{\alpha\phi_0} - A_1\alpha^2 H_0 t] \tag{21}$$

$$\chi(t) = \frac{1}{\beta}\ln[e^{\beta\chi_0} - A_1\beta^2 H_0 t] \tag{22}$$

$$a(t) = a_0\left(e^{\alpha\phi_0} - A_1\alpha^2 H_0 t\right)^{\frac{1}{2\alpha^2}}\left(e^{\beta\chi_0} - B_1\beta^2 H_0 t\right)^{\frac{1}{2\beta^2}} e^{H_0 t} \tag{23}$$

Ones again, applying our method, from equation (5) we derive

$$h_{12}(t) = \frac{\rho_{m0}\left(e^{\alpha\phi_0} - A_1\alpha^2 H_0 t\right)^{-\frac{n}{2\alpha^2}+1}\left(e^{\beta\chi_0} - B_1\beta^2 H_0 t\right)^{-\frac{n}{2\beta^2}+1} e^{-H_0 t}}{\alpha\beta A_1 B_1 H_0^2 a_0^n} \tag{24}$$

$$h_{12}(\phi,\chi) = \frac{\rho_{m0}\exp\left[\left(\alpha - \frac{n}{2\alpha}\right)\phi + \left(\beta - \frac{n}{2\beta}\right)\chi\right]}{\alpha\beta A_1 B_1 H_0^2 a_0^n \exp\left[\frac{n(e^{\alpha\phi_0} - e^{\alpha\phi})}{A_1\alpha^2}\right]} \tag{25}$$

$$h_{12}(\phi,\chi) = \frac{\rho_{m0}\exp\left[\left(\alpha - \frac{n}{2\alpha}\right)\phi + \left(\beta - \frac{n}{2\beta}\right)\chi\right]}{\alpha\beta A_1 B_1 H_0^2 a_0^n \exp\left[\frac{n(e^{\beta\chi_0} - e^{\beta\chi})}{B_1\alpha^2}\right]} \tag{26}$$

These solutions also determine the possible background dynamics of the universe at the modern era with considering the material fields as barotropic perfect liquid.

5. Conclusion

Thus, the work proposed a method for constructing exact solutions in the context of chiral cosmological models for both: cosmological models with two scalar fields and matter, also, for the inflation stage, which involves only scalar fields with nonzero curvature of FRW space as a special case of general solutions. It should also be noted that by specifying the components of the fields space tensor $h_{11} = h_{11}(\chi)$ and $h_{22} = h_{22}(\phi)$ it is possible to obtain more complicated models, based on known solutions, with cross potential interactions between scalar fields.

References

[1] Starobinsky A 1979 *Phys. Lett.* B **91** 99
[2] Guth A 1981 *Phys. Rev.* D **23** 347
[3] Albrecht A and Steinhardt P 1982 *Phys. Rev. Lett.* **48** 1220
[4] Mukhanov V, Feldman H and Brandenberger R 1992 *Phys. Rep.* **215** 203
[5] Perlmutter S *et al.* 1999 *Astrophys. J.* **517** 565
[6] Peebles P and Vilenkin A 1999 *Phys. Rep.* D **59** 063505
[7] Liddle A and Lyth D 1993 *Phys. Rep.* **231** 1
[8] Putter R and Linder E 2007 *Astropart. Phys.* **28** 263
[9] Fomin I 2015 (in Russian) *Herald of the Bauman Moscow State Tech. Univ. Nat. Sci.* **4** 37 (Original Russian title: Vestnik MGTU im. N.E. Baumana)
[10] Urena-Lopez M 2001 *Phys. Rev.* D **63** 6 06350
[11] Bohmer C *et al.* 2008 *Phys. Rev.* D **78** 2 023505
[12] Costa F *et al.* 2012 *Phys. Rev.* D **85** 107302
[13] Beesham A *et al.* 2009 *Class. Quant. Grav.* **26** 075017
[14] Chervon S *et. al.* 2015 *Russian Physics Journal* **58** 5 597
[15] Chervon S 2013 *Quantum Matter* **2** 71
[16] Ade P *et al.* 2016 *Astron. Astrophys.* **A13** 594
[17] Byrnes C and Tasinato G 2009 *JCAP* **16** 0908

Denaturation of collagen structures and their transformation under the physical and chemical effects

A. Ivankin[1], V. Boldirev[2], G. Fadeev[3], M. Baburina[4], A. Kulikovskii[5] and N. Vostrikova[6]

[1] Professor, Bauman Moscow State Technical University, Moscow, Russia
[2] Assistant Professor, Bauman Moscow State Technical University, Moscow, Russia
[3] Professor, Bauman Moscow State Technical University, Moscow, Russia
[4] Head Researcher, The V.M. Gorbatov All-Russian Meat Research Institute (VNIIMP), Moscow, Russia
[5] Head Researcher, VNIIMP, Moscow, Russia
[6] Head Laboratory, VNIIMP, Moscow, Russia

E-mail: aivankin@inbox.ru

Abstract. The process of denaturation of collagen structures under the influence of physical and chemical factors play an important role in the manufacture of food technology and the production of drugs for medicine and cosmetology. The paper discussed the problem of the combined effects of heat treatment, mechanical dispersion and ultrasonic action on the structural changes of the animal collagen in the presence of weak protonated organic acids. Algorithm combined effects of physical and chemical factors as a result of the formation of the technological properties of products containing collagen has been shown.

1. Introduction

The purpose of work - implementation of physical and chemical degradation of hard collagen structures in the presence of weak organic acids. As an object was the defatted collagen protein suspension characterized by the following complete amino acid composition (in g / 100g protein): Asp – 7.02; Thr – 2.12; Ser – 2.79; Glu – 13.31; Pro – 7.57; Gly – 10.53; Ala – 6.96; Cys – 0.04; Val – 4.13; Met – 1.76; Iley – 2.18; Ley – 5.00; Tyr – 4.56; Phe – 5.12; His – 8.13; Lys – 2.96; Arg – 8.27. Fractional composition of the soluble part of the protein, according to electrophoresis on 18% polyacrylamide gel was (fraction kD): 10 kD – 5%, 10 ... 17 – 8%, 17 ... 26 – 9%, 26 ... 43 – 14%, 43 ... 72 – 15%, 130 ... 72 – 16%, 130 ... 170 – 15%, more than 170 kDa – 18%. Co-content of free amino acids in the protein of 0.005%.

The concept of "food protein" animal includes the sum of a plurality of fractions of protein molecules with a molecular weight of several thousand to hundreds of thousands of different sequences with co-constitute amino acids, which are subject to degradation to smaller fragments to form intermediate poly- and oligopeptides and further fragmentation to free amino acids. Protein content of the animal tissue is typically approx. 20% and contains from 0.001% to 2% of free amino acids. Depending on the state raw protein content of free amino acids may exceed this level several times. In certain kinds of protein products medi-prophylactic action but which are protein hydrolysates, free amino acid content may be 50 – 100%.

Preliminary studies have shown that the way conventionally used the so-called hard collagens acid hydrolysis in the presence of strong mineral acids leads to total protein irreversible disintegration into constituent amino acids. The macromolecular structure ceases to exist. In some cases, it requires the implementation of a controlled degradation of collagen in the preparation of functional foods. This problem can be solved at the expense of the dosed physical and mechanical [1] and the heat [2] effects, as well as through the use of a weak organic acid, [3] as a suitable lactic acid.

2. Results

Hydro unit significantly affects the result. in terms of the effectiveness of the dispersing device to obtain a final protein precursor in the form of a gel with enhanced moisture-retaining properties, which is important for further production (Table 1).

Table 1: Effect of hydro module properties of the protein product

The value of the hydro unit (mass ratio of the raw material: liquid)	Viscosity gel, min	Moisture content of product, % wt.
1 : 1	No flow	< 85
1 : 10	30	90
1 : 20	5	97

An important factor affecting the speed and the result is an irreversible degradation of collagen is the effect of lactic acid concentration.

Table 2: Effect of the lactic acid concentration on the properties of the protein (1 hr, 20°C)

Lactic acid content in the mixture to be treated, % wt.	pH of gel	The proportion of free amino acids with respect to their total protein content,%
0.0028 (8% to the raw material)	2.8	0.02
0.0014 (4% to the raw material)	3.5	0.01
0.0007 (2% to the raw material)	4.0	0.005
0.0003 (1% to the raw material)	4.0	0.005

IOP Conf. Series: Journal of Physics: Conf. Series **918** (2017) 012010 doi:10.1088/1742-6596/918/1/012010

The system used by the hydrolytic treatment, as substances that contribute to mild hydrolysis of the components used in a weak organic acid - lactic acid, which is without any restrictions can be used in the preparation of technology products, since she is a natural component.

Data on the possible effects of lactic acid on the product properties are shown in Table 2.

It is seen from the data obtained that the lactic acid concentration in a certain range of 1...2% by weight. It has little effect on the value of the effectiveness of the hydrolysis treatment, as the content of free amino acids in the system is practically not increased. The same concentration is sufficient for effective pulverization feed structure and subsequent transfer to a homogeneous gel. Large concentrations of lactic acid to reduce the pH level of the system, which is already beginning to affect the pH of the finished product, in the case, for example adding to the product formulation of the composition at a level 10% of normal. Based on these considerations, the optimal amount of lactic acid added can be considered a level of 2% to the raw material.

Significant impact on the rate of protein degradation has a temperature. Table 3 summarized the major influence on the properties of the parameters derived from collagen degradation products. Also it contains the dispersion of the system to 200 microns, which is the threshold value of the food system, determining the absence of negative taste sensations.

Table 3: Effect of physical and chemical processing parameters on the properties of the protein products

Processing conditions	Protein fractions in the product, %				Free amino acids, %	The proportion of particles < 200 mkm, %
	< 10 kD	10 – 70 kD	70 – 170 kD	> 170 kD		
3 hours, hydro module 1:20, 20°C, 2% lactic acid	5	43	32	20	4.5	66
3 hours, hydro module 1:20, 60°C, 2% lactic acid	15	55	17	13	6.5	89
3 hours, hydro module 1:20, 20°C, 2% lactic acid, dispersing 5000 rpm	39	41	12	8	8.3	> 99
3 hours hydro module 1:10, 20°C, 2% lactic acid, 5000 rpm, ultrasound 200 W/ l	63	20	12	5	13.6	> 99

3. Conclusions

3.1. Varying the technical parameters of physico-chemical treatment of proteins with a rigid structure of collagen produces degraded protein structures.

3.2. The products with such characteristics can be used to produce formulations with the required functional components (bioavailability digestibility, moisture content, particle size distribution) properties.

Acknowledgments

The study was funded by a grant of the Russian Scientific Fund (project №15-16-00008).

References

[1] Fadeev G.N., Boldyrev V.S. and Sinkevich V.V. *Sonochemical transformations of chelate and clathrate structures in a low-frequency acoustic field* // Doklady Physical Chemistry. 2015. V. 462. № 2. P. 119-123

[2] Ivankin A N and Krasnoshtanova A A 2010 *Hydrolysis nano bio macromolecular systems* (Moscow: Forest University) 396 p.

[3] Neklyudov A D and Ivankin A N 2007 *Collagen: properties, preparation, application* (Moscow: Forest University) 336 p.

IOP Conf. Series: Journal of Physics: Conf. Series **918** (2017) 012011 doi:10.1088/1742-6596/918/1/012011

Determination of cracks in structural materials using the radio spectroscopy method

R Yu Gerasimov, Yu V Gerasimov and G N Fadeev
Bauman Moscow State Technical University, 5, 2ya Baumanskaya str., Moscow, 105005, Russia
E-mail: r_guerassimov@mail.ru

Abstract. The new method proposed in this research is based on the spectral registration of the microwave range radiation by means of a wideband scanning receiver. The experiments have proved its accuracy sufficient for the spectrum frequency determination. Sensitivity of 5-6 Hz/nm level is achieved. The obtained spectra are analyzed using a specialized hardware-software complex. The evidence from practice has shown that the method in question, apart from detecting the substances qualitative composition, helps determine the presence and propagation of cracks.

1. Introduction

The proposed method for determining the presence of substances in nanoscopic amounts is based on the registration of radio-wave emission spectra in the microwave band using a scanning wideband receiver [1]. This kind of radiation occurs as a result of excitation of the alternating high-frequency displacement and conduction currents in a testing specimen between the flexible plates of a capacitive sensor. The measuring results (the spectra obtained) are analyzed using a hardware-software complex. The universal character of this method allows determining temperature-induced change in water structure; estimating the character and quantitative composition of aqueous solutions of salts, as well as experimentally detecting heavy isotopes of hydrogen (deuterium and tritium) in water [2].

2. Measurements results

Samples with different size cracks and without them were analyzed. Below are the results from processing the experimental data obtained by determining the coordinates of the center of gravity by the spectral amplitude A_s defing the areas of the spectra planar representations:

Table 1. Spectral amplitudes of the test samples

Sample №	1	2	3
A_s	0.021641	0.021485	0.021146

Sample: № 1 without cracks; № 2 with 4 mm long cracks; № 3 with 17 mm long cracks.

As is seen, A_s index decreases with the emergence and propagation of cracks. The decline and the value of this index can serve as an indicator for the emergence and growth of cracks.

The accuracy of the data obtained is 0.01 Hz in frequency and 0.01 dB in amplitude [3, 4]. The values of parameter A_s were found using the formula

$$A_s = \Sigma(v_i A_i)/\Sigma v_i,$$

where A_i is the amplitude at the ith point, and v_i is the frequency at the ith point.

In order to analyze the crack width, one of the specimens was investigated with its crack extended. For this purpose, the sample was subjected to mechanical loading (from 50 to 100 Newton). Fig.1 shows the curve of the gravity center parameter.

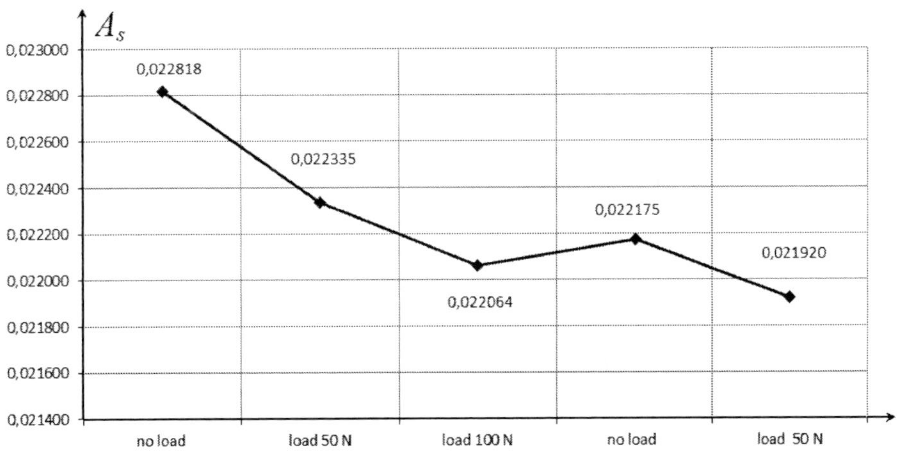

Figure 1. A_s parameter as function of load

As the graph shows, the gravity index curve falls, with load being applied and the crack width increasing. The decline of A_s does not linearly correlate with the load applied. On subsequent load removal, there occurs no return to the starting point, which may indicate the increased crack length.

The method under consideration allows estimating the nature and quantitative structure of aqueous solutions of salts [2]. Fig. 2 shows the processing results for the spectra of aqueous solutions of such salts as NaCl or LiCl in different concentrations. When processing the experimental data, the amplitude $A_i(v_i)$ for impurities was calculated according to the formula

$$A_i(v_i) = A_i(\text{sol})(v_i) - A_i(\text{water})(v_i),$$

where v_i is the characteristic frequency of the object under investigation.

The obtained data show the difference between the spectra of salt solutions and pure water. Each of the compounds is characterized by its own set of maxima, determined precisely according to the frequencies.

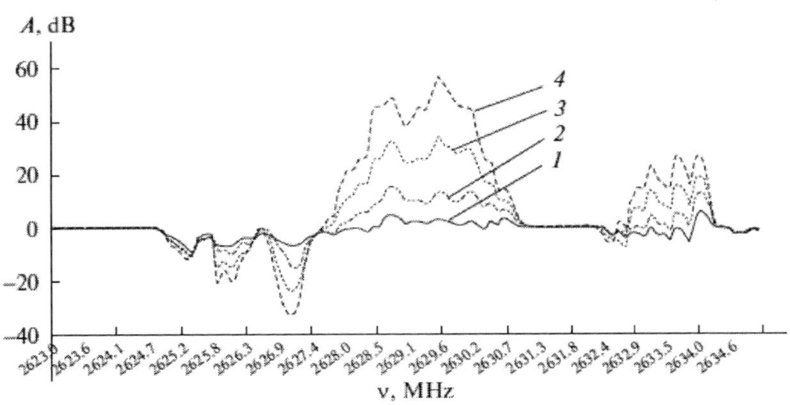

Figure 2. Results from processing the spectra of salt solutions with different concentrations: (*1*) 10% NaCl, (*2*) 15% NaCl, (*3*) 10% LiCl, (*4*) 15% LiCl

Comparing the spectra enabled us to quantify the content of the substances, as well as evaluate the differences in the qualitative composition of NaCl or LiCl solutions. The difference in the spectral amplitudes of the solutions with a different content of the same component (spectra *1* and *2* for NaCl, and spectra *3* and *4* for LiCl) helped determine the concentration of the solutes.

Conclusions.

1. It was verified experimentally that the method proposed could determine the presence of cracks, their size and growth.
2. The method of the microwave range radioscopy allows investigating both solid and liquid system by comparing the spectra of reference and the test samples.

REFERENCES

1. Gerasimov R Yu, Fadeev G N, Gerasimov Yu V and Kondrakova E A 2015 *Doklady RAN* **464**, Part 1, pp. 219-21.
2. Gerasimov R Yu , Fadeev G N,. Gerasimov Yu V and Kondrakova E A 2016 *Russ J of Phys Chem* A **90** No. 9 pp.1785-8
3. Bovenko V N and Gerasimov R Yu 2011 *Nonliner world* **6** pp 369–77
4. Gerasimov R Yu., Fadeev G N, Gerasimov Yu V and Kondrakova E A 2016, *J of Phys: Conf Ser* 731 (IOP Publishing) 012005

IOP Conf. Series: Journal of Physics: Conf. Series **918** (2017) 012012 doi:10.1088/1742-6596/918/1/012012

Estimating the railgun plasma piston rational parameters for the nanosatellite acceleration

A G Maslov[1], Y V Gerasimov[2] and V V. Korzhov[3]
[1]Assist. Professor, Bauman Moscow State Technical University, Moscow, Russian Federation
[2]Assoc. Professor, Bauman Moscow State Technical University, Moscow, Russian Federation
[3] Student, Bauman Moscow State Technical University, Moscow, Russian Federation

E-mail: maslov_art@bmstu.ru, drgerasimov.y@gmail.com

Abstract. The article examines the problem of finding rational parameters and composition of particles as well as optimizing plasma piston interaction with the launched object such as nano satellite. Optimization of plasma formation parameters is necessary for solving the following problems: implementation of particles into the launched object's surface with the formation of radiation porosity, increasing the object's specific volume, heating the object's construction materials up to the phase changing, heating the materials of the railgun, increasing the object's mass etc. These measures will increase the efficiency of the perspective rail systems for the nanosatellite acceleration. We introduce an algorithm for determining the rational parameters of plasma formation which allows perpendicular fall of plasma particles onto the surface in order to increase the nano satellite speed.

1. Introduction

Currently, there are many orbital satellites and clusters of satellites in use. They support systems of communication and Earth remote sensing. It helps to research processes in atmosphere and other important science-oriented problems. Satellite launching by means of modern rockets is extremely expensive and often causes heavy damage to the environment. That is why systems of pulse launch are being developed at present. Pulse systems design highly depends on the parameters of the projectile. In this article we use the parameters and the view of the nanosatellite from the article [1].

Electromagnetic launch using a rail accelerating system (railgun) holds much promise [2]. However, in case of launching nanosatellites the common practice of using conductor materials for the projectile has a number of weak points.

This problem can be solved by using different types of conductor armature, which accelerates the projectile. Usually metals such as cuprum and aluminium are used. But for various reasons it is not rational for the nanosatellites launching. Firstly, the efficiency of the railgun modern constructions is near 30%, so more than 60% of energy is wasted for the heating of the construction. Secondly, the efficiency of the systems is reduced due to the projectile mass increase. In addition, such systems have low reliability.

In our opinion, using a plasma piston for accelerating the nanosatellite is the most efficient way of launching, but there are various problems inherent to any interactions between the plasma and the solid. They are:

1. Implementation of plasma piston particles into the nanosatellite body and the launcher rail surface. That causes radiation-induced defects.
2. Increase of the nanosatellite body specific volume.
3. Heating of the nanosatellite materials up to the phase transition, heating of the launcher materials.

These effects efficiently reduce the plasma piston interaction with the nanosatellite. The level of such effects highly depends on plasma piston particles composition [4]. The equations of plasma piston state and rails metal are described in the articles [5-18].

Let us consider the multicomponent plasma formation in close volume interacting with the electromagnetic field to define the rational parameters of particles in the plasma piston. The trajectory of movement is close to the trajectory of a particle in the exponentially decreasing electromagnetic field, which is described in the article [19]. In this work we solve the problem of defining the rational parameters of the multicomponent plasma formation. The parameters rationalized in this article are the particle mass and charge. All other parameters are equal to the parameters in the process of a projectile electromagnetic acceleration by railgun for model and real experiments. Nevertheless, this problem appears in any interaction of plasma with a surface. Such optimization may be carried out for several reasons. Let us consider the extreme number of particles normal incidence onto the surface of the object, as a first approximation. It is equal to the parallelism of a particle's velocity vector to x axis. It occurs in different time for different types of trajectories. Let us analyse the problem of rationalisation in case of permanent electromagnetic field.

2. Rational parameters of plasma formation particles in case of permanent electromagnetic field

As it is well known, the particle's trajectory represents the following equations in case of permanent electromagnetic field:

$$\begin{cases} x(t) = -\dfrac{1}{2}\dfrac{U_0 md\cos\left(\dfrac{qk_{\text{м}}U_0 t}{md}\right)k_{\text{м}} + 2\sin\left(\dfrac{qk_{\text{м}}U_0 t}{md}\right)mdqRnS - 2tq^2U_0RnSk_{\text{м}} - mdU_0k_{\text{м}}}{q^2RnSk_{\text{м}}^2U_0} \\ z(t) = \dfrac{1}{2}\dfrac{md\left(U_0\sin\left(\dfrac{qk_{\text{м}}U_0 t}{md}\right)k_{\text{м}} - 2\cos\left(\dfrac{qk_{\text{м}}U_0 t}{md}\right)qRnS + 2qRnS\right)}{q^2k_{\text{м}}^2U_0RnS} \end{cases}$$

Hence, the form of the z-component of velocity is the following:

$$\dot{z} = \frac{1}{2}\frac{U_0\cos\left(\dfrac{qk_{\text{м}}U_0 t}{md}\right)k_{\text{м}} + 2\sin\left(\dfrac{qk_{\text{м}}U_0 t}{md}\right)qRnS}{qRnSk_{\text{м}}}$$

The time up to the normal incidence of a particle onto the surface was found from the equation $\dot{z} = 0$. It is equal to

$$t_{\text{н.п.}} = -\frac{\operatorname{arctg}\left(\dfrac{1}{2}\dfrac{U_0k_{\text{м}}}{qRnS}\right)md}{U_0k_{\text{м}}q} \tag{1}$$

Further we have to make a guess for the time of a particle's falling onto the surface in a real situation. Here we use the approximation expressed as:

$$t_{\text{н.п.}} = t_0 + Av, \tag{2}$$

t_0, A – constants.

This way:

IOP Conf. Series: Journal of Physics: Conf. Series **918** (2017) 012012 doi:10.1088/1742-6596/918/1/012012

$$t_0 + Av = -\frac{\operatorname{arctg}\left(\frac{1}{2}\frac{U_0 k_M}{qRnS}\right) md}{U_0 k_M q}$$

And after elementary conversion:

$$\operatorname{arctg}\left(\frac{1}{2}\frac{U_0 k_M}{qRnS}\right) md + \left(t_0 + Av\right) q k_M U_0 = 0 \tag{3}$$

This equation can be solved analytically, but it is possible to determine the particle's mass and charge with regard to the normal incidence using numerical values of the parameters from this formula.

It is possible to describe this sequence as the following algorithm:

Choosing the type of electromagnetic field variation → Estimating the particle's trajectory of movement → Calculating the time up to the normal incidence → Using linear approximation → Converting the equation to $Bq + f(q) = 0$ type. B – constant, $f(q)$ – some function → Choosing the particle fulfilling the equation.

3. Rational parameters of plasma formation particles in case of exponentially decreasing electromagnetic field

Let us develop the algorithm in case of exponentially decreasing electromagnetic field.

As shown in [19], the trajectory of the particle in this field is expressed as:

$$\begin{cases} x(t) = \left(\operatorname{Si}\left(\alpha\beta e^{-\frac{t}{RC}}\right) - \operatorname{Si}(\alpha\beta)\right)\left(\frac{RC}{k_M}\sin(\alpha\beta) + \frac{CU_0}{2\alpha mnS}\cos(\alpha\beta)\right) - \\ \quad -\left(\operatorname{Ci}\left(\alpha\beta e^{-\frac{t}{RC}}\right) - \operatorname{Ci}(\alpha\beta)\right)\left(\frac{CU_0}{2\alpha mnS}\sin(\alpha\beta) - \frac{RC}{k_M}\cos(\alpha\beta)\right) + \frac{t}{k_M}, \\ z(t) = \left(\operatorname{Si}\left(\alpha\beta e^{-\frac{t}{RC}}\right) - \operatorname{Si}(\alpha\beta)\right)\left(\frac{RC}{k_M}\cos(\alpha\beta) - \frac{CU_0}{2\alpha mnS}\sin(\alpha\beta)\right) - \\ \quad -\left(\operatorname{Ci}\left(\alpha\beta e^{-\frac{t}{RC}}\right) - \operatorname{Ci}(\alpha\beta)\right)\left(\frac{RC}{k_M}\sin(\alpha\beta) + \frac{CU_0}{2\alpha mnS}\cos(\alpha\beta)\right) \end{cases}$$

$\alpha = \frac{q}{m}$, $\beta = \frac{RCk_M U_0}{d}$.

Hence, the form of the velocity z-component is the following:

$$\dot{z} = -\frac{\sin\left(\alpha\beta e^{-\frac{t}{RC}}\right)\left(\frac{RC\cos(\alpha\beta)}{k_M} - \frac{1}{2}\frac{CU_0\sin(\alpha\beta)}{\alpha mnS}\right)}{RC} + \frac{\cos\left(\alpha\beta e^{-\frac{t}{RC}}\right)\left(\frac{RC\sin(\alpha\beta)}{k_M} + \frac{1}{2}\frac{CU_0\cos(\alpha\beta)}{\alpha mnS}\right)}{RC}$$

The time up to the normal incidence of a particle onto the surface was found from the equation $\dot{z} = 0$. It is equal to

$$t_{н.п.} = \ln\left(-\frac{\alpha\beta}{\operatorname{arctg}\left(\frac{2R\alpha mnS\,\operatorname{tg}(\alpha\beta)+U_0k_м}{-2R\alpha mnS+U_0k_м\,\operatorname{tg}(\alpha\beta)} \right)} \right) RC \quad (4)$$

We use the following linear approximation:

$$t_0 + Av = \ln\left(-\frac{\alpha\beta}{\operatorname{arctg}\left(\frac{2R\alpha mnS\,\operatorname{tg}(\alpha\beta)+U_0k_м}{-2R\alpha mnS+U_0k_м\,\operatorname{tg}(\alpha\beta)} \right)} \right) RC$$

And after the conversion we get:

$$\alpha\beta e^{-\frac{(t_0+Av)}{RC}} + \operatorname{arctg}\left(\frac{2R\alpha mnS\,\operatorname{tg}(\alpha\beta)+U_0k_м}{-2R\alpha mnS+U_0k_м\,\operatorname{tg}(\alpha\beta)} \right) = 0 \quad (5)$$

As (3) has no analytical solution for q, the equation (5) has no analytical solution for α, but it is possible to determine the function q(m) with regard to the normal incidence, using numerical values of the parameters from this formula. Thus the algorithm successfully works in case of generating the electromagnetic field by the high voltage capacitor.

However, the equation (5) has multiple roots, because of the function $\operatorname{arctg}\left(\frac{2R\alpha mnS\,\operatorname{tg}(\alpha\beta)+U_0k_м}{-2R\alpha mnS+U_0k_м\,\operatorname{tg}(\alpha\beta)} \right)$ is quickly oscillating. So it is rational to consider the minimization of the parameter $ù = \alpha\beta e^{-\frac{(t_0+Av)}{RC}} + \operatorname{arctg}\left(\frac{2R\alpha mnS\,\operatorname{tg}(\alpha\beta)+U_0k_м}{-2R\alpha mnS+U_0k_м\,\operatorname{tg}(\alpha\beta)} \right)$

4. Numerical calculations of rational parameters

In this experiment we use a system with the following parameters:

$$U_0 = 5\;KV$$

$$C = 8\;\mu F$$

$$R = 0.1\;Ohm$$

$$d = 30\;mm$$

$$k_м = 4\pi 10^{-6}\,\frac{S}{m}$$

$$n = 10^{29}\,\frac{1}{m^3}$$

$$S = 10^{-6}\,m^2$$

For choosing the ions with rational parameters the graph of parameter ϰ from the mass of ion was plotted for these experiments.

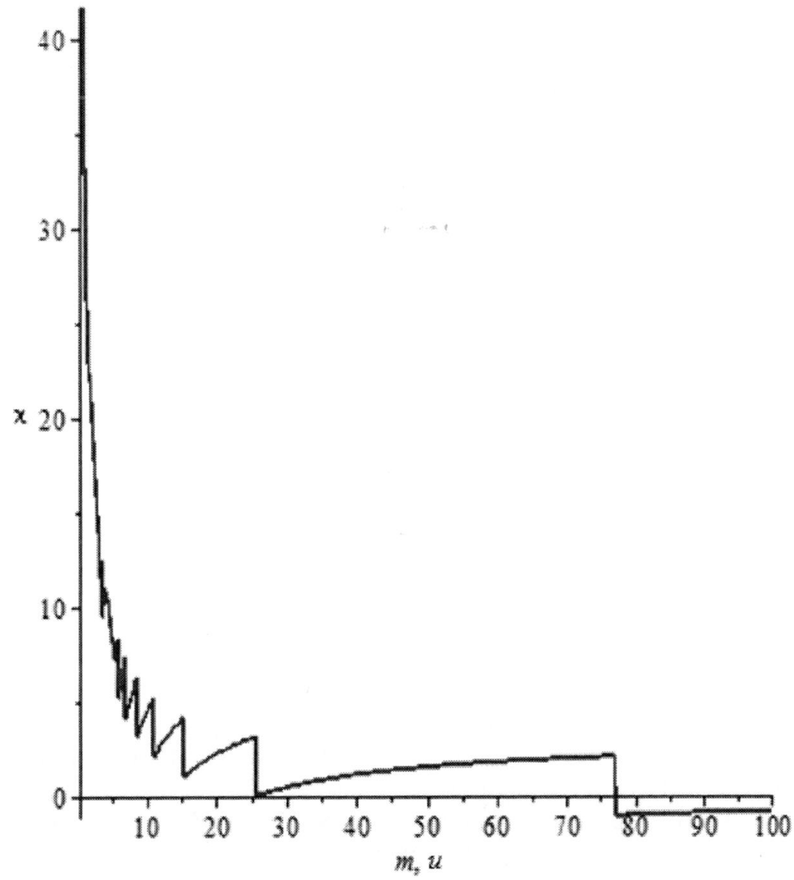

Figure 1. The plot of the parameter ϰ dependence on the mass of ion with the charge +1. Consequently, the heavy ion with the charge +1 is the most effective at the start of acceleration. It is Al+ for example in this experiment.

ϰ-parameter depends on the mass of ions in the plasma formation, but it is vanishing with the mass higher than 100 u. So, our model is the most effective in this region. This model is the most useful at the start of nanosatellite accelerating by the plasma formation, if the speeds of the nanosatellite and plasma formation are less than the speed of sound. Also, these examples show that rational parameters are different for different systems.

5. Rational parameters of experimental setup

An electromechanical scheme of the experimental setup is shown on fig. 2. AC current with 220 V and 50 Hz is regulated and transformed by TRESNSTELLTRAFO LTS 006 (10) to the battery charger МОСРЕНТГЕН М0СНХ ВС-20-10 (9) and controlled by the universal voltmeter В7-22А (11). 6 kV DC voltage from the battery charger is transmitted to the pulse capacitors ИК 5-200 У2 (7). It is controlled by the voltmeter В7-40-4 (12) with the voltage divider ДНВ-1000 (4). The charging time is 15 min. Then the charging scheme switches off. The gas-discharge arrester ИРТ-6 (8) is used for the process initiation. The speed of the accelerated object and plasma formation is registered by the pair of inductivity coils (3) with the resistance 50 Ohm and inductivity 3 μH connected to the grounded (1) oscilloscope С9-8 (2). The process is filmed by the digital camera and high-speed camera СФР. The rationalization of the plasma formation is considered at the initial stage.

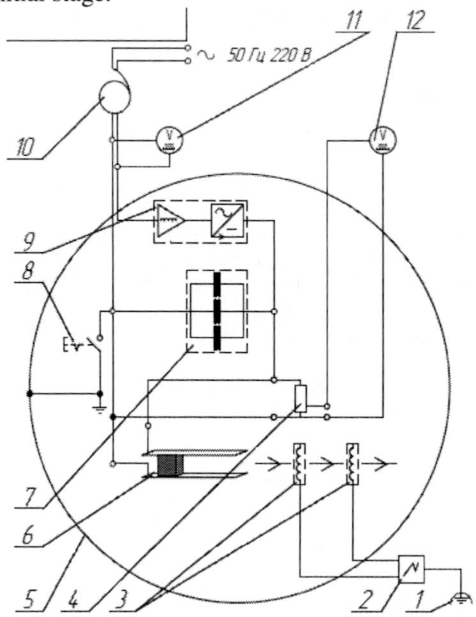

Figure 2. Electromechanical scheme of experimental setup. 1 – ground, 2 – oscilloscope С9-8, 3 – inductivity coils, 4 – voltage divider ДНВ-1000, 5 – wall of vacuum chamber, 6 – railgun, 7 – battery of capacitors ИК 5-200 У2, 8 – gas-discharge arrester ИРТ-6, 9 – battery charger МОСРЕНТГЕН М0СНХ ВС-20-10, 10 – laboratory transformer TRESNSTELLTRAFO LTS 006, 11 – voltmeter В7-22А, 12 – voltmeter В7-40-4.

Because the rational particle for the initial stage is Al^+ so the plasma piston formative charge consists of chemically pure aluminum and $C_{12}H_{22}O_{11}$ packed into the aluminum foil. According to the theory and parameters from [20] we can consider resonance of its eigen frequency with plasma frequency:

$$\omega_p = \sqrt{\frac{ne^2}{4\pi\varepsilon_0 m_e}} = \omega_i = \frac{\hbar}{2m}\vec{\eta}_i^{\,2}$$

If the interaction speed is about 2 km/s the Langmuir frequency should be:

$$\omega_p = 5.71\,10^{13}\ rad/s$$

So, the concentration of ionized aluminum particles in the plasma has to be about

$$n = \frac{4\pi\varepsilon_0 m_e \omega_p^{\,2}}{e^2} = 1.4\,10^{25}\ 1/m^3$$

It is possible to estimate the rational aluminum mass in the plasma formative charge with its volume V=5.4 10^{-7} m^3 and temperature of plasma formation T=7500 K, using the ionization equation:

$$m_{Al} = \mu_{Al}\left(\frac{nV}{\frac{(2m_e k_Б T)^{\frac{3}{4}}}{(2\pi\hbar)^{\frac{3}{2}}} e^{-\frac{E_a}{2k_Б T}}}\right) = 0.00015\ кг$$

The rational mass for other temperatures is shown on the fig. 3.

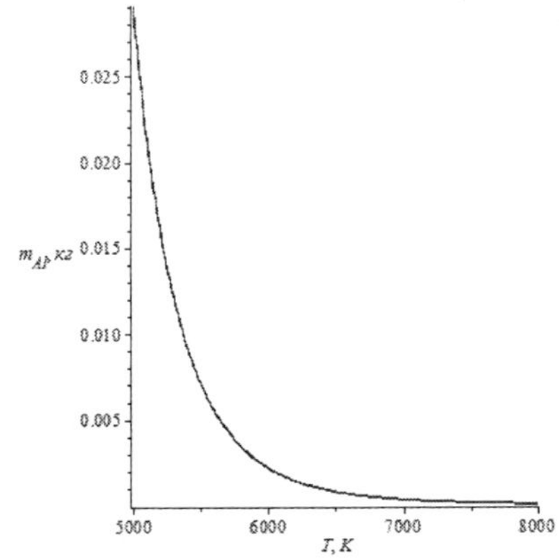

Figure 3. Plot of dependence of the rational mass of aluminum on the temperature of plasma formation

Finally, the optimal ratio between Al and $C_{12}H_{22}O_{11}$ is 1:1 for the total mass of the plasma formative charge of 0.32 g.

6. Conclusion

The rationalization model of plasma formation parameters is considered in the article. The main parameters that could be rationalized according to the parameters of the experimental setup and the interaction speed are ions of plasma formation specific charge for different stages and rational initial masses of unionized charge elements.

References

[1] Gerasimov Yu V, Karetnikov G K, Selivanov A B 2013 *Nano-satellites Stabilization under Condition of Pulse Launch in the Atmosphere and Space* J. Engineering Journal: Science and Innovation **8**

[2] Fionov A.S., Gerasimov Yu.V., Fillippenko A.V. 2007 *Options for the Development of Mass Acceleration Technology in Space* Proceedings of IV all-Russian conference

[3] Lehmann P., Reck B., Vo M.D., Behrens J. 2006 *Acceleration of a Suborbital Payload using an Electromagnetic Railgun* Proceedings of the 13-th International Symposium on Electromagnetic Launch Technology

[4] Vijayan T., Venkatramani N. 2003 *Heating and Transport of Metal Plasma in a Vacuum-Arc Rail Gun* IEEE Transactions on Plasma Science

[5] Fortov V.E., Ilkaev R.I., Arinin V.A. et al. 2007 *Phase Transition in a Strongly Nonideal Deuterium Plasma Generated by Quasi-Isentropical Compression at Megabar Pressures* Phys. Rev. Lett. **99** 185001
[6] Filinov V.S., Bonitz M., Fortov V.E. et al. 2005 *Coulomb Crystallization in Two-Component Plasmas* Phys. Rev. Lett. **95** 235006
[7] Eremets M.I., Trojan I.A. *Melting of Hydrogen at Megabar Pressures* 2007 21st AIRAPT and 45th EHPRG International Conference on High Pressure Science and Technology. Book of Abstracts/Eds. Angilella G.G., Pucci R., Siringo F. Catania
[8] Khomkin A., Shumikhin A. 2008 *Electrical Conductivity of Dense Metal Vapor Plasma* High Temperatures -High Pressures **37** 247.
[9] Boriskov G.V., Bykov A.I., Il"kaev R.I et al. 2005 *Shock Compression of Liquid Deuterium up to 109 GPa* Phys. Rev. B. **71** 092104
[10] Milyavskiy V.V., Utkin A.V., Zhuk A.Z. et al. 2005 *Shock Compressibility and Shock-Induced Phase Transitions of C_{60} Fullerite* Diamond & Related Materials **14** 1920.
[11] Pyalling A.A. 2010 *Semiempirical multiphase equation of state for hydrogen* High temperature **48** 163-169
[12] Kopyshev V.P., Medvedev A.B., Khrustalev V.V. 2006 *Equation of state of explosion products on the basis of a modified Van der Waals model* Combustion, Explosion, and Shock Waves. **42** 76-87.
[13] Verbitskaya O.V., Kuznetsova O.V., Mironova E.E., Sapozhnikov A.T., Sokolov V.P. 2007 *Integrated information and technological environment for development of equations of state* VANT **3-4** 37-45
[14] Khishchenko K.V., Fortov V.E., Lomonosov I.V. 2005 *Multiphase Equation of State for Carbon over Wide Range of Temperatures and Pressures* Int. J. Thermophys **26** 479.
[15] Jean Clérouin, Pierre Noiret. 2008 *Direct measurements and ab initio simulations for expanded fluid aluminum in the metal-nonmetal transition range* Phys. Rev. B. **78** 224203.
[16] Mazevet S., Ze´rah G. Ab 2005 *Initio Simulations of the K-Edge Shift along the Aluminum Hugoniot* Phys. Rev. Lett. **101** 155001.
[17] Gérald Faussurier, Christophe Blancard, Pier Luigi Silvestrelli 2009 *Evaluation of aluminum critical point using an ab initio variational approach* Phys. Rev. B. **79** 134202.
[18] Xianwen Ran, Yuying Yu, Hua Tan, Wenhui Tang 2008 *Behavior of aluminum shear modulus in solid-liquid mixed phase: Estimation with percolation theory* J. of Appl. Phys. **103** 103539.
[19] A.G. Maslov 2014 *Trajectory of charged particle in exponentially lowing electromagnetic field* Molodezhniy nauchno-technicheskiy vestnik **9**
[20] Y V Gerasimov, A G Maslov 2016 JPCS **731** 012006

Exact analytical formulae for linearly distributed vortex and source sheets influence computation in 2D vortex methods

K S Kuzmina[1,2], I K Marchevsky[1,2] and E P Ryatina[1,2]

[1] Applied Mathematics dep., Bauman Moscow State Technical University, 2-nd Baumanskaya st., 5, 105005 Moscow, Russia

[2] Institute for System Programming of the Russian Academy of Sciences, Alexander Solzhenitsyn st. 25, 109004 Moscow, Russia

E-mail: kuz-ksen-serg@yandex.ru, iliamarchevsky@mail.ru, evgeniya.ryatina@yandex.ru

Abstract. We consider the methodology of numerical schemes development for two-dimensional vortex method. We describe two different approaches to deriving integral equation for unknown vortex sheet intensity. We simulate the velocity of the surface line of an airfoil as the influence of attached vortex and source sheets. We consider a polygonal approximation of the airfoil and assume intensity distributions of free and attached vortex sheets and attached source sheet to be approximated with piecewise constant or piecewise linear (continuous or discontinuous) functions. We describe several specific numerical schemes that provide different accuracy and have a different computational cost. The study shows that a Galerkin-type approach to solving boundary integral equation requires computing several integrals and double integrals over the panels. We obtain exact analytical formulae for all the necessary integrals, which makes it possible to raise significantly the accuracy of vortex sheet intensity computation and improve the quality of velocity and vorticity field representation, especially in proximity to the surface line of the airfoil. All the formulae are written down in the invariant form and depend only on the geometric relationship between the positions of the beginnings and ends of the panels.

1. Introduction

Vortex methods of computational fluid dynamics (CFD) [1, 2, 3, 4, 5] that belongs to the class of Lagrangian particle methods are very suitable for a wide range of engineering applications: from estimating aerodynamical coefficients for aircraft and simulating aircraft trails to dealing with hydroelastic flow-induced oscillations of structural elements and problems of industrial aerodynamics. The flows that can be simulated using vortex methods are characterized by low subsonic Mach numbers, when the effects of flow compressibility can be neglected. Efficient parallel implementations of vortex methods, adapted for computations on both CPUs and GPUs, makes it possible to solve certain hydrodynamic and hydroelastic problems much faster in comparison with traditional mesh methods [6].

In the 2D case the most efficient modifications of vortex methods are able to simulate viscous incompressible flows that are described by the Navier — Stokes equations [4, 5, 7, 8, 9]. We consider the problem in the most general case, when there is an incident flow and an airfoil of an arbitrary shape, which can be movable and/or deformable.

Modifications of vortex methods to simulate 3D flows are also known, but there are fundamental differences between 2D and 3D flows. The main feature of 2D flows is that the vorticity vector is always perpendicular to the plane of the flow. It significantly simplifies the solution procedure for 2D problems. However, in a number of applications dealing with solving coupled aerohydrodynamic and aerohydroelastic problems a flat cross-sections approach is applicable, which reduces a 3D problem to a much simpler problem of simulating 2D flows around one or several cross-sections of the structure considered. Moreover, despite the differences between mathematical formulations of 2D and 3D problems and their properties, a number of basic approaches that proved to present an advantage when solving 2D problems may be successfully used in the 3D case in much the same manner. This fact, in turn, is a powerful stimulus for development of vortex methods.

2. Velocity field reconstruction

Vortex methods of simulating the flow around airfoils are based on the principle that was discovered by Professor N.E. Zhukovsky. According to this principle, it is possible to replace a stationary airfoil by an attached vortex layer placed on its surface line [10]. Conformal mapping technique can be used to find the vortex sheet intensity for airfoils of primitive shapes; such solutions can serve as benchmarks for verification of the numerical schemes being developed.

When solving the Navier — Stokes equations, Zhukovsky's classical ideas remain correct, but such vortex sheet is free instead of being attached, and according to Lighthill's approach [11], it can be considered as a result of the vorticity flux acting on the airfoil surface line K during an infinitesimal time period. It means that the vorticity concentrated in this vortex sheet with the intensity $\gamma(\boldsymbol{r})$, $\boldsymbol{r} \in K$ is not connected with the airfoil and, after being generated on its surface line, it becomes part of the vortex wake and moves in the flow according to the governing equations.

If the airfoil is movable and/or deformable, we should introduce in addition to the free vortex sheet mentioned above, *i.e.* attached vortex sheet with the intensity γ^{att} and attached source sheet with the intensity q^{att} [12, 13]. For simplicity, we assume that the velocity $\boldsymbol{V}_K(\boldsymbol{r})$ is known for an arbitrary point on the airfoil surface line K, so the intensities of the attached sheets can be found explicitly:

$$\gamma^{att}(\boldsymbol{r}) = \boldsymbol{V}_K(\boldsymbol{r}) \cdot \boldsymbol{\tau}(\boldsymbol{r}), \qquad q^{att}(\boldsymbol{r}) = \boldsymbol{V}_K(\boldsymbol{r}) \cdot \boldsymbol{n}(\boldsymbol{r}), \qquad \boldsymbol{r} \in K,$$

where $\boldsymbol{\tau}(\boldsymbol{r})$ and $\boldsymbol{n}(\boldsymbol{r})$ are tangent and normal unit vectors to the airfoil surface line, respectively.

In order to compute the flow velocity for an arbitrary point of the flow domain, we use a generalized Biot — Savart law [14]:

$$\boldsymbol{V}(\boldsymbol{r}) = \boldsymbol{V}_\infty + \frac{1}{2\pi} \int_S \frac{\boldsymbol{\Omega}(\boldsymbol{\xi}) \times (\boldsymbol{r} - \boldsymbol{\xi})}{|\boldsymbol{r} - \boldsymbol{\xi}\,|^2} dS_\xi + \frac{1}{2\pi} \oint_K \frac{\boldsymbol{\gamma}(\boldsymbol{\xi}) \times (\boldsymbol{r} - \boldsymbol{\xi})}{|\boldsymbol{r} - \boldsymbol{\xi}\,|^2} dl_\xi + \\ + \frac{1}{2\pi} \oint_K \frac{\boldsymbol{\gamma}^{att}(\boldsymbol{\xi}) \times (\boldsymbol{r} - \boldsymbol{\xi})}{|\boldsymbol{r} - \boldsymbol{\xi}\,|^2} dl_\xi + \frac{1}{2\pi} \oint_K \frac{q^{att}(\boldsymbol{\xi})(\boldsymbol{r} - \boldsymbol{\xi})}{|\boldsymbol{r} - \boldsymbol{\xi}\,|^2} dl_\xi. \qquad (1)$$

Here S is the flow domain; attached and free vortex sheet intensity vectors and the vorticity vector are

$$\boldsymbol{\gamma}^{att}(\boldsymbol{\xi}) = \gamma^{att}(\boldsymbol{\xi})\,\boldsymbol{k}, \qquad \boldsymbol{\gamma}(\boldsymbol{\xi}) = \gamma(\boldsymbol{\xi})\,\boldsymbol{k}, \qquad \boldsymbol{\Omega}(\boldsymbol{\xi}) = \Omega(\boldsymbol{\xi})\,\boldsymbol{k}, \qquad (2)$$

respectively, where $\boldsymbol{k}$ is a unit vector orthogonal to the flow plane; on the airfoil surface $\boldsymbol{n}(\boldsymbol{r}) \times \boldsymbol{\tau}(\boldsymbol{r}) = \boldsymbol{k}$.

The flow velocity limit value on the surface line of the airfoil cannot be computed directly by using (1), because $\boldsymbol{V}(\boldsymbol{r})$ has a jump discontinuity there [1]. The following formula should be used to compute it:

$$\boldsymbol{V}_{-}(\boldsymbol{r}) = \boldsymbol{V}(\boldsymbol{r}) - \frac{\gamma(\boldsymbol{r}) - \gamma^{att}(\boldsymbol{r})}{2}\boldsymbol{\tau}(\boldsymbol{r}) + \frac{q^{att}(\boldsymbol{r})}{2}\boldsymbol{n}(\boldsymbol{r}), \qquad \boldsymbol{r} \in K. \tag{3}$$

3. Integral equation

In order to solve the problem of flow simulation around an airfoil we need to find the distribution of free vortex sheet intensity over the surface line. We can do it by satisfying the boundary condition on the surface line:

$$\boldsymbol{V}_{-}(\boldsymbol{r}) = \boldsymbol{V}_K(\boldsymbol{r}), \qquad \boldsymbol{r} \in K. \tag{4}$$

It can be shown that the equalities between normal components of velocities

$$\boldsymbol{V}_{-}(\boldsymbol{r}) \cdot \boldsymbol{n}(\boldsymbol{r}) = \boldsymbol{V}_K(\boldsymbol{r}) \cdot \boldsymbol{n}(\boldsymbol{r}), \qquad \boldsymbol{r} \in K, \tag{5}$$

or between their tangent components

$$\boldsymbol{V}_{-}(\boldsymbol{r}) \cdot \boldsymbol{\tau}(\boldsymbol{r}) = \boldsymbol{V}_K(\boldsymbol{r}) \cdot \boldsymbol{\tau}(\boldsymbol{r}), \qquad \boldsymbol{r} \in K, \tag{6}$$

are two equivalent sufficient conditions for satisfying the equality (4).

However, the resulting integral equations with respect to the unknown $\gamma(\boldsymbol{r})$, which can be derived from (5) and (6) after substituting the expression (3) for the limit value of velocity in those, have quite different properties.

In the framework of "classical" implementations of vortex methods, the boundary condition is used in form (5), and the governing integral equation, which is a first kind equation in this case, can be written down as

$$\frac{1}{2\pi}\oint_K Q_n(\boldsymbol{r}, \boldsymbol{\xi})\gamma(\boldsymbol{\xi})\, dl_\xi = f_n(\boldsymbol{r}), \qquad \boldsymbol{r} \in K, \tag{7}$$

where the kernel $Q_n(\boldsymbol{r}, \boldsymbol{\xi})$ is an unbounded (singular) one of Hilbert-type

$$Q_n(\boldsymbol{r}, \boldsymbol{\xi}) = \frac{\boldsymbol{k} \times (\boldsymbol{r} - \boldsymbol{\xi})}{|\boldsymbol{r} - \boldsymbol{\xi}|^2} \cdot \boldsymbol{n}(\boldsymbol{r}) = -\frac{\boldsymbol{\tau}(\boldsymbol{r}) \cdot (\boldsymbol{r} - \boldsymbol{\xi})}{|\boldsymbol{r} - \boldsymbol{\xi}|^2},$$

and the right-hand side $f_n(\boldsymbol{r})$ is a known function which depends on the shape of the airfoil, incident flow velocity, velocity of the airfoil itself and vorticity distribution in the flow domain:

$$f_n(\boldsymbol{r}) = -\frac{1}{2}q^{att}(\boldsymbol{r}) - \boldsymbol{n}(\boldsymbol{r}) \cdot \Bigg(\boldsymbol{V}_\infty - \boldsymbol{V}_K(\boldsymbol{r}) + \frac{1}{2\pi}\int_S \frac{\boldsymbol{\Omega}(\boldsymbol{\xi}) \times (\boldsymbol{r} - \boldsymbol{\xi})}{|\boldsymbol{r} - \boldsymbol{\xi}|^2} dS_\xi + \\ + \frac{1}{2\pi}\oint_K \frac{\boldsymbol{\gamma}^{att}(\boldsymbol{\xi}) \times (\boldsymbol{r} - \boldsymbol{\xi})}{|\boldsymbol{r} - \boldsymbol{\xi}|^2} dl_\xi + \frac{1}{2\pi}\oint_K \frac{q^{att}(\boldsymbol{\xi})(\boldsymbol{r} - \boldsymbol{\xi})}{|\boldsymbol{r} - \boldsymbol{\xi}|^2} dl_\xi \Bigg). \tag{8}$$

An approximate numerical solution of equation (7) requires specific quadrature formulae; in the simplest case, those can be discrete vortex type quadrature formulae [1], which permit us to select the principal value of the corresponding singular integral in the Cauchy sense.

If we construct the numerical scheme of our vortex method on the basis of a "tangent" approach (6), the integral equation transforms into the second kind equation

$$\frac{1}{2\pi}\oint_K Q_\tau(\boldsymbol{r}, \boldsymbol{\xi})\gamma(\boldsymbol{\xi})\, dl_\xi - \frac{1}{2}\gamma(\boldsymbol{r}) = f_\tau(\boldsymbol{r}), \qquad \boldsymbol{r} \in K, \tag{9}$$

where the kernel

$$Q_\tau(\boldsymbol{r}, \boldsymbol{\xi}) = \frac{\boldsymbol{k} \times (\boldsymbol{r} - \boldsymbol{\xi})}{|\boldsymbol{r} - \boldsymbol{\xi}|^2} \cdot \boldsymbol{\tau}(\boldsymbol{r}) = \frac{\boldsymbol{n}(\boldsymbol{r}) \cdot (\boldsymbol{r} - \boldsymbol{\xi})}{|\boldsymbol{r} - \boldsymbol{\xi}|^2}$$

is now a uniformly bounded function when K is a smooth curve, and unbounded only in proximity to angle points or sharp edges of a non-smooth airfoil surface line K, and the right-hand side has the following form:

$$f_\tau(\boldsymbol{r}) = -\frac{1}{2}\gamma^{att}(\boldsymbol{r}) - \boldsymbol{\tau}(\boldsymbol{r}) \cdot \Bigg(\boldsymbol{V}_\infty - \boldsymbol{V}_K(\boldsymbol{r}) + \frac{1}{2\pi} \int_S \frac{\boldsymbol{\Omega}(\boldsymbol{\xi}) \times (\boldsymbol{r} - \boldsymbol{\xi})}{|\boldsymbol{r} - \boldsymbol{\xi}|^2} dS_\xi + \\ + \frac{1}{2\pi} \oint_K \frac{\gamma^{att}(\boldsymbol{\xi}) \times (\boldsymbol{r} - \boldsymbol{\xi})}{|\boldsymbol{r} - \boldsymbol{\xi}|^2} dl_\xi + \frac{1}{2\pi} \oint_K \frac{q^{att}(\boldsymbol{\xi})(\boldsymbol{r} - \boldsymbol{\xi})}{|\boldsymbol{r} - \boldsymbol{\xi}|^2} dl_\xi \Bigg). \quad (10)$$

Results of numerical experiments show that the "tangent" approach, which corresponds to numerical solving the equation (9), makes it possible to raise the accuracy of vortex sheet intensity computation significantly. Nevertheless, in some cases a "normal" approach (7) can also be useful.

We should note that both equations (7) and (9) have an infinite set of solutions. In order to select a physically plausible unique solution, an additional condition should be added, which in most cases corresponds to given value of the integral from solution over the surface line:

$$\oint_K \gamma(\boldsymbol{\xi}) dl_\xi = \Gamma. \quad (11)$$

4. Problem discretization

When solving a problem using vortex methods, the vortex wake is normally simulated by discrete vortex-type singularities (vortex elements):

$$\Omega(\boldsymbol{r}, t) = \sum_{w=1}^{n} \Gamma_w \tilde{\delta}(\boldsymbol{r} - \boldsymbol{r}_w). \quad (12)$$

Here n is the number of vortex elements, Γ_w and $\boldsymbol{r}_w$ are intensities and positions of vortex elements, respectively, and $\tilde{\delta}$ is a two-dimensional Dirac delta function. Taking into account (2) and (12), we obtain for the term, which corresponds to the integral over the flow domain in right-hand side functions (8) and (10)

$$\frac{1}{2\pi} \int_S \frac{\boldsymbol{\Omega}(\boldsymbol{\xi}) \times (\boldsymbol{r} - \boldsymbol{\xi})}{|\boldsymbol{r} - \boldsymbol{\xi}|^2} dS_\xi = \sum_{w=1}^{n} \frac{\Gamma_w}{2\pi} \frac{\boldsymbol{k} \times (\boldsymbol{r} - \boldsymbol{r}_w)}{|\boldsymbol{r} - \boldsymbol{r}_w|^2}.$$

The simplest way to discretize the surface line K of the airfoil is to approximate it by a polygon consisting of N rectilinear segments K_i, which we call hereinafter "panels". Let us denote panel lengthes by L_i, their normal and tangent unit vectors by $\boldsymbol{n}_i$ and $\boldsymbol{\tau}_i$, respectively, $i = 1, \ldots, N$. Now all the integrals over the surface line in the previous formulae can be written down as sums of integrals over separate panels.

We then consider the intensity distributions of the free and attached vortex sheets and source sheet over the surface line to be piecewise-linear functions:

$$\gamma(\boldsymbol{r}) = \sum_{i=1}^{N} \Big(\gamma_i \phi_0^i(\boldsymbol{r}) + \delta_i \phi_1^i(\boldsymbol{r}) \Big), \qquad \gamma^{att}(\boldsymbol{r}) = \sum_{i=1}^{N} \Big(\gamma_i^{att} \phi_0^i(\boldsymbol{r}) + \delta_i^{att} \phi_1^i(\boldsymbol{r}) \Big),$$

$$q^{att}(\boldsymbol{r}) = \sum_{i=1}^{N} \Big(q_i^{att} \phi_0^i(\boldsymbol{r}) + p_i^{att} \phi_1^i(\boldsymbol{r}) \Big), \qquad \boldsymbol{r} \in K.$$

Here $\phi_j^i(\boldsymbol{r})$, $i = 1, \ldots, N$, $j = 0, 1$, is a set of basis functions:

$$\phi_0^i(\boldsymbol{r}) = \begin{cases} 1, & \boldsymbol{r} \in K_i, \\ 0, & \boldsymbol{r} \notin K_i; \end{cases} \qquad \phi_1^i(\boldsymbol{r}) = \begin{cases} \dfrac{(\boldsymbol{r} - \boldsymbol{c}_i) \cdot \boldsymbol{\tau}_i}{L_i}, & \boldsymbol{r} \in K_i, \\ 0, & \boldsymbol{r} \notin K_i, \end{cases} \tag{13}$$

where $\boldsymbol{c}_i$ is the center of the i-th panel, and the unit tangent vector $\boldsymbol{\tau}_i$ is directed from the beginning to the end of the i-th panel. So, the basis function ϕ_0^i is constant over the panel, while the basis function ϕ_1^i varies linearly from $-1/2$ to $1/2$ along the i-th panel.

Different approaches can be used to construct a numerical solution $\gamma(\boldsymbol{r})$ over the whole airfoil surface line, as well as to approximate $\gamma^{att}(\boldsymbol{r})$ and $q^{att}(\boldsymbol{r})$. In order to find unknown coefficients in the representation of the solution, we use a Galerkin-type approach: the residual of equations (7) or (9) should be orthogonal to projection functions $\{\psi_k(\boldsymbol{r})\}$.

A number of possible ways to numerical solution of the equation (9) are briefly described below.

(i) In the simplest case a piecewise-constant discontinuous approximation can be considered, so $\delta_i = 0$, $\delta_i^{att} = 0$ and $p_i^{att} = 0$ for all $i = 1, \ldots, N$. So γ_i^{att} and q_i^{att} should be chosen as average values of the corresponding intensities over the panels:

$$\gamma_i^{att} = \frac{1}{L_i} \int_{K_i} \gamma^{att}(\boldsymbol{r}) dl_\xi \approx \boldsymbol{V}_K(\boldsymbol{c}_i) \cdot \boldsymbol{\tau}_i, \qquad q_i^{att} = \frac{1}{L_i} \int_{K_i} q^{att}(\boldsymbol{r}) dl_\xi \approx \boldsymbol{V}_K(\boldsymbol{c}_i) \cdot \boldsymbol{n}_i.$$

As a result, the unknown distribution depends only on N unknown values γ_i, which can be found from conditions of orthogonality to a set of projection functions $\{\psi_k(\boldsymbol{r})\} = \{\phi_0^k(\boldsymbol{r})\}_{k=1}^N$.
This approximation makes it possicle to achieve second-order accuracy with respect to the average values γ_i of the free vortex sheet intensity over the panels and first-order accuracy in the L_1 norm.
Publication [15] presents a description of this approach and its comparison with less accurate ones, as well as all the necessary formulae for computations.

(ii) Approximations of the solution $\gamma(\boldsymbol{r})$ and known distributions $\gamma^{att}(\boldsymbol{r})$ and $q^{att}(\boldsymbol{r})$ remain the same as in the previous case, but unknown values γ_i, $i = 1, \ldots, N$ can be found from collocation-type conditions, which means that the residual of equations (7) or (9) is equal to zero only at certain selected points on the surface line, as a rule, at the centers of the panels.
This modification is equivalent to the previous one, if $\{\psi_k(\boldsymbol{r})\} = \{\delta(\boldsymbol{r} - \boldsymbol{c}_k)\}_{k=1}^N$, where δ is a Dirac delta-function:

$$\int_{K_i} g(\boldsymbol{r}) \psi_i(\boldsymbol{r}) dl_\xi = g(\boldsymbol{c}_i), \qquad i = 1, \ldots, N.$$

(iii) In the most complicated case, piecewise-linear discontinuous approximation are considered, so the resulting solution depends on $2N$ unknown values γ_i and δ_i, $i = 1, \ldots, N$, which can be found from the same approach with the only difference that now the set of projection functions consists of functions $\{\phi_0^i(\boldsymbol{r})\}_{i=1}^N$ and $\{\phi_1^i(\boldsymbol{r})\}_{i=1}^N$ together. This method provides the second-order accuracy with respect to the average values of the free vortex sheet intensity γ_i and to the vortex sheet intensity distribution in the L_1 norm.
Paper [16] presents a description of this method and the results of solving test problems for airfoils of simple shapes (elliptical and the Zhukovsky wing).

(iv) A "finite element-type" approach provides a continuous piecewise-linear solution approximation. The basis functions in this case are first-order linear shape functions, which can be easily constructed from (13):

$$\hat{\phi}^i(\boldsymbol{r}) = \begin{cases} \frac{1}{2}\phi_0^{i-1}(\boldsymbol{r}) + \phi_1^{i-1}(\boldsymbol{r}), & \boldsymbol{r} \in K_{i-1}, \\ \frac{1}{2}\phi_0^i(\boldsymbol{r}) - \phi_1^i(\boldsymbol{r}), & \boldsymbol{r} \in K_i, \qquad i = 1, \ldots, N, \quad K_0 \equiv K_N, \end{cases}$$

and finally

$$\gamma(\boldsymbol{r}) = \sum_{i=1}^{N} \hat{\gamma}_i \hat{\phi}^i(\boldsymbol{r}),$$

where unknown values $\hat{\gamma}_i$ correspond to the intensities of the free vortex sheet at the beginnings of the panels. In order to determine them, as in the FEM, we select projection functions to be equal to the shape functions: $\{\psi_k(\boldsymbol{r})\} = \{\hat{\phi}^k(\boldsymbol{r})\}_{k=1}^N$.
If the exact solution is continuous, this modification provides second-order accuracy, same as the previous one, but the dimension of the problem is twice as small and is equal to the number of panels N. But if the airfoil has sharp edges or angle point, it leads to incorrect behavior of the numerical solution in proximity to such points.
Note, that in the framework of the FEM approach Dirac delta functions can also be used as projection ones.

(v) A "hybrid" schemes involves a piecewise-linear solution, which is discontinuous only at N_d chosen points, and continuous at all others. The number of unknown values here is $N + N_d$, all basis functions can be expressed through ϕ_0^i and ϕ_1^i. Projection functions again coincide with the basis ones.
Numerical experiments show that this approach remains correct and provides second-order accuracy for airfoils with angle points and sharp edges when they are marked as discontinuity points. This scheme may be called "cost-efficient" because the dimension of the resulting matrix is $N + N_d$, which is much smaller than $2N$.

Some other modifications are also possible for solving both equations (7) and (9) numerically, but the main principles of constructing those remain the same.

The easiest way to take into account the additional condition (11) is to add one extra "regularization" variable R to all the equations in the resulting linear system and solve it together with the linear equation, which follows from (11) after discretization. Another possible approach is to find pseudo-solution of a non-regularized linear system.

5. Analytical computation of integrals

In order to compute matrix and right-hand side coefficients for discrete analogues of the equations (7) and (9) according to numerical schemes described in the previous section or some other schemes that are constructed using Galerkin-type approach with piecewise-constant and piecewise-linear basis and projection functions, it is necessary to compute the following integrals (note that $\phi_0^i \equiv 1$ over the i-th panel):

$$\boldsymbol{I}_j^0(\boldsymbol{r}) = \int_{K_j} \boldsymbol{Q}(\boldsymbol{r}-\boldsymbol{\xi})dl_\xi, \qquad \boldsymbol{I}_j^1(\boldsymbol{r}) = \int_{K_j} \boldsymbol{Q}(\boldsymbol{r}-\boldsymbol{\xi})\phi_1^j(\boldsymbol{\xi})dl_\xi,$$

$$\boldsymbol{I}_{ij}^{00} = \int_{K_i} dl_r \int_{K_j} \boldsymbol{Q}(\boldsymbol{r}-\boldsymbol{\xi})dl_\xi, \qquad \boldsymbol{I}_{ij}^{01} = \int_{K_i} dl_r \int_{K_j} \boldsymbol{Q}(\boldsymbol{r}-\boldsymbol{\xi})\phi_1^j(\boldsymbol{\xi})dl_\xi,$$

$$\boldsymbol{I}_{ij}^{10} = \int_{K_i} dl_r \int_{K_j} \boldsymbol{Q}(\boldsymbol{r}-\boldsymbol{\xi})\phi_1^i(\boldsymbol{r})dl_\xi, \qquad \boldsymbol{I}_{ij}^{11} = \int_{K_i} dl_r \int_{K_j} \boldsymbol{Q}(\boldsymbol{r}-\boldsymbol{\xi})\phi_1^i(\boldsymbol{r})\phi_1^j(\boldsymbol{\xi})dl_\xi.$$

We write down these integrals with a "vector kernel"

$$\boldsymbol{Q}(\boldsymbol{r}-\boldsymbol{\xi}) = \frac{\boldsymbol{r}-\boldsymbol{\xi}}{|\boldsymbol{r}-\boldsymbol{\xi}|^2},$$

in order to have the possibility to adapt them easily to computing the integrals in equations (7) and (9), taking into account the expressions (8) and (10) for their right-hand sides. It should be noted that the kernel $\boldsymbol{Q}$ is anti-symmetric:

$$\boldsymbol{Q}(\boldsymbol{r}-\boldsymbol{\xi}) = -\boldsymbol{Q}(\boldsymbol{\xi}-\boldsymbol{r}),$$

so all other necessary integrals can be expressed through the ones considered.

5.1. Computation of integrals $\boldsymbol{I}_j^0(\boldsymbol{r})$ and $\boldsymbol{I}_j^1(\boldsymbol{r})$

Figure 1 shows the j-th panel of the airfoil and the point $\boldsymbol{r}$ and introduces several auxiliary vectors for computation of the integrals $\boldsymbol{I}_j^0(\boldsymbol{r})$ and $\boldsymbol{I}_j^1(\boldsymbol{r})$: vectors $\boldsymbol{s}$ and $\boldsymbol{p}$ join the beginning and the end of the j-th panel to the point $\boldsymbol{r}$, respectively.

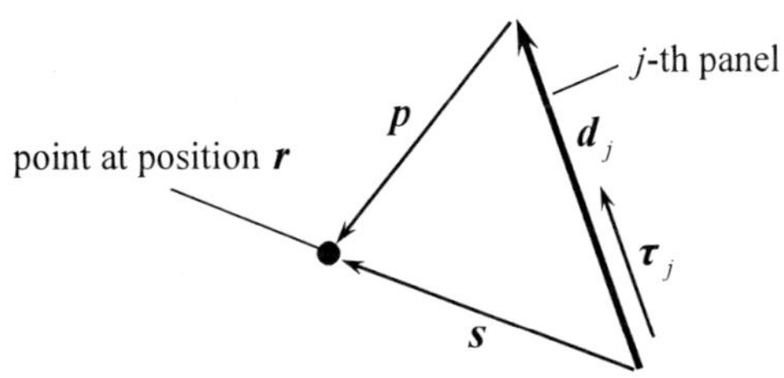

Figure 1. The j-th panel and auxiliary vectors

Employing these vectors after some transformations it is possible to write down the following formulae:

$$\boldsymbol{I}_j^0(\boldsymbol{r}) = \int_{K_j} \boldsymbol{Q}(\boldsymbol{r}-\boldsymbol{\xi})dl_\xi = \alpha\boldsymbol{u}_0 \times \boldsymbol{k} + \lambda\boldsymbol{u}_0,$$

$$\boldsymbol{I}_j^1(\boldsymbol{r}) = \int_{K_j} \boldsymbol{Q}(\boldsymbol{r}-\boldsymbol{\xi})\phi_1^j(\boldsymbol{\xi})dl_\xi = \alpha\boldsymbol{u}_1 \times \boldsymbol{k} + \lambda\boldsymbol{u}_1 - \boldsymbol{\tau}_j.$$

Here we denote

$$\boldsymbol{u}_0 = \boldsymbol{\omega}(\boldsymbol{\tau}_j, \boldsymbol{\tau}_j, \boldsymbol{\tau}_j) = \boldsymbol{\tau}_j, \quad \boldsymbol{u}_1 = \frac{1}{2|\boldsymbol{d}_j|}\boldsymbol{\omega}(\boldsymbol{p}+\boldsymbol{s}, \boldsymbol{\tau}_j, \boldsymbol{\tau}_j),$$
$$\alpha = \angle(\boldsymbol{p}, \boldsymbol{s}), \quad \lambda = \ln\frac{|\boldsymbol{s}|}{|\boldsymbol{p}|},$$

where

$$\boldsymbol{\omega}(\boldsymbol{a}, \boldsymbol{b}, \boldsymbol{c}) = (\boldsymbol{a}\cdot\boldsymbol{b})\,\boldsymbol{c} + (\boldsymbol{a}\times\boldsymbol{b})\times\boldsymbol{c}, \qquad \angle(\boldsymbol{p}, \boldsymbol{s}) = \mathrm{Arg}\big(\boldsymbol{p}\cdot\boldsymbol{s} + i(\boldsymbol{p}\,\boldsymbol{s}\,\boldsymbol{k})\big). \tag{14}$$

5.2. Computation of integrals $\boldsymbol{I}_{ij}^{00}$, $\boldsymbol{I}_{ij}^{01}$, $\boldsymbol{I}_{ij}^{10}$ and $\boldsymbol{I}_{ij}^{11}$

Figure 2 shows the i-th and j-th panels and introduces auxiliary vectors similarly to the previous case: vectors $\boldsymbol{d}_i$ and $\boldsymbol{d}_j$ are codirectional with the i-th and j-th panels; vectors $\boldsymbol{s}_1$ and $\boldsymbol{s}_2$ join the beginning of the j-th panel with the end and the beginning of the i-th panel, respectively;

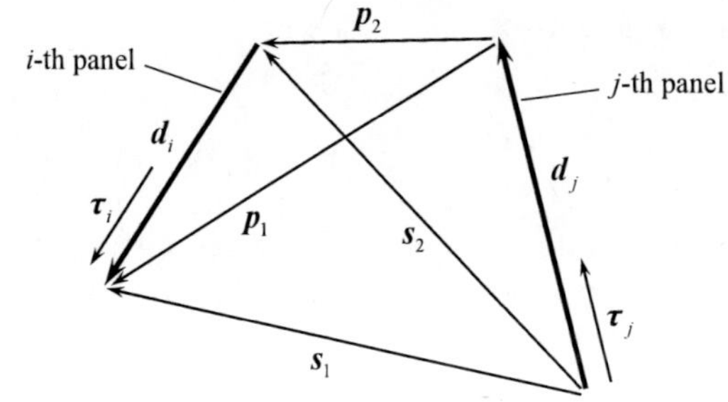

Figure 2. The i-th and j-th panels and auxiliary vectors

vectors $\boldsymbol{p}_1$ and $\boldsymbol{p}_2$ join the end of the j-th panel with the end and the beginning of the i-th panel, respectively.

These vectors enabled us to derive the following formulae after a number of transformations performed in the *Wolfram Mathematica* computer algebra system:

$$\boldsymbol{I}_{ij}^{00} = \int_{K_i} dl_r \int_{K_j} \boldsymbol{Q}(\boldsymbol{r}-\boldsymbol{\xi})dl_\xi = (\alpha_1\boldsymbol{v}_1 + \alpha_2\boldsymbol{v}_2 + \alpha_3\boldsymbol{v}_3) \times \boldsymbol{k} + (\lambda_1\boldsymbol{v}_1 + \lambda_2\boldsymbol{v}_2 + \lambda_3\boldsymbol{v}_3),$$

$$\boldsymbol{I}_{ij}^{01} = \int_{K_i} dl_r \int_{K_j} \boldsymbol{Q}(\boldsymbol{r}-\boldsymbol{\xi})\phi_1^j(\boldsymbol{\xi})dl_\xi =$$
$$= \big((\alpha_1+\alpha_3)\boldsymbol{v}_4 + \alpha_3\boldsymbol{v}_5\big) \times \boldsymbol{k} + \big((\lambda_1+\lambda_3)\boldsymbol{v}_4 + \lambda_3\boldsymbol{v}_5\big) + \frac{1}{2}|\boldsymbol{d}_j|\boldsymbol{\tau}_i,$$

$$\boldsymbol{I}_{ij}^{10} = \int_{K_i} dl_r \int_{K_j} \boldsymbol{Q}(\boldsymbol{r}-\boldsymbol{\xi})\phi_1^i(\boldsymbol{r})dl_\xi =$$
$$= \big((\alpha_1+\alpha_3)\boldsymbol{v}_6 + \alpha_4\boldsymbol{v}_7\big) \times \boldsymbol{k} + \big((\lambda_1+\lambda_3)\boldsymbol{v}_6 + \lambda_4\boldsymbol{v}_7\big) - \frac{1}{2}|\boldsymbol{d}_i|\boldsymbol{\tau}_j,$$

$$\boldsymbol{I}_{ij}^{11} = \int_{K_i} dl_r \int_{K_j} \boldsymbol{Q}(\boldsymbol{r}-\boldsymbol{\xi})\phi_1^i(\boldsymbol{r})\phi_1^j(\boldsymbol{\xi})dl_\xi =$$
$$= \big((\alpha_1+\alpha_3)\boldsymbol{v}_8 + \alpha_4\boldsymbol{v}_9 + \alpha_3\boldsymbol{v}_{10}\big) \times \boldsymbol{k} + \big((\lambda_1+\lambda_3)\boldsymbol{v}_8 + \lambda_4\boldsymbol{v}_9 + \lambda_3\boldsymbol{v}_{10}\big) +$$
$$+ \frac{1}{12}\big(|\boldsymbol{d}_j|\boldsymbol{\tau}_i + |\boldsymbol{d}_i|\boldsymbol{\tau}_j - 2\boldsymbol{\omega}(\boldsymbol{s}_1, \boldsymbol{\tau}_i, \boldsymbol{\tau}_j)\big).$$

Here we have introduced the following notations, taking into account (14):

$$\alpha_1 = \angle(\boldsymbol{s}_2, \boldsymbol{s}_1), \quad \alpha_2 = \angle(\boldsymbol{s}_2, \boldsymbol{p}_1), \quad \alpha_3 = \angle(\boldsymbol{p}_1, \boldsymbol{p}_2), \quad \alpha_4 = \alpha_2 + \alpha_3 = \angle(\boldsymbol{s}_2, \boldsymbol{p}_2)$$

$$\lambda_1 = \ln\frac{|\boldsymbol{s}_1|}{|\boldsymbol{s}_2|}, \quad \lambda_2 = \ln\frac{|\boldsymbol{p}_1|}{|\boldsymbol{s}_2|}, \quad \lambda_3 = \ln\frac{|\boldsymbol{p}_2|}{|\boldsymbol{p}_1|}, \quad \lambda_4 = \lambda_2 + \lambda_3 = \ln\frac{|\boldsymbol{p}_2|}{|\boldsymbol{s}_2|},$$

$$\boldsymbol{v}_1 = \boldsymbol{\omega}(\boldsymbol{s}_1, \boldsymbol{\tau}_i, \boldsymbol{\tau}_j), \quad \boldsymbol{v}_2 = -\boldsymbol{\omega}(\boldsymbol{d}_i, \boldsymbol{\tau}_i, \boldsymbol{\tau}_j), \quad \boldsymbol{v}_3 = \boldsymbol{\omega}(\boldsymbol{p}_2, \boldsymbol{\tau}_i, \boldsymbol{\tau}_j),$$

$$\boldsymbol{v}_4 = -\frac{1}{2|\boldsymbol{d}_i|}\big(((\boldsymbol{s}_1+\boldsymbol{s}_2)\cdot\boldsymbol{\tau}_i)\boldsymbol{\omega}(\boldsymbol{s}_1, \boldsymbol{\tau}_i, \boldsymbol{\tau}_j) - |\boldsymbol{s}_1|^2\boldsymbol{\tau}_j\big), \quad \boldsymbol{v}_5 = \frac{|\boldsymbol{d}_j|}{2|\boldsymbol{d}_i|}\boldsymbol{\omega}(\boldsymbol{s}_1+\boldsymbol{p}_2, \boldsymbol{\tau}_i, \boldsymbol{\tau}_i),$$

$$\boldsymbol{v}_6 = \frac{1}{2|\boldsymbol{d}_j|}\big(((\boldsymbol{p}_1+\boldsymbol{s}_1)\cdot\boldsymbol{\tau}_j)\boldsymbol{\omega}(\boldsymbol{s}_1, \boldsymbol{\tau}_i, \boldsymbol{\tau}_j) - |\boldsymbol{s}_1|^2\boldsymbol{\tau}_i\big), \quad \boldsymbol{v}_7 = -\frac{|\boldsymbol{d}_i|}{2|\boldsymbol{d}_j|}\boldsymbol{\omega}(\boldsymbol{s}_1+\boldsymbol{p}_2, \boldsymbol{\tau}_j, \boldsymbol{\tau}_j),$$

$$\boldsymbol{v}_8 = \frac{1}{12|\boldsymbol{d}_i||\boldsymbol{d}_j|}\Big(2(\boldsymbol{s}_1\cdot\boldsymbol{\omega}(\boldsymbol{s}_1-3\boldsymbol{p}_2, \boldsymbol{\tau}_i, \boldsymbol{\tau}_j))\boldsymbol{\omega}(\boldsymbol{s}_1, \boldsymbol{\tau}_i, \boldsymbol{\tau}_j) - |\boldsymbol{s}_1|^2(\boldsymbol{s}_1-3\boldsymbol{p}_2)\Big) - \frac{1}{4}\boldsymbol{\omega}(\boldsymbol{s}_1, \boldsymbol{\tau}_i, \boldsymbol{\tau}_j),$$

$$\boldsymbol{v}_9 = -\frac{1}{12}\frac{|\boldsymbol{d}_i|}{|\boldsymbol{d}_j|}\boldsymbol{\omega}(\boldsymbol{d}_i, \boldsymbol{\tau}_j, \boldsymbol{\tau}_j), \quad \boldsymbol{v}_{10} = -\frac{1}{12}\frac{|\boldsymbol{d}_j|}{|\boldsymbol{d}_i|}\boldsymbol{\omega}(\boldsymbol{d}_j, \boldsymbol{\tau}_i, \boldsymbol{\tau}_i).$$

If $i = j$, then the corresponding integrals are singular, and their limit values, which are required to compute matrix coefficients and right-hand sides, are equal to zero:

$$\boldsymbol{I}_{ii}^{00} = \boldsymbol{I}_{ii}^{01} = \boldsymbol{I}_{ii}^{10} = \boldsymbol{I}_{ii}^{11} = \mathbf{0}.$$

For neighboring panels, when $i = j + 1$ ($\boldsymbol{p}_2 = \mathbf{0}$, $\boldsymbol{s}_1 \neq \mathbf{0}$), coefficients α_3 and λ_3 vanish. For the other neighboring case, when $j = i + 1$ ($\boldsymbol{s}_1 = \mathbf{0}$, $\boldsymbol{p}_2 \neq \mathbf{0}$), coefficients α_1 and λ_1 should be equal to zero.

6. Conclusion

We derive exact analytical formulae for computing the integrals that arise in 2D vortex methods during simulating the flow around airfoils in hydrodynamic and coupled hydroelastic problems. These formulae are suitable for arbitrary airfoils when their surface lines are approximated by rectilinear panels. The distribution of known or unknown intensities of vortex or source sheets over the panels can be piecewise-constant or piecewise-linear, and either collocation-type or finite element-type methods can be used for solving boundary integral equations. Implementing a similar approach with higher-order basis functions (in particular, piecewise-quadratic) is unreasonable since it would not lead to achieving higher accuracy. Further improvement of the algorithm is possible only by taking into account the curvilinearity of the panels, for example, by using a least squares approach [17].

Usage of the derived formulae makes it possible to raise significantly the accuracy of computing the vortex sheet intensity and to improve the quality of velocity and vorticity field representation, especially, in proximity to the surface line of the airfoil.

Acknowledgement

This work is supported by the Russian Science Foundation (grant 17-79-20445).

References

[1] Lifanov I 1996 *Singular Integral Equations and Discrete Vortices* (Utrecht: VSP)
[2] Cottet G-H and Koumoutsakos P 2000 *Vortex Methods: Theory and Practice* (Cambridge: CUP)
[3] Lewis R 2005 *Vortex Element Methods for Fluid Dynamic Analysis of Engineering Systems* (Cambridge: CUP)
[4] Dynnikova G 2003 *Fluid Dynamics* **38** 670
[5] Dynnikova G 2004 *Doklady Physics* **49** 648
[6] Kuzmina K, Marchevsky I and Moreva V 2015 *Procedia Computer Science* **66** 73
[7] Chorin A 1973 *J. Fluid Mech.* **57** 785
[8] Degond P and Mas-Gallic S 1989 *Math. Comp.* **53** 485
[9] Ogami Y and Akamatsu T 1991 *Computers & Fluids* **19:3–4** 433
[10] Tokaty G 1994 *A History and Philosophy of Fluid Mechanics* (Dover: Courier Corporation)
[11] Lighthill M 1963 *Boundary Layer Theory* ed J Rosenhead (Oxford: OUP) Introduction pp 54–61
[12] Anronov P, Guvernuk S and Dynnikova G 2006 *Vortex Methods for Computation of Unsteady Hydrodynamic Loads* (Moscow: Moscow State University Press)
[13] Kuzmina K and Marchevsky 2015 *Proc. 4th Int. Conf. on Particle-Based Methods – Fundamentals and Applications, PARTICLES 2015 (Barcelona)* (Barcelona: Int. Center for Numerical Methods in Engineering (CIMNE)) p 680
[14] Saffman P 1992 *Vortex Dynamics* (Cambridge: CUP)
[15] Moreva V and Marchevsky I 2012 *ECCOMAS 2012 – European Congress on Computational Methods in Applied Sciences and Engineering, e-Book Full Papers (Vienna)* (Barcelona: Int. Center for Numerical Methods in Engineering (CIMNE)) p 5952
[16] Kuzmina K, Marchevskii I and Ryatina E 2017 *Russian Aeronautics* **60**
[17] Kuzmina K and Marchevsky I 2016 *Herald of the Bauman Moscow State Technical University, Series Natural Sciences* **69** 93

Features of surface phase formation during case-hardening of iron- and titanium-based alloys

B E Vintaikin[1], A V Kamynin[2,3], V S Kraposhin[2], A E Smirnov[1], K V Terezanova[1], S A Cherenkova[1] and V I Sheykina[1]

[1]Bauman Moscow State Technical University, Russia, 105005, Moscow, 2nd Baumanskaya street, 5/1
[2]Spetsmagnit JSC, Russia, 127238, Moscow, Dmitrov highway, 58
[3]NUST MISiS, Russia, 119049, Moscow, Leninsky Prospekt, 4

E-mail: vintaikb@mail.ru; kseniya-t94@yandex.ru

Abstract. The article provides a detailed analysis of formation features for surface phases in technical iron and Cr20-Ni80 alloy samples that undergo case-hardening at a temperature of 850°C for 2, 4 and 6 hours of saturation in two different environments: acetylene, and molten salt consisting of sodium tetraborate and amorphous boron. We carried out an X-ray phase analysis to determine the phase structure of surface material layers that formed as a result of the case-hardening process. We discovered that after carburising it was possible to detect Fe_3C and Fe-α phases on the surface of technical iron samples, and after boriding we found FeB, Fe_2B and Fe_3B phases; we noted a lack of characteristic Fe-α and Fe-γ peaks on the X-ray diffraction pattern. We detected many different phases in the Cr20-Ni80 alloy after the same type of case-hardening. Titanium oxides appeared after case-hardening of titanium in air at 800°C. We provide data on surface structure of samples subjected to vacuum carburising: over a 2 to 6 hour interval, the layer thickness is a parabolic function of time. When carrying out electrolysis-free liquid boriding, increasing exposure time from 2 to 6 hours alters the thickness of the strengthened layer only slightly, so, when carrying out case-hardening, it is less efficient to increase saturation time in molten salt containing sodium tetraborate and amorphous boron.

1. Introduction

Some of the most important material properties are durability and wear resistance, since they affect potential industrial applications of these materials. Today there exist many methods to improve performance parameters of materials, for example, alloying or surface hardening. However, alloying is a very uneconomic method; moreover, certain properties of the materials to be strengthened sometimes make it unfeasible [1-3].

During operation, the surface of parts bears the main load, and, therefore, phase structure of a surface and its properties are greatly important for operability. Therefore, there is a significant interest in methods that make it possible to improve the properties of machine parts by only affecting their surface [4, 5].

There are a good number of methods dedicated to altering surface structure and properties. However, many of them have certain shortcomings, for example, they require expensive equipment. One of the most optimum methods is diffusion hardening of alloy surfaces during heat treatment [6-11].

Altering surface layer properties of parts by employing case-hardening to form diffusion layers with reinforcing phases is a promising way of manufacturing parts that are strong, with a hard surface but preserving the properties of their core material. Developing non-destructive testing methods for these processes is an urgent matter.

2. Results

In this study, we used X-ray diffractometry to perform phase analysis of surface layers of several iron-, titanium- and nickel-based alloys, subjected to different types of case-hardening. We investigated samples of various chemical compositions after carburising, boriding and oxidation.

We investigated three samples of technical iron after vacuum carburising in acetylene. Saturation time varied from 2 to 6 hours at a temperature of 850° C. Our research demonstrated that after processing, sample surface featured Fe_3C and Fe-α phases.

Analysis of X-ray data (figure 1) showed that increasing saturation time from 2 to 4 hours altered the ratio of Fe_3C to Fe-α (the amount of Fe_3C, cementite, went up); increasing saturation time up to 6 hours didn't change the phase ratio significantly. This result means that when the saturation time was 4 hours or more, the surface part of the diffusion layer at the X-ray radiation penetration depth (about 20 microns) was not significantly altered.

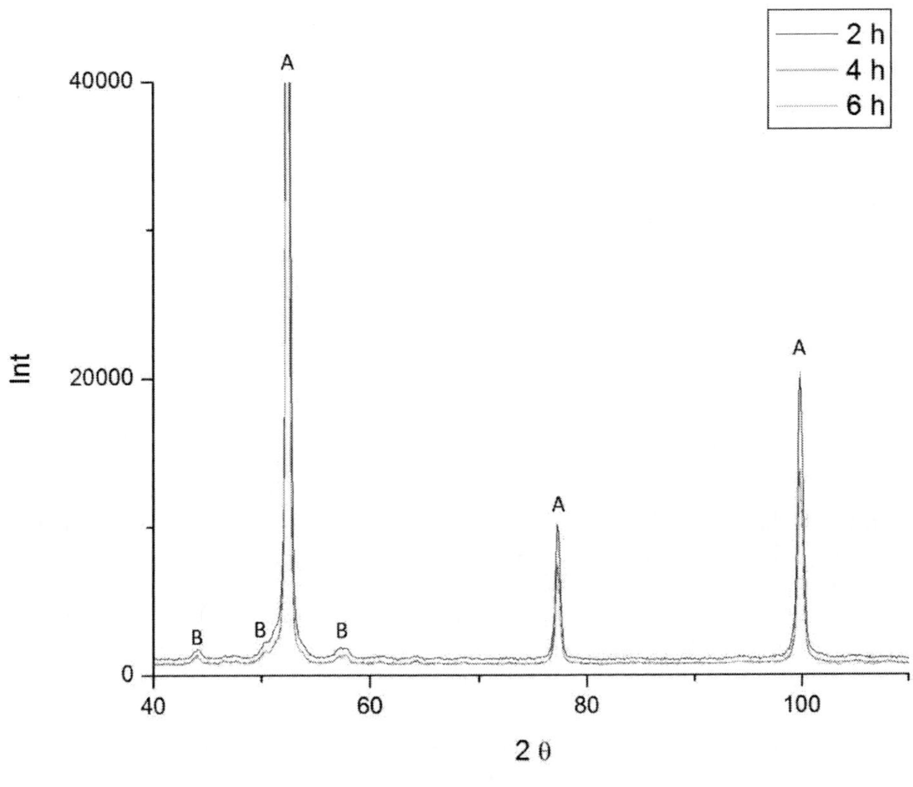

Figure 1. X-ray diffraction data for technical iron samples after carburising. Case-hardening time varied from 2 to 6 hours; A- Fe-α, B- Fe_3C.

We carried out electrolysis-free liquid boriding of technical iron samples in a melt of sodium tetraborate and amorphous boron at a temperature of 850 °C for 2, 4 and 6 hours.

X-ray phase analysis of the sample surface showed a lack of characteristic Fe-α and Fe-γ peaks (figure 2). There are primarily iron boride lines: FeB, Fe_2B and Fe_3B.

IOP Conf. Series: Journal of Physics: Conf. Series **918** (2017) 012014 doi:10.1088/1742-6596/918/1/012014

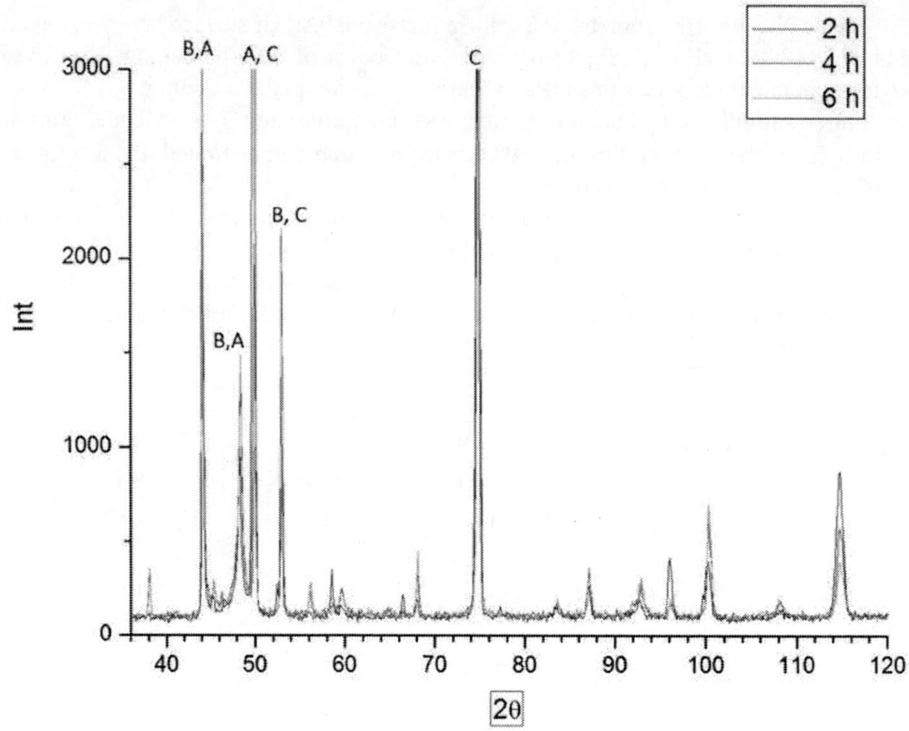

Figure 2. X-ray diffraction data for several technical iron samples after boriding. Three samples were subjected to this type of case-hardening for 2, 4 and 6 hours respectively; C - Fe_2B, B - FeB, A - Fe_3B.

Analysis of intensity variation of characteristic boride peaks shows that increasing boriding time increases the intensity of FeB lines, while the characteristic Fe_2B peak intensity reduces. At the same time, we recorded relatively weak Fe_3B peaks for all samples.

The results obtained demonstrate that a continuous boride layer forms on the iron surface. It is over 20 microns thick, and this layer shields the iron matrix. Comparison of peak intensity for FeB and Fe_2B shows that as saturation develops, the FeB layer thickens. Metallographic analysis results confirm this. Since relatively weak Fe_3B lines are present in the X-ray diffraction data for all saturated samples, we may assume that this phase builds up on the sample surface in the form of a thin layer or as isolated particles.

We detected several weak Fe_2O_3 lines, which may be explained by the surface oxidising lightly during boriding.

We carried out an X-ray phase analysis of the Cr20-Ni80 alloy sample after boriding (figure 3). The chemical composition of this sample is as follows: Cr – 20%, Ni – 72-78.8%, Fe – up to 1%, C – up to 0.06%, Si – 1-1.5%, Mn – up to 0.6%, Ti and Al up to 0.2%, Zn – 0.2-0.5%. After the case-hardening we discovered a variety of nickel, chrome and iron borides, as well as intermetallic compounds: Ni_2B, Fe_3B, FeB, Cr_2B, Ni_2Cr_3, Cr_2B_3.

IOP Conf. Series: Journal of Physics: Conf. Series **918** (2017) 012014 doi:10.1088/1742-6596/918/1/012014

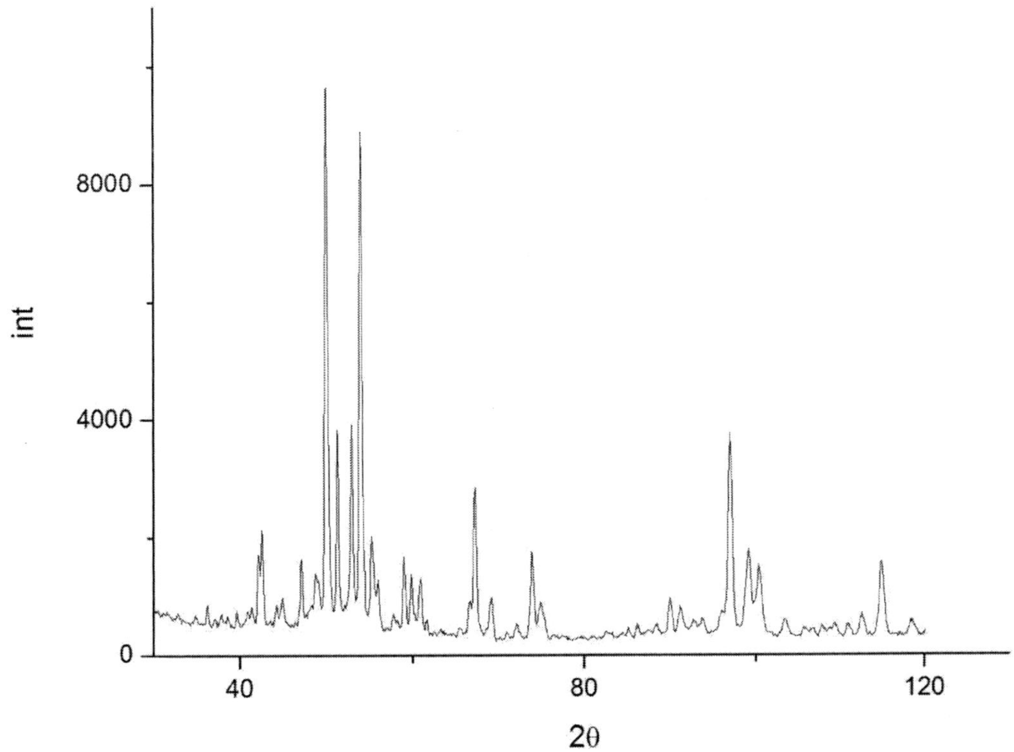

Figure 3. X-ray diffraction data of a Cr20-Ni80 sample after boriding.

We also investigated two samples of a titanium (PT-1M) alloy. These samples were oxidized at a temperature of 800°C for 12 hours, the first one in air and the second one in graphite filling. X-ray analysis results allowed us to detect the main phase structure of these samples; when the surface oxidises, higher (TiO_2) and lower titanium dioxides are formed, which have high hardness and wear resistance (figure 4). Characteristic titanium peaks show that the oxide film thickness is small.

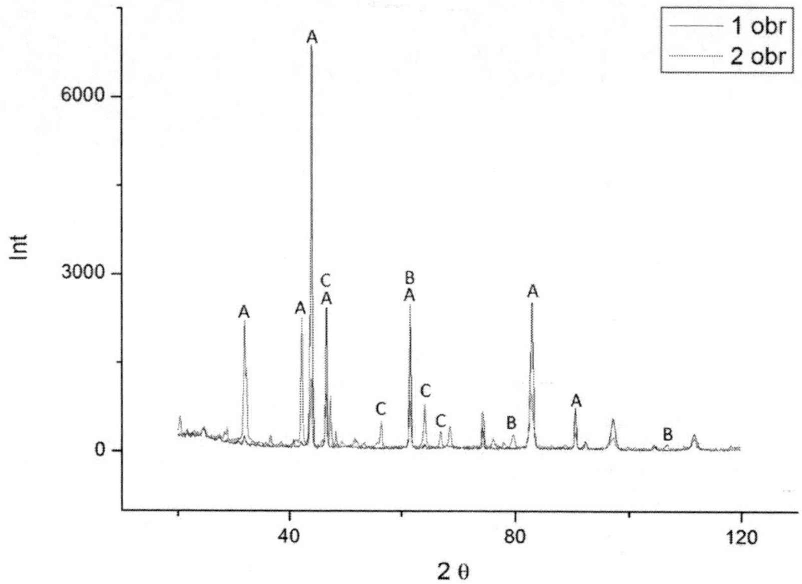

Figure 4. X-ray diffraction data of both PT1-M alloy samples; A - TiO, C - TiO_2, B - Ti_2O_3.

We found that the technique we applied in order to determine phase structure of thin surface layers of iron-, titanium- and nickel-based alloys has sufficient sensitivity to reliably record formation of carbides, borides, oxides and intermetallic compounds of various chemical compositions.

3. Conclusion

We analysed X-ray diffraction data of our iron- and titanium-based samples that we subjected to various types of case-hardening. As a result, we detected the presence of the following reinforcing phases:

1. We found the Fe_3C phase forming in the technical iron samples after vacuum carburising at 850 °C. When we increased case-hardening time from 2 to 4 hours, the amount of the Fe_3C phase grew significantly; a 6 hour case-hardening duration doesn't provide a significant increase in the amount of this phase as compared to a 4 hour case-hardening duration.
2. Our research revealed the presence of such phases as Fe_2B, FeB in the technical iron samples after liquid boriding at 850 °C. With an increase in processing time from 2 to 6 hours, we observed an increase in the FeB phase; when increasing case-hardening time from 4 to 6 hours, we did not find the FeB phase to grow significantly.
3. We carried out X-ray diffraction analysis of a Cr20-Ni80 sample after boriding. After case-hardening we observed a variety of phases forming: Ni_2B, Fe_3B, FeB, $C_{r2}B$, Ni_2Cr_3, Cr_2B_3.
4. After oxidation in air, we located TiO_2 phases in a PT1-M alloy sample; after oxidation in graphite, we detected the TiO_2, Ti_2O_3 and TiO phases in another PT1-M alloy sample.

References

[1] Malkova N Yu 2007 *Uspekhi sovremennogo estestvoznaniya - Advances in current natural sciences* **12** 106-7

[2] Krukovich M G 2012 *Novye materialy i tekhnologii v mashinostroenii - New materials and technologies in mechanical engineering* **15** 69-76
[3] Evdokimov V D, Klimenko L P and Evdokimova A N 2005 *Tekhnologiya uprochneniya mashinostroitelnykh materialov: Uchebnoe posobie-spravochnik* [Mechanical engineering material hardening technology: textbook and reference book] (Odessa, Nikolaev: Petro Mohyla Black Sea State University) p 352
[4] Ryzhov N M, Smirnov A E, Fakhurtdinov R S and Mordovin A I 2008 *Metal Science and Heat Treatment* **50**(1-2) 63-7
[5] Ryzhov N M, Smirnov A E, Fakhurtdinov R S, Mulyakaev L M and Gromov V I 2004 *Metal Science and Heat Treatment* **46**(5-6) 230-5
[6] Isayenkova M G, Perlovich Yu A, Skrytnyy V I, Sokolov N A and Yaltsev V N *Rentgenovskaya difraktometriya* [X-ray diffractometry] (Moscow: MEPhI) p 60
[7] Guryev A M, Lygdenov B D and Vlasova O A 2008 *Fundamental research* [Fundamentalnye issledovaniya] **8** 48-50
[8] Grigorienko D O and Byk M V 2015 *Austrian Journal of Technical and Natural Sciences* **3-4** 64-7
[9] Balandin Yu A 2006 *Fizikokhimiya Poverkhnosti i Zashchita Materialov - Protection of Metals and Physical Chemistry of Surfaces* **2** 150-3
[10] Shipko A A, Rudenko S P, Valko A L and Chichin A N 2016 *Lite i metallurgiya - Foundry production and metallurgy* **2**(83) 104-9
[11] Barabash M A, Kolmykov D V, Goncharov A N and Kolmykov V I 2009 *Uprochnyayushchie tekhnologii i pokrytiya - Strengthening Technologies and Coatings* **5** 44-7

IOP Conf. Series: Journal of Physics: Conf. Series **918** (2017) 012015 doi:10.1088/1742-6596/918/1/012015

Fractal and topological sustainable methods of overcoming expected uncertainty in the radiolocation of low-contrast targets and in the processing of weak multi-dimensional signals on the background of high-intensity noise: A new direction in the statistical decision theory

A. A. Potapov[1, 2, 3]

[1]Chief Researcher, V.A. Kotelnikov Institute of Radio Engineering and Electronics, Russian Academy of Sciences, 11, Mokhovaya str., 125009, Moscow, Russia
[2]Jinan University (JNU), College of Information Science and Technology / College of Cyber Security, Guangzhou, 510632, China
[3]Cooperative Chinese-Russian Laboratory of Informational Technologies and Signals Fractal Processing, Guangzhou, 510632, China

E-mail: potapov@cplire.ru

Abstract. The main purpose of this work is to interpret the main directions of radio physics, radio engineering and radio location in "fractal" language that makes new ways and generalizations on future promising radio systems. We introduce a new kind and approach of up-to-date radiolocation: fractal-scaling or scale-invariant radiolocation. The new topologic signs and methods of detecting the low-contrast objects against the high-intensity noise background are presented. It leads to basic changes in the theoretical radiolocation structure itself and also in its mathematical apparatus. The fractal radio systems conception, sampling topology, global fractal-scaling approach and the fractal paradigm underlie the scientific direction established by the author in Russia and all over the world for the first time ever.

1. Introduction

The intensive development of the modern radar technology establishes new demands to the radiolocation theory [1, 2]. Some of these demands do not refer to the theory basis and contribute to the precision improvement and the development of new calculation methods. The other ones are fundamental and related to the basis of the radiolocation theory. The last demands are the most important both in theory and in practice.

The entire current radio engineering is based on the classical theory of an integer calculation. Thus, an extensive area of mathematical analysis which is called the fractional calculation and deals with derivatives and integrals of a random (real or complex) order as well as the fractal theory has been historically "outboard" (!) [2-14]. At the moment the integer measures (integrals and derivatives with integer order), Gaussian statistics, Markov processes etc. are mainly and habitually used everywhere in the radio physics, radio electronics and processing of multidimensional signals. It is worth noting that the Markov processes theory has already reached its satiation and researches are conducted at the level of abrupt complication of synthesized algorithms. The radar systems should be considered with relation to the open dynamical systems. The improvement of classical radar detectors of signals and its mathematical support have basically reached their saturation and limit. It forces to look for fundamentally new ways of solving the problem of increasing the sensitivity or the range of coverage for various radio systems.

Nowadays it is absolutely clear that the application of scale invariance ideas – "scaling" along with the set theory, fractional measure theory, general topology, measure geometrical theory and dynamical systems theory reveals big opportunities and new prospects in the processing of multidimensional signals in the related scientific and engineering fields.

In this paper the presented alternative solutions for actual problems of modern radiolocation are based on the ideas and methods of the new scientific fundamental direction "Fractal Radio Physics and Fractal Radio Electronics: Designing of Fractal Radio Systems". This direction was initiated by Professor A. Potapov in 1980 at the IREE RAS (Kotelnikov Institute of Radio Engineering and Electronics of the Russian Academy of Sciences, Moscow, Russia) and it is currently widely developed in his works and acknowledged in the Scientific World.

2. "Fractal" Conception in Radiolocation

In general terms a radar image (RI) can always be presented as a set of elements X_k, whose values are proportional to the scattering cross-section (SCS) of a k-th element of the radar resolution [6-10]. In Figure 1a the RI of the terrain which was obtained at the wavelength $\lambda = 8.6$ mm from a helicopter is shown. In Figure 1b the RI of the same terrain region which was obtained by a radar at the wavelength $\lambda \approx 30$ cm is shown. Both images are two-dimensional with gray level proportional to SCS.

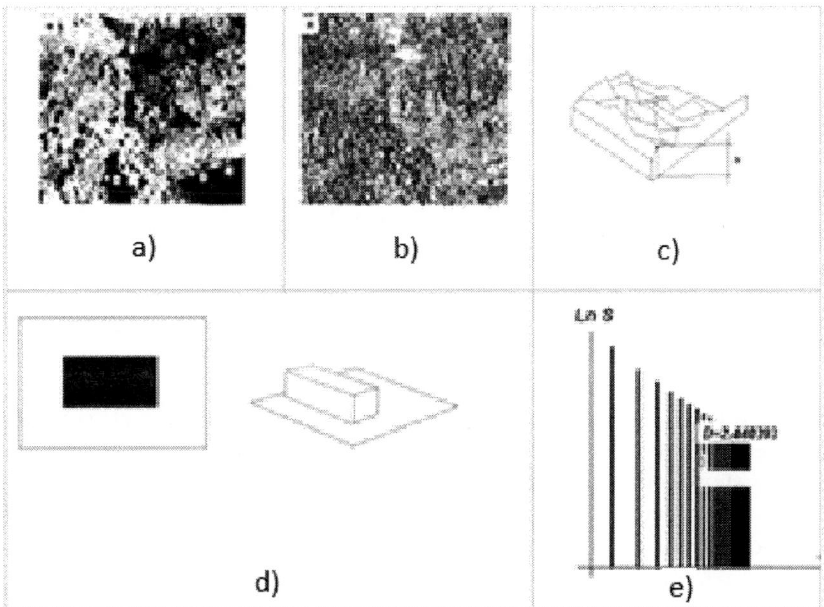

Figure 1. a - d) Examples explaining the matter of fractal processing, e) fractal space signature

Let us suppose that for every RI a surface (Figure 1c) with the height h which is proportional to the gray level is built. Let us suppose that we need to measure the square of the resulting surface. On the RI which corresponds to $\lambda \approx 30$ cm the square will be less than for the RI on $\lambda \approx 8.6$ mm since the smaller wavelength is the more terrain details can be recognized.

A probing electromagnetic wave is some kind of a "ruler" in this case. Due to that an increasingly finer structure of time-spatial signals or wave fields begins to have an effect.

If we have a RI which was obtained at even shorter waves then its square will be bigger. By decreasing the wavelength λ we will get increasing values of the squares. Then the question arises: what is the square of the surface which the RI was obtained from in reality? If the surface is covered with simple objects, for example a rectangular eminence (Figure 1d), and the sizes of this eminence are much higher than the wave length then the squares of the objects on the RI will be approximately equal for short and long waves. Then we would answer the mentioned question by calculating the number of resolution elements covering the object. The surface area S in this case would be equal to:

$$S \equiv S(\lambda) = N(\lambda)\delta(\lambda)\,, \qquad (1)$$

where $\delta(\lambda)$ - the square of the resolution element of the radar; $N(\lambda)$ - the number of the resolution elements required to cover the object, λ - the wavelength of the radar. As it was already noted, for a simple object (Figure 1d) value $S(\lambda) = const$.

For the RI on Figure 1a and Figure 1b one can build a dependence $S(\lambda) = f(\lambda)$ and assuming that $\delta(\lambda) = K(\lambda)$, where *K* is a known function then one can build a dependence $S(\lambda) = f(\delta)$. It happens that the measured square *S* is described well by the formula

$$S(\lambda) = k\lambda^{-D} . \tag{2}$$

Then just using the logarithm we can calculate the parameter *D*. The dependence which determines a fractal signature $D(t, f, \vec{r}\ )$ of a RI by itself is shown on Figure 1e. This dependence describes a space fractal cepstrum of an image (this conception was introduced by the author in 1990s).

The fractional parameter *D* is called the Hausdorf-Besikovitch dimension or the fractal dimension [3,5,7,8]. For the RIs of the objects with a simple geometric form (rectangles, circles, smooth curves) this dimension coincides with the topological one that is equal to 2 for the two-dimensional RIs and it is determined by the slope of straight lines (2) in binary logarithmic coordinates. However, the value of *D* for the majority of real coverings images and meteorological formations turns out to be higher than the topological dimension $D_0 = 2$ that emphasizes its complexity and random nature.

3. Textural and Fractal Measures in Radio Physics and Radiolocation

A radar along with the observation objects and radio waves propagation medium forms a space-time radar-location probing channel. During the process of radio location the useful signal from the target is a part of the general wave field which is created by all reflecting elements of the observed fragments of the target surrounding background, that is why in practice the signals from these elements form an interfering component. It is worthwhile to use the texture conceptions to create radio systems for the landscape real inhomogeneous images automatic detecting [6-8]. A texture describes spatial properties of the earth covering images regions with locally homogenous statistical characteristics [15]. Target detecting and identification occurs in the case when the target shades the background region at that changing integral parameters of the texture.

Many natural objects such as soil, flora, clouds and so on reveal fractal properties in certain scales [5-10]. Today the analysis of natural textures is undergone by significant changes due to the use of metrics taken from the fractal geometry. After the texture they introduced a conception of fractals that is based on the fractional measure theory for fundamentally different approach to solving modern radio location problems. The fractal dimension *D* or its signature in different regions of the surface image is a measure of texture i.e. properties of spatial correlation of radio waves scattering from the corresponding surface regions. At the very beginning the author initiated a detailed research of the texture conception during the process of the radio location of the earth coverings and objects against its background. Further on a particular attention was paid to the development of textural methods of objects detecting against the earth coverings background with low ratios of signal/background (see for example [6-10,14,16-19] and references).

4. Textural and Fractal Measures in Radio Physics and Radiolocation

The author for the first time ever has shown that the fractal processing excellently suits for solving the problem of processing the low-contrast images and detecting superweak signals in high-intensity noise when the modern radars do not practically function [2, 6-11, 14, 16, 17]. When using the fractal approach, as it was pointed out above, it is natural to focus attention on the description and also on the processing of radio physical signals (fields) exceptionally in the fractional measure space using the hypothesis of the scaling and universal distributions with "heavy tails" or stable distributions [7, 8, 20].

The author has developed a fractal classification which was personally approved by B. Mandelbrot [9, 10] in the USA in 2005. It is presented on Figure 2 where the fractal properties are described on the assumption that D_0 is the topological dimension of the space of embeddings.

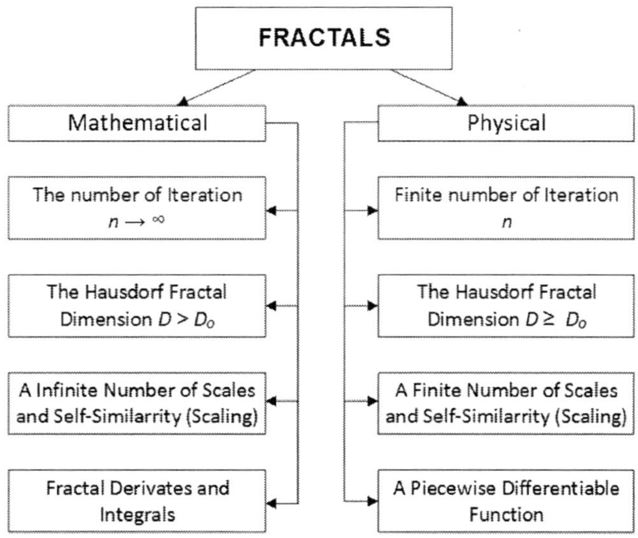

Figure 2. The author's classification of fractal sets and signatures

The textural and fractal digital methods developed by the author and his followers (Figure 3) allow to partially overcome a prior uncertainty in the radar problems using the geometry or the sampling topology (one- or multidimensional) [6, 16]. At that topological peculiarities of the sampling become very important as opposed to the average realizations which have different behavior.

It turns out that the concepts of fractal signatures and fractal spectra are very helpful for measurements. For example, these concepts are effectively applied to solve the problems of detecting the low-contrast targets and weak signals in the presence of intense non-Gaussian (!) interference. The methods of fractal processing should take into account the scaling effect of real radio signals and electromagnetic fields. The introduction of a fractional measure and scaling invariants necessitates the predominant use of power-series probabilistic distributions. These distributions result from the feedback that amplifies the events. Note that, for the distributions with the heavy tails, the sample means are unstable and carry little information because the law of large numbers cannot be applied in this situation.

IOP Conf. Series: Journal of Physics: Conf. Series **918** (2017) 012015 doi:10.1088/1742-6596/918/1/012015

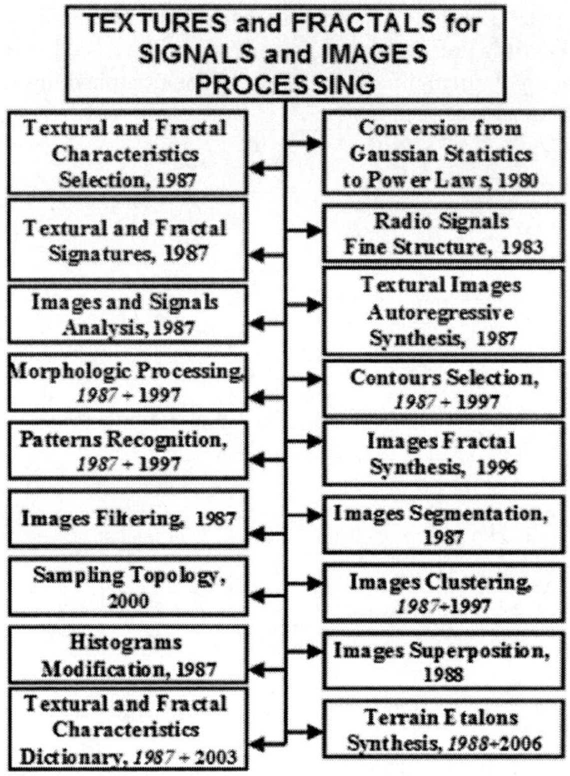

Figure 3. Textural and fractal methods for processing low-contrast images and super weak signals in high-intensity noise

Figure 4 shows the general view of distributions with the fractal dimension D. While fractal processing the signals in noise it is shown that at the signals in noise ratio (SNR) q_0^2 = +10 dB we precisely measure the statistics of a signal. With the reduction of value aside negative values (for example, q_0^2 = -3 dB) there is a displacement of maximum final fractal distributions aside the values fractal dimensions of the noise or cluttering.

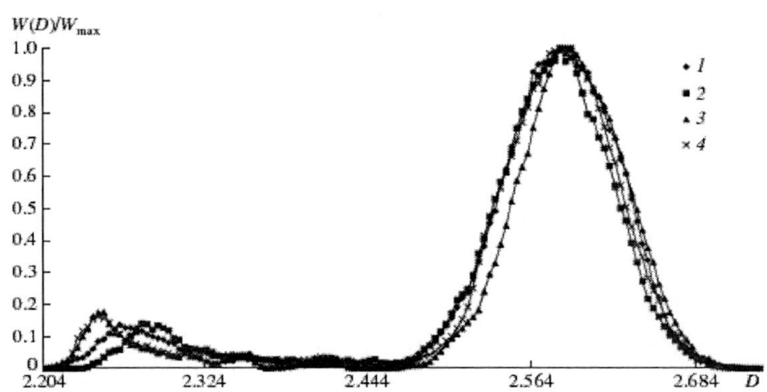

Figure 4. Empirical fractal distributions with heavy tails for images observed in the presence of an intense Gaussian noise: (1 and 3) scene **A**, (2 and 4) scene **B**, (1 and 2) $q_0^2 = -10$ dB, and (3 and 4) $q_0^2 = -20$ dB

Thus, always in a vicinity of value fractal dimensions of a useful component there is "a heavy tail" fractal distribution, reaching the stable size, about 20 %. The given tendency is kept at much smaller values, equal SNR q_0^2 = -10 dB and q_0^2 = -20 dB, as shown in Figure 4.

Figure 5 shows the selected results of the low-contrast objects fractal nonparametric filtering. Aircraft images were masked by an additive Gaussian noise. In this case, the SNR was -3 dB. It is seen in the figures that all the desired information is hidden in the noise. The optimum mode of filtering the necessary contours or objects is chosen by the operator using the spatial distribution of fractal dimensions D of a scene. This distribution is determined automatically and is shown in the right panel of the computer display [8-11, 13, 21].

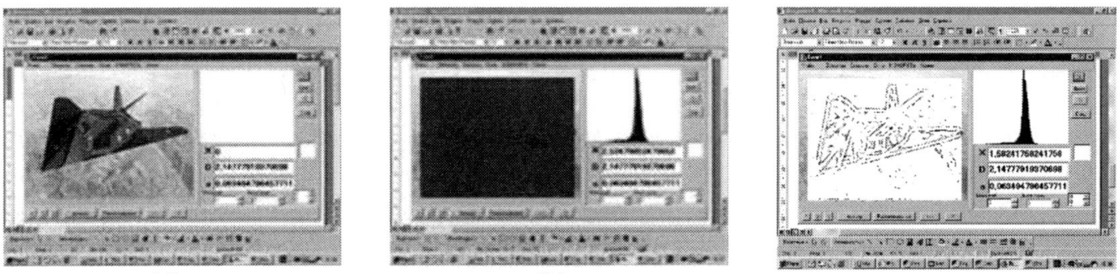

(a) (b) (c)

Figure 5. a) Source image, b) source image and noise q_0^2 = -3 dB, c) results of fractal filtration image

High sensitivity of estimating the non-integral dimension functionals to the presence of a continuous contour in images suggests a large potential of the objects contours fractal filtering in strong interference (Figure 6). The observation was made using a ground-based telescope, the distance between the telescope and objects being about 800 km. These data are presented in the book [11]. None of the modern methods of digital processing can provide comparable objects resolution!

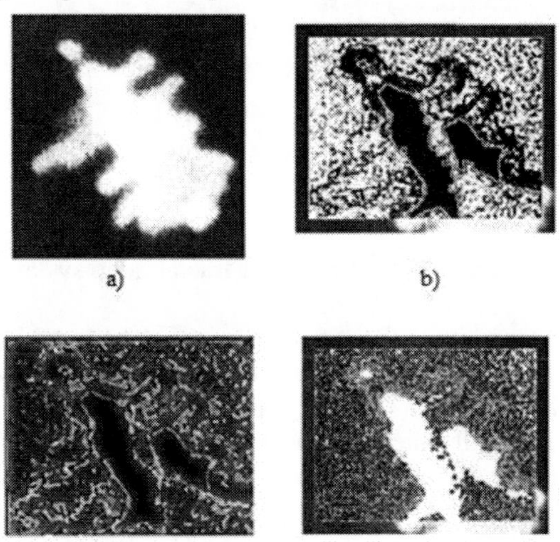

Figure 6. a) The initial image of space complex at the time of joining "Shuttle" – "Peace", b–d) the results of the fractal processing (targets clustering) for various values of the threshold D of topological fractal nonparametric detector

This concept can be widely applied to solving the modern problems of radar, correlation-extremely navigation, artificial intelligence and dynamic systems. The algorithms developed by the authors for calculating the fractal signatures are efficient over an extremely wide range of physical sizes of the characteristic image details and provide the detection estimation for the scaling effects, including even those masked by the noise.

5. Innovative Fractal-Scaling Technologies: Creation, Development and Application of Fractal Methods for Radiolocation Tasks and Developing the Foundations of Fractal Element Base

A critical distinction between the author's proposed fractal and scaling methods and classical ones is due to fundamentally different (fractional) approach to the main components of a physical signal. It allowed us to come to the new level of informational structure of the real non-Markov signals and fields. Thus, this is a *fundamentally new radio engineering*. For 35 years of scientific researches the global fractal and scaling method designed by the author has justified itself in many applications. We labeled all of this briefly and expressively – "The Fractal Paradigm" [9, 22-27].

The fractal geometry is a huge and genius merit of mathematician B. Mandelbrot [5]. But its radio physical/radio engineering implementation is a merit of the Russian (now it is international) scientific school of fractal methods and fractional operators under the supervision of Professor A.A. Potapov (V.A. Kotelnikov IREE RAS). In the current situation all the attempts to belittle their meaning and to reckon only with the classical knowledge endure the intellectual fiasco. A triad of specified problems in the general "Fractal Analysis and Synthesis" therefore creates the basis for "Fractal Radio Systems" (2005) – Figure 7, "Global Fractal–Scaling Method" (2006) and "Fractal Paradigm" (2011) [2,6-14,16-19,21-28].

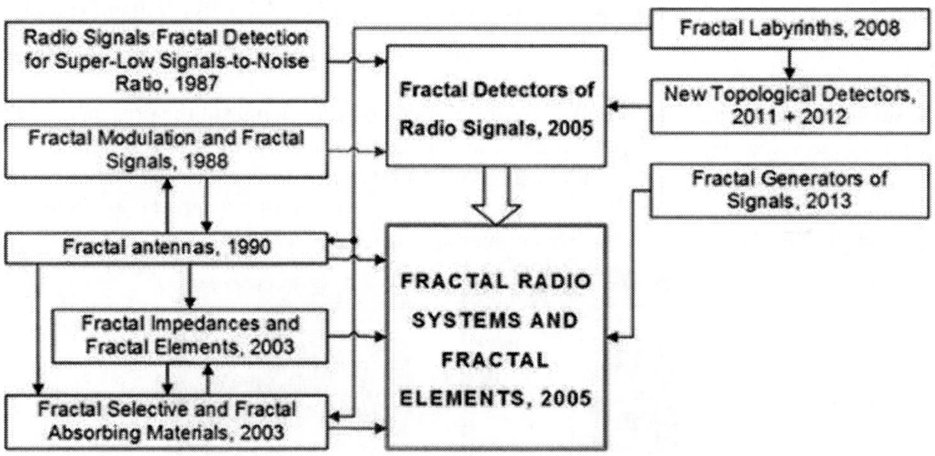

Figure 7. Author's conception of fractal radio systems and devices

The work fulfilled in Kotelnikov IREE of RAS by the author and his apprentices is based on the theoretical and experimental results in the scheduled introduction of the fractals, fractional integration-differentiation and the scaling effects in radio physics, radio engineering, and some contiguous scientific directions (Figure 8). We have published a sufficiently large number of works for each direction using the data from Figure 8 in Russia and abroad.

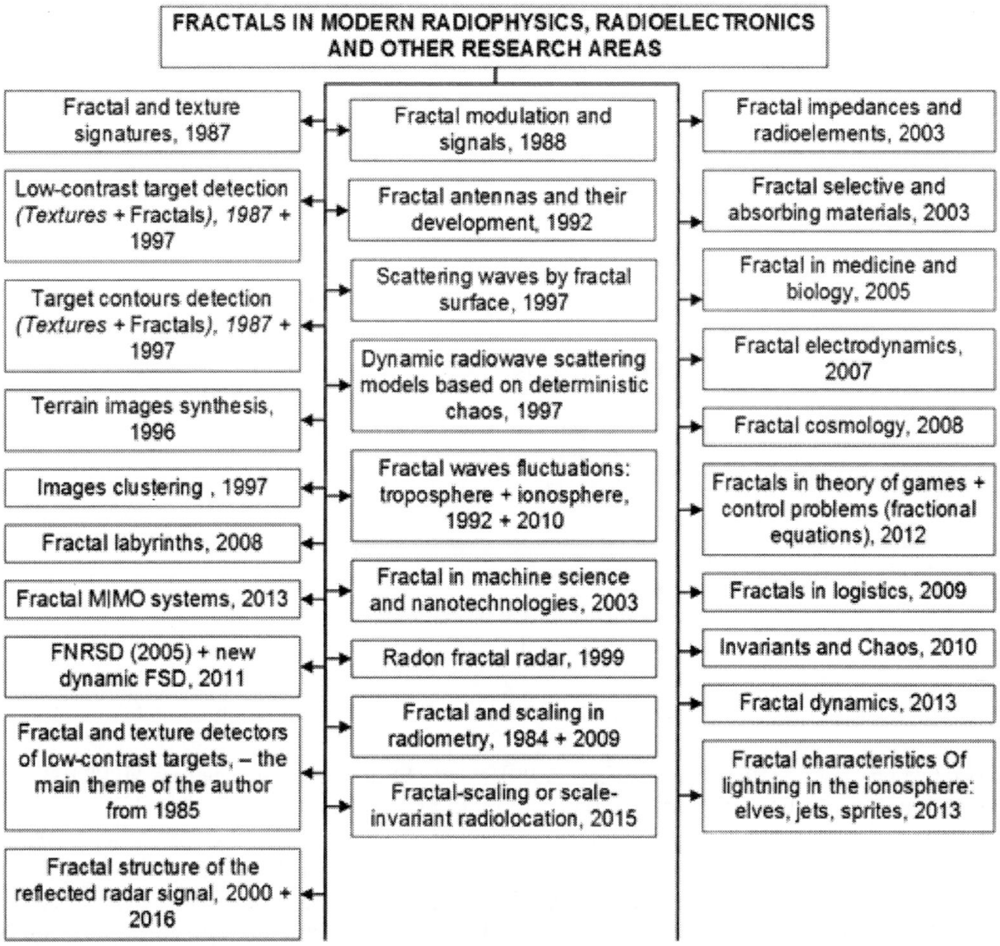

Figure 8. Sketch of author's development of new information technologies based on fractals, fractional operators and scaling effects

6. Fractal-Scaling or Scale-Invariant Radiolocation and Fractal MIMO-Radars as a New Kind of Radio System

For further instantiation of the problems regarding the weak radar signals detection, we believe the initial information to come from a variety of radio systems in the form of a one-dimensional signal and / or a radar image (RI) - Figure 9.

The simplified scheme of primary radio systems and the investigation of radar image and one-dimensional signal in millimeter wave band (MWB) were introduced by the author much earlier.

Currently the fractal radar, the MIMO-radar and the fractal MIMO radar as well as unmanned aerial vehicles (UAVs) are added to the scheme in Figure 9. The concept of fractal radar is presented in [7-11, 13, 14, 26-29], the concept of fractal MIMO-radar is considered in [7-11, 13, 26-29].

In general, the technology of MIMO systems implies that each wireless device involved in the data exchange has several spatially distributed receiving and transmitting antennas. The basic idea of fractal MIMO-radars is using fractal antennas and fractal detectors [9, 13, 18, 19, 23-29]. The capability of fractal antennas to work on several frequencies simultaneously or to radiate broadband sounding signal provides a sharp increase in the number of degrees of freedom that defines many of the important advantages of this type of radiolocation and significantly expands adaptation possibilities.

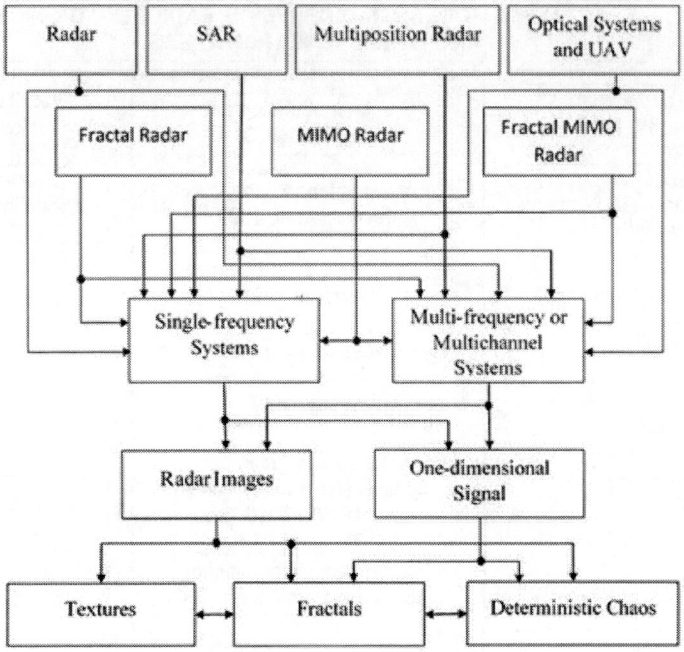

Figure 9. Radio systems of initial information

To represent these specifics in [9, 13, 26-29] a new term "fractal-frequency MIMO systems (FF MIMO)" was introduced, which reflects their physical properties much better. The MIMO technologies related to the spatial multi-channel systems provide great opportunities for the application of the author's global fractal-scaling method for signal processing, various algorithms and technologies of fractal detectors at all stages of the synthesis of information MIMO systems.

The idea of a fractal radar station (Figure 9 and Figure 10) is based on the concept of fractal radio systems developed by the author - Figure 7.

7. Postulates of Fractal Radar

Fractal radar (Figure 10) defined in [7-11, 13, 14, 16-19, 26-29] is based on four main postulates:

1) Intelligent signal processing is based on the theory of fractional measure, scaling effects and fractional operator's theory;

2) Hausdorff dimension or fractal dimension *D* of a signal or a radar image is directly connected with the topological dimension;

3) Robust non-Gaussian probability distributions of the fractal dimension of the processed signal;

4) "Maximum topology with a minimum of energy" for the received signal. It allows using the advantages of fractal scaling information processing more effectively.

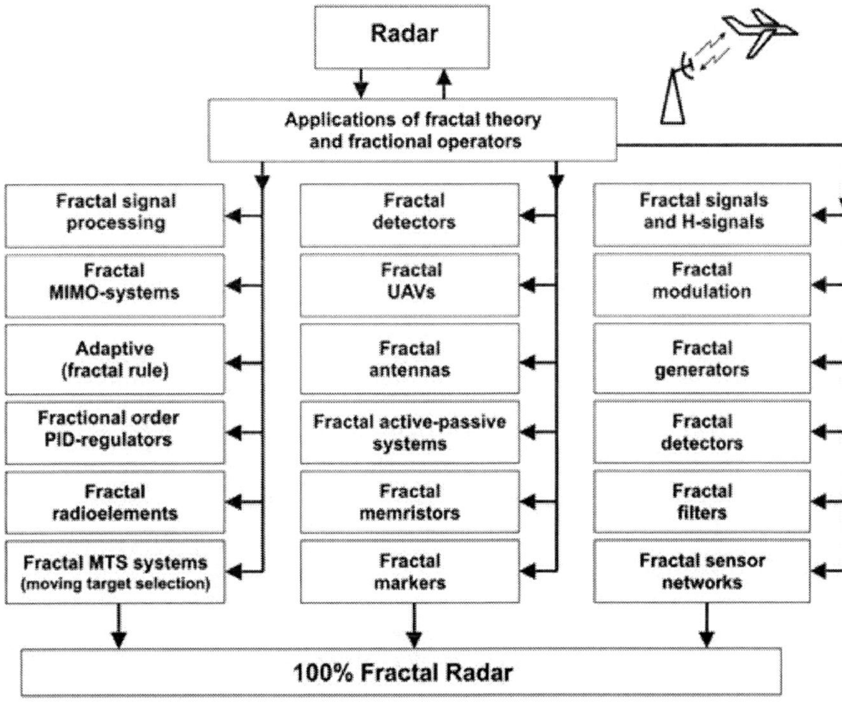

Figure 10. The points of application of fractals, scaling and fractional operators in a classical radar for proceeding to a fractal radar

The key point of fractal approach is to focus on describing and processing of radar signal (fields) exclusively in the space of fractional measure with the use of the scaling hypothesis and distributions with heavy-tailed or stable distributions (non-Gaussian). The methods of fractal-scaling processing of signals, wave fields and images are in a broad sense based on the pieces of information which is not usually taken into account and irretrievably lost if the classical methods of processing are applied. This work is concerned with the main area of radio physics – radiolocation and it aims to ascertain what is done and has to be done in this field on the basis of the fractal theory. The investigations carried out showed the correctness of the path chosen by the author (since 1980) to improve the radiolocation technique.

It is necessary to think about the processing of the input signals with a low threshold at high levels of false alarm and then changing to a low level of false alarms. Moreover, *the false alarm probability is never measured in the real time.* In principle, we need new metric and new parameters of radar detection.

8. Postulates of Cognitive Radar

Cognitive radar defined in [30] is based on 3 main postulates:

1) intelligent signal processing, which builds on learning through interactions of the radar with the surrounding environment;

2) feedback from the receiver to the transmitter, which is a facilitator of intelligence;

3) preservation of the information content of radar returns, which is realized by the Bayesian approach to the target detection through tracking.

We focus on future possibilities of radars with the special emphasis on the fractal, chaos and synergistic ideas.

9. New Features and Topological Methods for the Detection of Low-Contrast Targets

All the currently existing methods and features of detecting the low-contrast targets against the background of high-intensity reflections from the sea, ground and meteorological formations are presented on Figure 11. The correlations between various signs and methods are also shown.

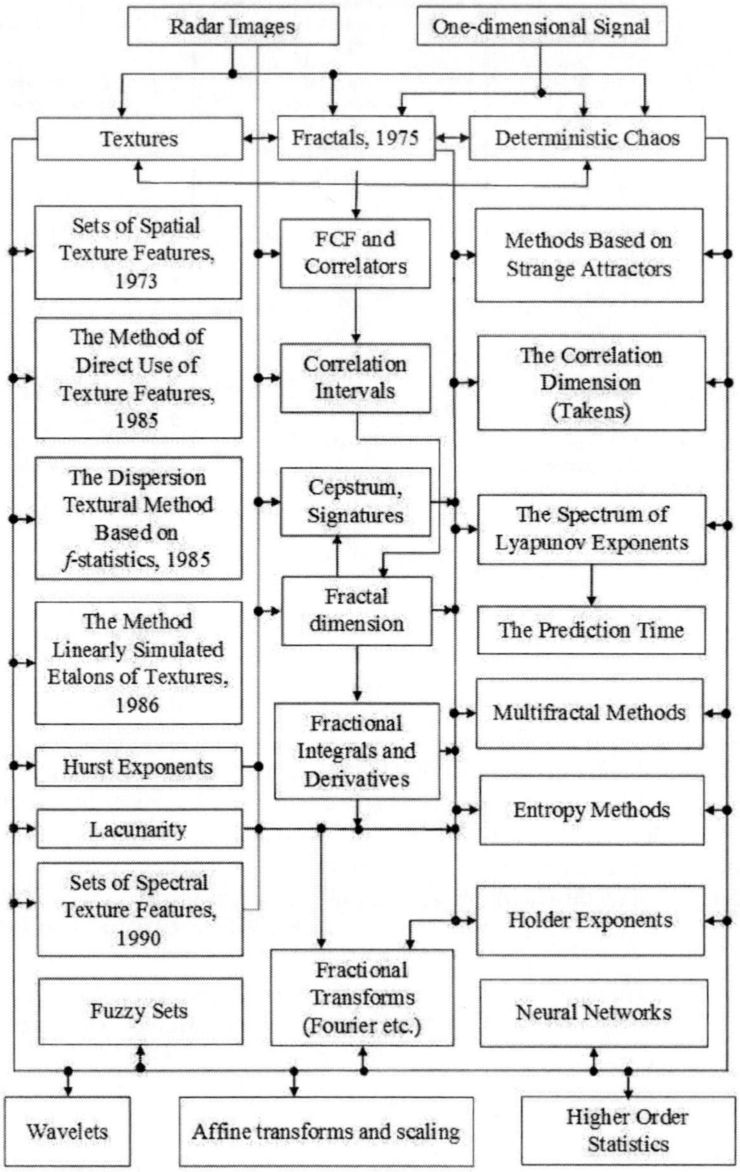

Figure 11. Topological features and methods for detecting the low-contrast targets against the background of high-intense noise

The introduction of the textural features ensemble concept to the US in 1973 [15] made it possible for the author in the 1980-ies to be the first to calculate complete ensembles of 28 textural features and to conduct their detailed synchronic analysis for the real (optical aerial photography (OAP) and radar images within the MMW range at a wavelength of 8.6 mm) as well as for the synthesized textures based on the autoregressive models depending on the season [6].

IOP Conf. Series: Journal of Physics: Conf. Series **918** (2017) 012015 doi:10.1088/1742-6596/918/1/012015

For a long time the works on the study of radar images of the land cover at MMW using textural information have actually been carried out only in Russia and are of interest so far (especially now) [9,13,14]. After calculating the ensembles of textural features based on the optical and radar images, in 1985-1986 the author proposed the methods and algorithms for the detection of low-contrast targets against the background of high-intense noise. Those included the method of the direct use of textural features (1985), the dispersion method based on *f*-statistics (1985) and the detection method using linearly simulated patterns, i.e. textures (1986) [6].

The created methods of detection are quite valid at low signals. To the author's knowledge, no textural methods for detecting the low-contrast target have been proposed abroad. Moreover, an important advantage of the textural methods of processing is the possibility to neutralize the speckles at coherent images of the Earth's surface obtained by SAR.

The methods of deterministic chaos are widespread; they are shown in the right column of Figure 11. It should be only noted that the algorithms of radar detection of the low-contrast targets against the background of woodlands for the radar at a wavelength of 2.2 mm were tested by the experimenters in 2001. It was the first time when a strange attractor was reconstructed. It controlled the radar scattering of millimeter radio waves. Its dynamic and geometric characteristics were measured; D fractal dimensions, depending on the value of m embedding dimension were calculated. The most accurate estimate of D can be obtained at the breakpoint of $D(m)$ convex curve, at that paying no attention to the reduction in scale ratio above and below.

The calculation of the correlation integral $C(r)$ was conducted using the F. Takens' theorem on the sampling out of 50000 counts which corresponds to the angle of incidence of an electromagnetic wave $\theta = 50^0$. Based on the found maximum Lyapunov exponent $\lambda_1 > 0.6$ bit/s it has been shown that when measuring the current conditions with an accuracy of up to 1 bit we lose all predictive power over time during 1.7 seconds. Therefore, *the prediction interval* of the echoed signal intensity is by about 8 times greater than the classical correlation time τ ($\tau \approx 210$ ms at a wind speed of 3 m/sec). The prediction interval provides an opportunity to estimate roughly the amplitude of further samples in the sample collection and, as noted by the author, it can be used in the radar practice. The calculations of Hurst exponent H showed that in two out of three cases the scattering process of millimeter waves by woodlands corresponds to the persistent process with $H > 0{,}5$, i.e. to the process with the maximal rank.

The obtained results along with the family of fractal distributions underlie the new dynamical model of signals scattered by the plant coverings. The proposed model of electromagnetic waves scattering by the earth coverings has a fundamental difference from the existing classical models. It has a finite number of degrees of freedom, describes processes of the non-Gaussian scattering and introduces the interval of prediction of the received radar signal intensity and its fractal features into consideration. We particularly note that taking into account the fractality of the earth coverings allows to describe the earth coverings scattering indicatrixes more precisely than the classical models which are used now.

10. Fractal Scaling Topological Detectors of Low-Contrast Targets

Currently great interest is evinced in various fractal and scaling methods (Figure 11). Those fractal investigations started almost simultaneously in Russia, the USA and China in the 1980s. [7-11, 13, 14, 16-19, 26-29, 31-36].

Fractal dimension D or its signature $D(t, f, \vec{r})$ in different regions of the surface image is at the same time a texture measure that is the properties of the space correlation of the radio waves scattering by the corresponding surface regions. At the same time the texture determines *lacunarity* (Figure 11) which uses the statistics of the second order for fractal images [7, 8]. The lacunarity is small for a dense texture and large when the texture is coarse-grained. Lacunarity Λ (by Mandelbrot [5]) is defined by the formula

$$\Lambda = \langle((M/\langle M\rangle) - 1)^2\rangle. \tag{3}$$

Here M is the "mass" of the fractal formation, $\langle M\rangle$ is the expected "mass", brackets <…> mean the data ensemble averaging. The consideration of lacunarity as a sign of objects detection was conducted by the author in 1997. The introduction of the fractional measure and the scaling invariants requires working with exponential probabilistic distributions.

The main principles of the fractal detector were discovered and proposed by the author as early as in 1980s and getting out (also for the first time ever - see Figure 7 and Figure 10) to the functioning prototype of *the fractal nonparametric detector of radar signals* (FNDRS) was done in 2003-2005. It was demonstrated in the USA on the ISTC project with Central Design Bureau "Almaz" and IREE RAS in 2005 and it earned highly positive feedback from many specialists [8-10, 31, 32]. The high stability of our proposed original algorithms of fractal-and-scaling detection was shown.

Some initial versions of radar fractal detectors generalized structures are presented on Fig. 12. A schematic view of the probable detector is given on Figure 12a. A set of textural or fractal signs ξ is determined on the received radio signal or the image. Then in a threshold device the decision on signal presence $\mathbf{H_1}$ or its absence $\mathbf{H_0}$ is generated at the threshold value $\mathbf{\Pi}$ and a certain level of probability of a false alarm **F**. The values of fractal dimension *D*, Hurst exponents $0 \le H \le 1$ for multidimensional surfaces, Holder exponents, lacunarity values and so on can be used as signs ξ. The Hurst exponent equals

$$H = 3 - D \tag{4}$$

for the RI and

$$H = 2 - D \tag{5}$$

for the one-dimensional signal.

A structural enlarged pattern of a radar signals fractal detector is shown on Figure 12b. It includes a contours filter and a calculator of *fractal cepstrums*. The further specification of the structural pattern of FNDRS is given on Figure 12c. The input signal (RI, one-dimensional sampling) comes to the input transducer. Its designation is preliminary preparing of the sampling under analysis. This preparation can include either a forced noise contamination (in case of low resolution of an analogous-and-digital transducer of a radar) or for example a contrast compression - in case of the sampling with the high dynamic range.

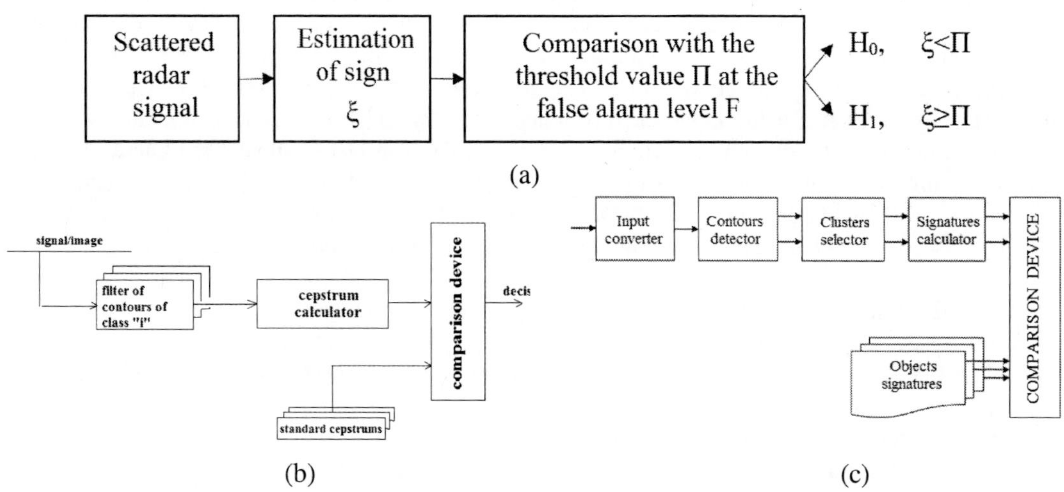

Figure 12. a) Initial and b, c) detailed structures of the first fractal detectors

A detector on the basis of the Hurst exponent (Figure 13) works using one or several search frequencies of the radar [29]. The Hurst exponent *H* reflects irregularity of the fractal object – (4) and (5). The less exponent *H* is the more irregular the fractal object becomes. So, the Hurst exponent gets higher when an object appears.

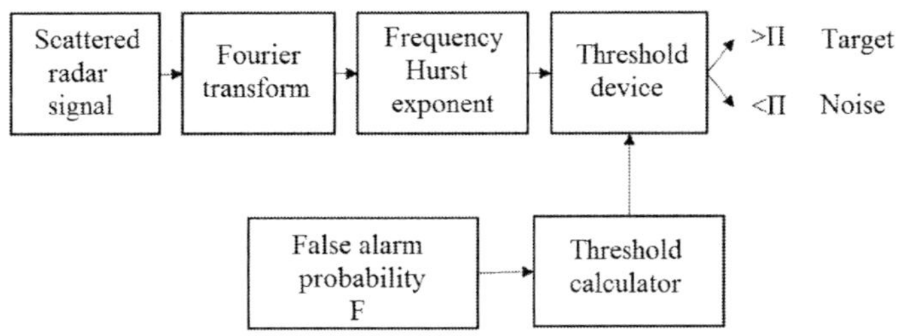

Figure 13. Fractal detector based on the Hurst exponent

On Figure 14 there is a scheme of fractal detector with the autoregressive estimation of the power spectrum of the interference from the Earth surface [29]. The autoregressive model represents a linear model of prediction which estimates the power spectrum of the interference from the surface and forms its autocorrelation matrix. The autoregressive equation describes a relation between the current and preceding counts of the sampled stochastic process. Earlier, in the 1980s, we were resolving the problem of autoregression on the basis of the canonical system of Yule-Walker equations with the transform of brightness histograms [6]. Thus, in the detector on Figure 14 the real fractal properties of the power spectrum on the basis of the autoregressive spectral estimation which are applied for detecting the low-contrast objects are used.

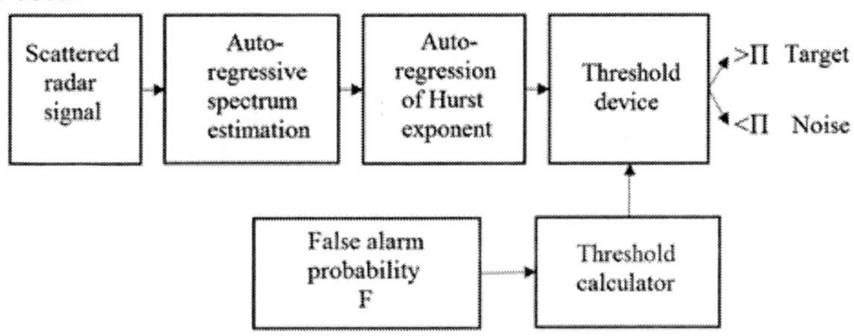

Figure 14. Fractal detector with autoregressive estimation of the interference spectrum and the Hurst exponent

11. New Directions in the Theory of Statistical Solutions and Testing Statistical Hypotheses

The fast development of the fractal theory in radar and radio physics led to establishing a new theoretical direction in modern radar. It can be described as «*Statistical theory of fractal radiolocation*». This direction includes (at least at the initial stage) the following fundamental questions [1-5, 7-13, 20, 37]:

1). The theory of the integer and fractional measure.
2). Caratheodory construction in the measure theory.
3). Hausdorff measure and Hausdorff-Besicovitch dimension.
4). The theory of topological spaces.
5). The dimension theory.
6). The line from the mathematician's point of view.
7). Non-differentiable functions and sets.
8). Fundamentals of the theory of probability.
9). Stable probability distributions.
10). The theory of fractional calculus.
11). The classical Brownian motion.
14). Anomalous diffusion.
15). The main criteria for statistical decision theory in radar.
16). Wave propagation in the fractal random media.
17). Wave scattering generalized Brownian surface.
18). Wave scattering surface on the basis of non-differentiable functions.
19). Difractals.
20). Cluster analysis.
21). Theory and circuitry of fractal detectors.
22). Fractal-scaling or scale-invariant radar.
23). The multi-radar.
24). MIMO radar.

12). Generalized Brownian motion.
13). Fractal sets.
25). Cognitive radar.

***

This list of studied questions, of course, is supposed to be expanded and refined in the future. The author has been dealing with it for nearly 40 years of his scientific career.

12. Fractal Processing of Medical Images

It is known that X-ray and fluorographic images are very important for express diagnostics of such diseases as numerous fractures and initial stages of malignant tumors. One of the methods is X-ray mammography making possible to diagnose at the earliest stage of the disease. To solve the problem of isolating suspicious areas on X-ray images, we chose a series of images with accumulations of calcinates. The information about the possible size of such formations helps to identify them against the background of other elements of the image. In fractal processing, the scales K and M varied from 2 to 10. Correspondingly, different slices were obtained for estimating D – Figure 15. The suspicious regions are in Figure 15c.

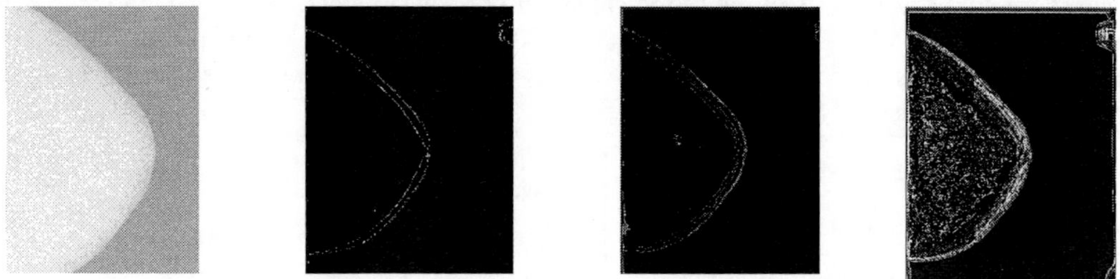

(a) (b) (c) (d)

Figure 15. Images obtained for different values of the fractal dimension estimate D ($D_1 > D_2 > D_3$): a) initial X-ray image, b) – $D = D_1$, c) $D = D_2$, d) $D = D_3$

Figure 16 shows the results of the processing another X-ray image. The accumulations of calcinates in Figure 16a are marked by the squares.

(a) (b)

Figure 16. An example of selecting suspicious areas in the image: a) the initial image, b) the result of the fractal processing

Figure 17 shows the final results of solving the problems of image clustering based on the value of the fractal dimension D and the tasks of extracting the contours of the image.

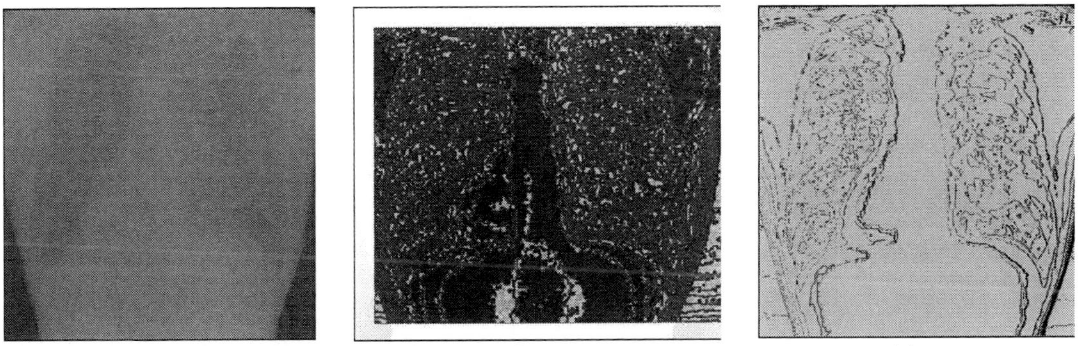

(a) (b) (c)

Figure 17. a) An example of solving the problem of clustering an X-ray image, b) from the value of estimating the fractal dimension *D*, and c) image contours extraction

The results of the fractal processing show a reliable improvement in the quality of the processed medical images. The proposed methods are verified by the authors on various real medical images. It should be noted that the algorithms used are characterized by modest computational requirements and high stability.

13. Conclusions

The author created, developed and applied the fractal-scaling methods for radiolocation problems and forming the foundations of fractal element base [6-14, 16-19, 21-29, 31-36, 38, 39]. For the first time ever approaches to the development of a fractal radar and a fractal MIMO- radar were considered. The author emphasizes that the synthesis of topological (fractal, textural, chaotic, etc.) detectors makes for a fresh look at the problem of detecting the super weak actual signals. As a result of that, the author's discovery in the 1980-ies takes the meaning of the *generalized detection*. Thus, pure energy and pure topological detectors are not contrary to each other and they do not duplicate but complement one another.

Due to the topological detectors it is possible to see the process of energy detection in a new light and to find some essential faults in it. Consequently, topological detection becomes not less, if not more, valuable for the theory and practice than energy detection. The theory of topological detection is formulated in [7-11, 13, 14, 16-19, 26-29, 31-36, 38, 39]. It is especially necessary for the purpose of reexamining the former theory and in that way producing new results that are not available to the traditional concepts of radiolocation.

Thus, the topological detection opens the door to a radically new field of statistical decision theory and provides an opportunity to correct the ideas in this field and even to create new ones, which is of great theoretical and practical importance. The sufficiently detailed reasoning reported here should contribute to a better understanding of proposed by the author fundamentally new interpretation of the problem of radar (and other kinds of) detection. The proposed theory has much in common with the cognitive radar.

Thus, during more than 35 years, almost from scratch, fundamental bases of the theory that will be applied in the following decades were formed. It is not the results and the specific solutions that are the most valuable, but the solution method and the approach to it. The created method is presented in [7-11, 13, 14, 16-19, 26-29, 31-36, 38, 39].

The author raised the foregoing problems as early as in 1980 and for more than 35 years he has been successfully working on their solution and development. Careful bibliographic studies show complete and absolute world priority of the author in all "fractal" fields of radiolocation and radio-physics (the list of the author's works in cooperation with students has about 830 publications, including 23 monographs).

IOP Conf. Series: Journal of Physics: Conf. Series **918** (2017) 012015 doi:10.1088/1742-6596/918/1/012015

14. Acknowledgment
This work was supported in part by the project of International Science and Technology Center No. 0847.2 (2000 – 2005, USA), Russian Foundation for Basic Research (projects №№ 05-07-90349, 07-07-07005, 07-07-12054, 07-08-00637, 11-07-00203), and also was supported in part by the project "Leading Talents of Guangdong Province", № 00201502 (2016 – 2020) in the JiNan University (China, Guangzhou).

15. References:
[1] Skolnik M 2008 *Radar Handbook* (New York: McGraw-Hill Co)
[2] Bunkin B V, Potapov A A, Reutov A P *et al* 2003 *Aspects of Perspective Radiolocation* (Moscow: Radiotekhnika)
[3] Rogers C A 1970 *Hausdorff Measures* (London: Cambridge University Press)
[4] Oldham K B and Spanier J 1974 *The Fractional Calculus* (New York: Academic Press)
[5] Mandelbrot B 1983 *The Fractal Geometry of Nature* (San Francisco: W.H. Freeman and Co)
[6] Potapov A A 1994 *Synthesis of Images of Earth Coverings in Optical and Millimeter Waves Ranges: Dissertation of Doctor of Physics and Mathematics: (01.04.03 - Radio Physics)* (Moscow: IREE RAS)
[7] Potapov A A 2002 *Fractals in Radiophysics and Radar* (Moscow: Logos)
[8] Potapov A A 2005 *Fractals in Radiophysics and Radiolocation: Sample Topology* (Moscow: Universitetskaya kniga)
[9] Potapov A A 2016 *Chaos Theory, Fractals and Scaling in the Radar: A Look from 2015* in: ed C Skiadas *The Foundations of Chaos Revisited: From Poincaré to Recent Advancements* (Switzerland, Basel: Springer) chapter 12 pp 195–218
[10] Potapov A A 2006 *Fractals and Chaos as the Basis of New Breakthrough Technologies in Modern Radio Systems* in: Kronover R *Fractals and Chaos in Dynamic Systems* (Moscow: Technosphere) pp 374–479
[11] Potapov A A, Gulyaev Yu V, Nikitov S A, Pakhomov A A and German V A 2008 *Newest Images Processing Methods* ed A A Potapov (Moscow: FIZMATLIT)
[12] Podosenov S A, Potapov A A, Foukzon J and Men'kova E R 2015 *Nonholonomic, Fractal and Linked Structures in Relativistic Continuous Medium, Electrodynamics, Quantum Mechanics and Cosmology*: In three volumes ed A A Potapov (Moscow: LENAND)
[13] Potapov A A, Wu Hao, Foukzon J, Podosenov S A and Men'kova E R 2016 *Fractals in Radar, Fields, Control and Low-Contrast Target Detection*, ed. A A Potapov (China) in press
[14] Potapov A A 2009 "The Textures, Fractal, Scaling Effects and Fractional Operators as a Basis of New Methods of Information Processing and Fractal Radio Systems Designing" *Proc. SPIE* **7374** pp 73740E-1–3740E-14
[15] Haralick R M, Shanmugan K and Dinstein I 1973 "Textural Features for Image Classification" *IEEE Trans* **SMC–3** no 6 pp 610–621
[16] Potapov A A 2004 "Topology of Sample" *Non-Linear World (Moscow)* **2** no 1 pp. 4–13
[17] Potapov A A 2007 "The Theory of Functionals of Stochastic Backscattered Fields" *J. Communications Technology and Electronics* **52** no 3 245–292
[18] Potapov A A and German V A 1998 "Detection of Artificial Objects with Fractal Signatures" *Pattern Recognition and Image Analysis* **8** no 2 pp. 226–229
[19] Potapov A A and German V A 2001 "Fractals, Fractal Target Selection and Fractal Antennas" *Proc. 1st Int. Workshop on Mathematical Modeling of Physical Processes in Inhomogeneous Media (Mexico)* pp 44–46
[20] Gnedenko B V and Kolmogorov A N 1954 *Limit Distributions for Sums of Independent Random Variables* (London: Addison–Wesley)

[21] Potapov A A and German V A 2004 "Methods of measuring the fractal dimension and fractal signatures of a multidimensional stochastic signal" *J. Communications Technology and Electronics* **49** no 12 pp. 1370–1391

[22] Potapov A A 2012 *The Fractal Method and Fractal Paradigm in Modern Natural Science* (Voronezh: Publishing and Polygraphic Centre "Nauchnaya kniga")

[23] Potapov A A 2008 "Fractal Models and Methods on the Basis of the Scaling in Fundamental and Applied Problems of the Modern Physics" *Irreversible Processes in Nature and Engineering* ed V S Gorelik and A N Morozov (Moscow: N. E. Bauman MSTU and P. N. Lebedev Physical Institute RAS) issue II pp 5–107

[24] Potapov A A 2013 "Fractal Paradigm and Fractal-Scaling Methods in Fundamentally New Dynamic Fractal Signal Detectors" *Proc. the Eighth Int. Kharkov Symposium on Physics and Engineering of Millimeter and SubMillimeter Waves-MSMW'13 (Kharkov)* pp 644–647

[25] Potapov A A 2013 "The Global Fractal Method, Fractal Paradigm and the Fractional Derivatives Method in Fundamental Radar Problems and Designing of Revolutionary Radio Signals Detectors" *Zbornik radova Konferencije MIT - Matematicke i informacione tehnologije, (Kosovska Mitrovica Prirodno-matematicki fakultet Ulverziteta u Pristini Serbia)* pp 539–552

[26] Potapov A A 2015 "Fractal Radar: Towards 1980 – 2015" *Proc. of CHAOS 2015 Int. Conf. (Paris, Henri Poincaré Institute)* pp 559–573

[27] Potapov A A 2016 "The Fractal-Scaling Radiolocation: Formation History 1980–2015" *Chaotic Modeling and Simulation* **3** pp 317–331

[28] Potapov A A 2016 "New Conception of Fractal Radio Device with Fractal Antennas and Fractal Detectors in the MIMO Systems" *Book of abstracts 9th int. conf. (CHAOS' 2016) on Chaotic Modeling, Simulation and Applications* (London, Senate House, University of London) p 85

[29] Potapov A A 2016 "Analysis and Synthesis of Topological Radar Detectors of Low-Contrast Targets Against the Background of High Intensity Noise as a New Branch of Radiolocation and the Theory of Statistical Solutions" *Eurasian Physical Technical Journal* **13** no 2(26) pp 12–23

[30] Haykin S 2006 "Cognitive Radar: A Way of the Future" *IEEE Signal Processing Magazine* **23** pp 30–40

[31] Potapov A A 2003 "New Information Technology in Radar Detection of Low-Contrast Targets Based on Probabilistic Texture and Fractal Features *J. of Communications Technology and Electronics* **48** no 9 pp 1012–1029

[32] Gulyaev Yu V, Nikitov S A, Potapov A A and German V A 2006 "Concepts of Scaling and Fractal Dimension in the Design of a Fractal Detector of Radio Signals" *J. Communications Technology and Electronics* **51** no 8 pp 909–916

[33] Potapov A A, Il'yin E M, Chigin E P and German V A 2005 "Development and Structure of the First Etalon Dictionary of Fractal Properties of Target Classes" *Electromagnetic Phenomena* **5** no 2(15) pp. 107–142

[34] Potapov A A 2016 "Fractality and scaling problems in radio location and radio physics with new methods of detection of low-contrast targets against a background of high intensity noise" *Proc. of XV Int. Academic Congress "Fundamental and Applied Studies in the Modern World" (Oxford, Oxford University Press)* pp 314–322

[35] Potapov A A, German V A and Pahomov A A 2016 "Processing of Images Obtained From Unmanned Aerial Vehicles in the Regime of Flight over Inhomogeneous Terrain with Fractal-Scaling and Integral Method" *Proc. CIE Int. Conf. on Radar "Radar 2016" (China, Guangzhou)* pp 585–587

[36] Potapov A A 2016 "Strategic Directions in Synthesis of New Topological Radar Detectors of Low-Contrast Targets against the Background of High-Intensity Noise from the Ground, Sea and Precipitations" *Proc. CIE Int. Conf. on Radar "Radar 2016" (China, Guangzhou)* pp 692–696

[37] Lehmann E L and Romano J P 2004 *Testing statistical hypotheses* (New York: Springer)

[38] Potapov A A 2017 "Diffractals at frequency 36 GHz which are observed at radar scattering of an electromagnetic wave by a fractal surface and wave catastrophes in fractal randomly

inhomogeneous media" *Proc. XIII Int. Conf. "Zababakhin Scientific Talks" Dedicated to 100th anniversary of academician E I Zababakhin (Snezhinsk)* pp 137–138

[39] Potapov A A 2017 "On the fractal structure of high-altitude storm discharges in the ionosphere: elves, jets and sprites" *Proc. XIII Int. Conf. "Zababakhin Scientific Talks" Dedicated to 100th anniversary of academician E I Zababakhin (Snezhinsk)* pp 335–337

IOP Conf. Series: Journal of Physics: Conf. Series **918** (2017) 012016 doi:10.1088/1742-6596/918/1/012016

Implementation of plasma ions in pistons as a cause of nanosatellite hull defects

A G Maslov[1], Y V Gerasimov[2] and O P Plotnikova[3]
[1]Assist. Professor, Bauman Moscow State Technical University, Moscow, Russian Federation
[2]Assoc. Professor, Bauman Moscow State Technical University, Moscow, Russian Federation
[3] Student, Bauman Moscow State Technical University, Moscow, Russian Federation
E-mail: maslov_art@bmstu.ru, drgerasimov.y@gmail.com

Abstract. The article deals with estimating strength losses during the pulse launch of a nanosatellite in the railgun barrel and their allocation in the process of hydrogen plasma ions irradiation. The authors take into account that the plasma piston consists of charged and non-charged particles, so there will be both bosons and fermions. That is why we use modified quantum statistics. As a first estimate we use the Kinchin and Pease theory of radiation defects and our assumption of the defects even distribution in the nanosatellite hull crystal. The article describes estimations of strength losses time variation.

1. Introduction

The strength losses of a nanosatellite hull result from the implementation of plasma ions in pistons. This problem appears during the nanosatellite pulse launch [1-6]. The pulse launch looks to be more effective for the launch of different numbers of nanosatellites compared to the rocket launch which has a number of problems. Different systems of a pulse launch are proposed. The authors consider the electromagnetic railgun system using a plasma piston. As the load upon the nanosatellite reaches the yield point it can cause the appearance of defects and extending cracks.

The interaction between the materials that the plasma piston and the nanosatellite hull are made of is complicated and multi-parametrical. It is possible to make decomposition on the following tasks:

1. The definition of plasma formation micro-parameters and macro-parameters;
2. The modification of quantum distribution taking into account additional conditions and the degree of ionization;
3. The number of defects in nanosatellite hull materials and the strength losses calculation;

The authors estimate that the plasma piston consists of the hydrogen plasma, the material of the nanosatellite hull is structural steel, the temperature (T) of ions rises from 300 K up to 3000 K during the process of acceleration.

2. Determining relations and parameters

The plasma piston particles energy distribution is equal to the quantum statistics from the article [7]

$$\frac{(n_b+1)(n_f-1)}{n_b n_f}=C, C=\mathrm{e}^{\frac{E+\frac{4}{3}\pi P r^3-\mu;}{kT(t)}} \qquad (1)$$

where n_b is the energy distribution of bosons, n_f is the energy distribution of fermions, μ is the chemical potential, t – time, r – radius of plasma formation particles, P – pressure, k – the Boltzmann constant.

According to [7] it is necessary to obtain the relation between bosons and fermions. The Saha ionization equation (Saha-Langmuir equation) is used as this relation:

$$\frac{n_i}{n}=\frac{1}{2}\frac{2^{1/4}\left(m_e kT(t)\right)^{3/4}\mathrm{e}^{-\frac{1}{2}\frac{Ea}{kT(t)}}}{(\pi\hbar)^{3/2}\sqrt{\frac{N}{V}}} \tag{2}$$

where n is the volume concentration of all particles of plasma piston, n_i – volume concentration of ions, V – volume of system, m_e – mass of electron, Ea – energy of ionization, $\hbar$ – Planck constant over two pi.

The formula for density of states (Z):

$$Z=\pi\sqrt{2}\left(\frac{m_e}{\pi^2\hbar^2}\right)^{3/2}\sqrt{E} \tag{3}$$

The number of defects (N_{def}) in the material of a nanosatellite hull is calculated according to Kinchin and Pease theory [9]:

$$N_{def}=\frac{1}{2}\frac{N_{im}E}{E_d} \tag{4}$$

where N_{im} – the number of implemented ions, E_d – formation of dislocation energy.

The estimation of strength losses is equal to the variation of yield point (σ) by changing the averaged square of atoms layer (S):

$$\sigma_{def}=\frac{\sigma_0 S_{def}}{S_0} \tag{5}$$

The function of temperature rise is equal to [10]

$$T=T_{\max}\left(1-\mathrm{e}^{-\frac{t}{\tau}}\right)+300 \tag{6}$$

where $T_{\max}$ – the maximum temperature in the process, τ – the discharge time of capacitors and 300 K – the initial temperature.

3. Energy distribution and calculations

Because

$\frac{n_i}{n}=\frac{n_f}{n_f+n_b}$ as $n_c=n_f+n_b=\frac{N_f+N_b}{Z}=\frac{N}{Z}$ and $n_i=\frac{N_i}{V}$, $N_i=N_f$

the Saha ionization equation is expressed as

$$\frac{n_f}{n_c}=A,\ A=\frac{1}{2}\frac{2^{1/4}\left(m_e kT(t)\right)^{3/4}\mathrm{e}^{-\frac{1}{2}\frac{Ea}{kT(t)}}}{(\pi\hbar)^{3/2}\sqrt{\frac{N}{V}}} \tag{7}$$

After substituting (7) with (1) and solving the square equation

$$n_c=\frac{1}{2}\frac{2A-1-\sqrt{-4A^2C+4AC+1}}{A(AC+A-C-1)} \tag{8}$$

the second root of this equation has no physical sense because it is negative. For calculating the number of particles with energy we have to integrate the distribution function (8) plotted on Figure 1 and DOS over energy and time:

$$N_{im} = V\left(\int_{0}^{t_{\max}} \int_{E_d}^{E_{\max}} n_c\left(E,t\right) Z\left(E\right) \mathrm{d}E \, \mathrm{d}t \right) \tag{9}$$

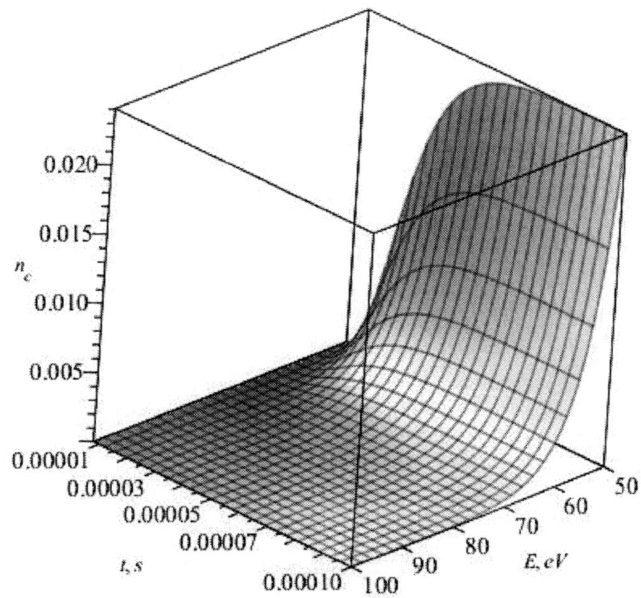

Figure 1. Distribution of particles with energy in time

The calculations are made for the following parameters:

$$E_a = 13.6 eV$$

$$t_{\max} = \frac{1}{10000} s$$

$$E_d = 50 eV \tag{10}$$

$$\tau = 10 \mu s$$

$$P = 250 MPa$$

$$r = 5.29 \cdot 10^{-11} m$$

For these parameters the numerical calculation gives

$$N_{im} = 5.192610^{19} \tag{11}$$

So, according to (4) $N_{def} = 5.4523 \cdot 10^{20}$

the new yield point is $\sigma = 0.9998\,\sigma_0$. On the one hand, cumulative strength losses within the bounds 0.02% are truly low and may not cause a failure. On the other hand, when 2 defects from $5.4523\,10^{20}$ are close enough to make a line the defect (micro crack) can occur. In this case the formation of extending crack is possible. While the shockwave passes through the material with the intensity up to the values from the articles [11-13] the failure is also possible.

To estimate the strength losses with other times Figure 2 was plotted based on the results of the numeric modeling.

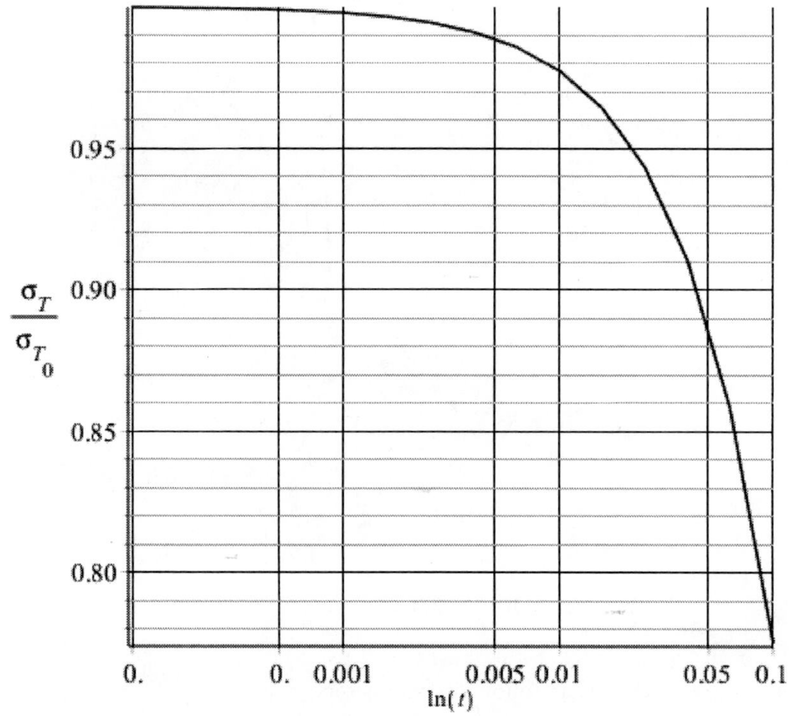

Figure 2. Strength losses in time

4. Mechanism of the material destruction

Previously we made an assumption that the defects are distributed in the satellite volume evenly and considered the volumetric strength loss. But a more accurate model should include the mechanism of ions penetration which allows figuring out how the material will be destructed in reality.

For this purpose the Bethe-Bloch theory of ionization energy loss was chosen:

$$f = \frac{dE}{dx} = -\frac{4\pi n_e z^2 e^4}{m_e v^2}\left[\ln\frac{2m_e v^2}{I} - \ln\left(1-\beta^2\right) - \beta^2\right] \tag{12}$$

z is particle charge, e – electron charge, v – velocity of the particle, $\beta = v/c$ - the relativistic factor, I – ionization energy.

Only the kinetic energy is considered. M is the particle mass.

$$E = \frac{Mv^2}{2} \tag{13}$$

After conversion the formula takes the following form:

$$f = \frac{dE}{dx} = -\frac{4\pi n_e z^2 e^4 M}{m_e E}\left[\ln\frac{4m_e E}{IM} - \ln\left(1 - \frac{2E}{Mc^2}\right) - \frac{2E}{Mc^2}\right] \tag{14}$$

After integrating we obtain relation between the energy and the penetration depth:

$$R = \int_0^E \frac{1}{f(E)} dE$$

The calculations were made numerically. The result is presented in the graphic form.

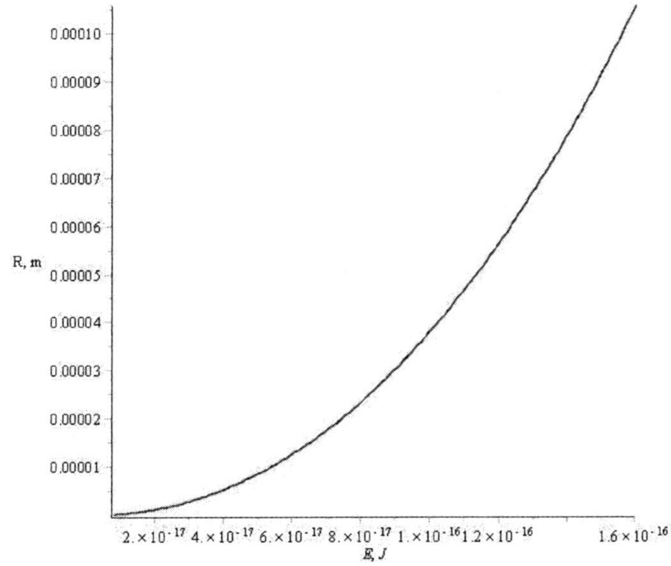

Figure 3. Function of penetration depth

Then by using the energy distribution function the average value was obtained:

$$R_{mid} = 2.147 \cdot 10^{-7}\ m$$

The Gauss distribution of penetration depth with $\sigma = \left|R_{mid} - R\left(E_{mid}\right)\right| = 2.349 \cdot 10^{-9}$ m has the following form:

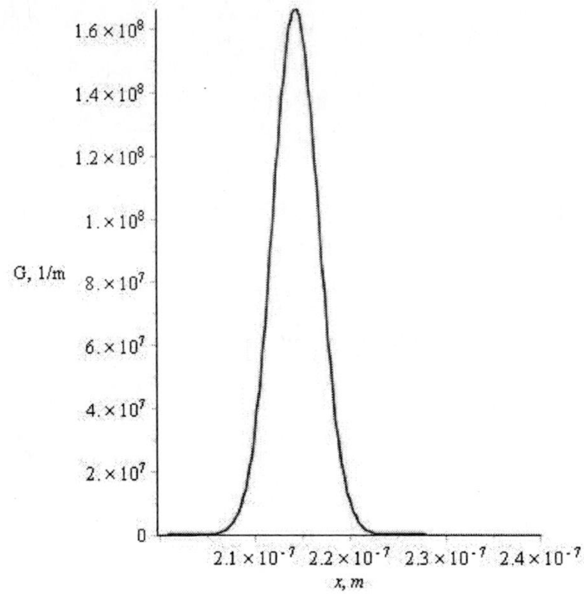

Figure 4. Gauss distribution for penetration depth

It allows accepting that all particles will be able to interact only with the 0.2 micrometers layer of the hull surface.

Previously the amount of Kinchin-Pease defects was calculated. The volume of the material destroyed during the interaction in the experiment being considered (structural steel, 10x10 mm square in cut) can be estimated as:

$$V_D = 6.41 \cdot 10^{-9}\ m$$

Total destroyed layer thickness in the assumption that the defects are formed independently amounts to 64.13 mkm, which is much more than 0.2 mkm from the previous calculations. We assume the worst variant, when all the particles with sufficient energy form their dislocations. In this case we conclude the following: the whole depth of destruction is determined by the number of Kinchin-Pease particles which knock out every molecule of the steel surface. In each moment only a 0.2 micrometer layer is destroyed, and this layer moves into the depth of the material until all the particles with sufficient energy make the interaction. In our experiment the whole depth of destruction amounted approximately 64 micrometers.

5. Conclusion

Volumetric strength loss can be easily ignored. But the surface destruction should be taken into account because microcracks are formed on the surface and this type of damage is the most dangerous in case of pulse load. Such strength loss does not make the pulse launch of nanosatellites impossible, but it must be noticed while constructing especially for composite hulls of nanosatellite, which are more subjected to failure by formation of microcracks.

One of the solutions is to use the protective wafer, which will be dropped after the satellite leaves the railgun bore.

To sum up, we have to say that:

1. The number of defects in the crystal of the nanosatellite hull appearing due to the penetration of plasma piston ions into the railgun pulse launcher was calculated.

2. The form of the damage and its quantitative estimation was obtained.
3. The volumetric strength loss from plasma formation ions penetration is negligible.

References

[1] Gerasimov Yu V, Karetnikov G K, Selivanov A B, Fionov A S 2013 *EVALUATION OF RELATIVE FINAL MASS OF A NANOSATELLITE DELIVERED TO THE NEAR-EARTH SPACE USING A PULSED LAUNCHER AND A PULSED CORRECTING THRUSTER* Herald of BMSTU. Series Mechanical Engineering 3.

[2] Fionov A S, Gerasimov Yu V, Fillipenko A V 2007 *Options for the Development of Mass Acceleration Technology in Space* Proceedings of IV all-Russian conference

[3] Lehmann P, Reck B, Vo M D, Behrens J 2006 *Acceleration of a Suborbital Payload using an Electromagnetic Railgun* Proceedings of the 13-th International Symposium on Electromagnetic Launch Technology

[4] Vijayan T, Venkatramani N 2003 *Heating and Transport of Metal Plasma in a Vacuum-Arc Rail Gun* IEEE Transactions on Plasma Science

[5] Landen D and Satapathy S 2007 *Eddy Current Effects in the Laminated Containment Structure of Railguns* IEEE Transactions on Magnetics **1**

[6] Reznikov B I, Bobashev S V, Zhukov B G, Kurakin R O, Ponyaev S A, Rozov S I 2010 *Parameters of Erosion Carbon Plasma in the Railgun Bore* J. of Technical Physics **12** 45

[7] Gerasimov Yu V, Maslov A G. 2015 *Multicomponent Quantum Particles System Statistics* Herald of the Bauman Moscow State Technical University, Series Natural Science **1** 69

[8] Artsimovich L A, Sagdeev R Z: *Plasma Physics for Physicists*, Moskow, Atomizdat, P 28

[9] Pavlov P V, Chochlov A F 2000 *Solid State Physics,* Moscow, Vusshaya Shkola, P 128

[10] Duvall G E, Graham R A 1977 *Phase Transitions under Shock-Wave Loading* Rev. Mod. Phys. V 49 № 3 P 523.

[11] Kanel G I, Fortov V E, Razorenov S V 2007 *Shock Waves in Condensed Matter Physics* Uspekhi Fizicheskikh Nauk T 177 № 8 P 809

[12] Nellis W J, Mitchell A C, Young D A 2003 *Equation-of-State Measurements for Aluminum, Copper and Tantalum in the Pressure Range 80-440 GPa (0.8-4.4 Mbar)* J. Appl. Phys. V 93(1) P 304

IOP Conf. Series: Journal of Physics: Conf. Series **918** (2017) 012017 doi:10.1088/1742-6596/918/1/012017

Light scattering from polar and pseudoscalar modes in doped $LiNbO_3$ and $LiTaO_3$ monocrystals

V S Gorelik[1,2], N V Sidorov[3], A I Vodchits[4], B P Gorshunov[5,6] and A Yu Pyatyshev[2,7]

[1]Lebedev Physical Institute of the Russian Academy of Sciences, Moscow, Russia
[2]Bauman Moscow State Technical University, Moscow, Russia
[3]Tananaev Institute of Chemistry and Technology of Rare Elements and Mineral Raw Materials of RAS, Apatity, Russia
[4]Stepanov Institute of Physics of National Academy of Sciences of Belarus, Minsk, Belarus
[5]Moscow Institute of Physics and Technology, Dolgoprudny, Russia
[6]Prokhorov General Physics Institute of the Russian Academy of Sciences, Moscow, Russia
[7]JSC "RPC "Istok" named after Shokin", Fryazino, Russia

E-mail: gorelik@sci.lebedev.ru

Abstract. The paper deals with spontaneous scattering in $LiNbO_3$ and $LiTaO_3$ crystals we synthesized to contain different dopants (Mg^{2+}, Zn^{2+}, Gd^{3+}, B^{3+}, Cu^{+} and others). We investigated photoinduced light scattering patterns in doped $LiNbO_3$ crystals. We recorded Raman spectra at different light scattering geometries and temperatures. We detected pseudoscalar modes in the Raman spectra of doped $LiNbO_3$ and $LiTaO_3$ crystals.

1. Introduction and background

Ferroelectrics $LiNbO_3$ and $LiTaO_3$ have very high spontaneous polarization values and go through the transition into the paraelectric phase at very high temperatures [1, 2]. A lack of an inversion centre in the point group symmetry (C_{3V}) and high values of nonlinear optical coefficients in these crystals make it possible to use single crystals of $LiNbO_3$ and $LiTaO_3$ to generate optical harmonics and parametrically convert electromagnetic radiation from the visible spectral range to the far infrared, including the terahertz region of the spectrum [3-5]. All properties of $LiNbO_3$ and $LiTaO_3$ crystals depend on different types of defects: crystalline lattice disorder and impurities. In this paper we present the results of investigating Raman scattering by polar and pseudoscalar modes in $LiNbO_3$ and $LiTaO_3$ crystals containing various impurities and defects.

2. Experimental technique

We synthesized $LiNbO_3$ and $LiTaO_3$ crystals doped with various ions (Mg^{2+}, Zn^{2+}, Gd^{3+}, B^{3+}, Cu^{+} and others). We investigated Raman spectra in pure (without additional defects) and doped crystals. Figure 1 shows several $LiNbO_3$ crystals doped with Cu^{+}. The samples we grew were oriented single crystals 10x7x5 mm in size, whose polar axes were perpendicular to the largest surface of the single crystal. We investigated Photoinduced Light Scattering (PLS) patterns in our $LiNbO_3$ crystals doped with Gd,

Mg, Fe, Ta and Zn. As a source of excitation radiation we used the second optical harmonics of a YAG:Nd^{3+} laser (λ=532 nm) with the average power of 160 mW. We found that the PLS patterns vary along the polar axis.

Figure 1. Photographs of single crystal lithium niobate samples: pure single crystal (left), a single crystal of lithium niobate doped with a 0.043% mass fraction of copper Cu^{+} ions (centre) and a single crystal of lithium niobate doped with a 0.015% mass fraction of copper Cu^{+} ions (right).

3. Results and discussion

Figure 2 shows a small optical distortion that takes place in the first group of doped $LiNbO_3$ crystals. In this case the shape of the PLS spot does not significantly depend on exposure time.

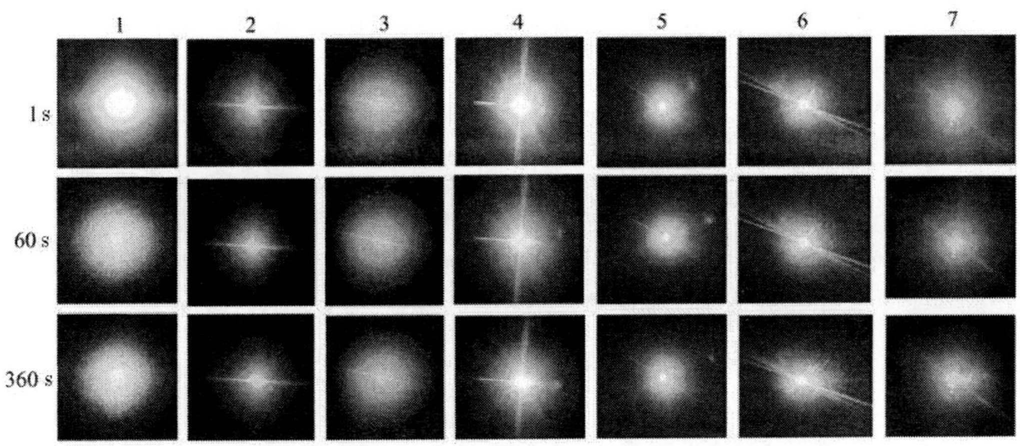

Figure 2. Temporal dynamics of Photo-induced Light Scattering (PLS) in the first group of $LiNbO_3$ crystals: 1. $LiNbO_3$, 2. $LiNbO_3$:Gd(0.51%), 3. $LiNbO_3$:Mg(0.078%), 4. $LiNbO_3$:Fe(0.0036%):Mg(0.86%), 5. $LiNbO_3$:Gd(0.23%):Mg(0.75%), 6. $LiNbO_3$:Ta(1.13%):Mg(0.011%), 7. $LiNbO_3$:Zn(0.03%).

For the second group of samples (see figure 3) temporal variations of PLS spots in doped $LiNbO_3$ crystals were larger: we can see the PLS spots elongating along the polar axis P_s.

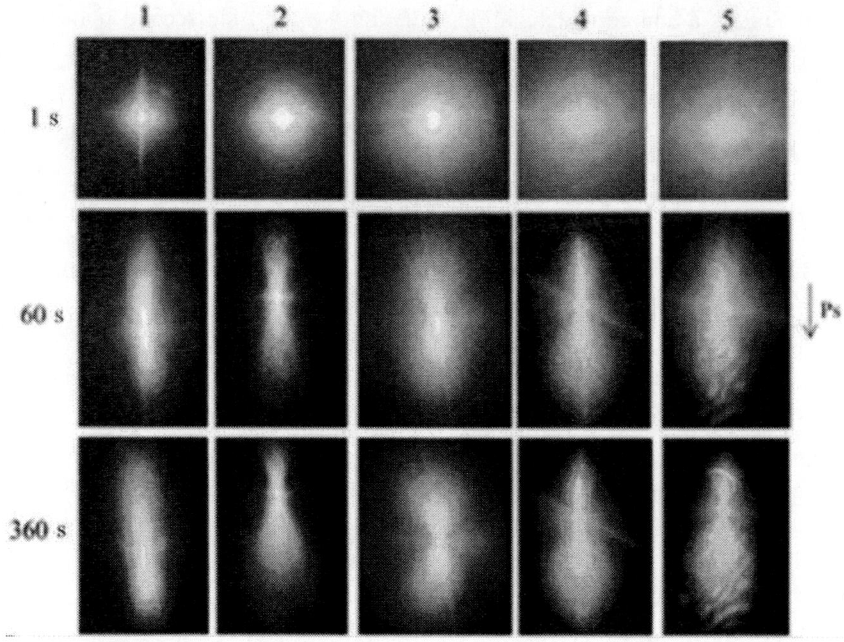

Figure 3. Temporal dynamics of Photo-induced Light Scattering(PLS) in the second group of $LiNbO_3$ crystals: 1) $LiNbO_3$, 2) B (0.12%), 3) Cu (0.015%), 4) Zn (0.52%), 5) Zn (0.62%). P_s is the spontaneous polarization vector

At last we observed a significant change in the PLS spot for the third group of samples (see figure 4): the shape of the PLS spot became asymmetric and elongated.

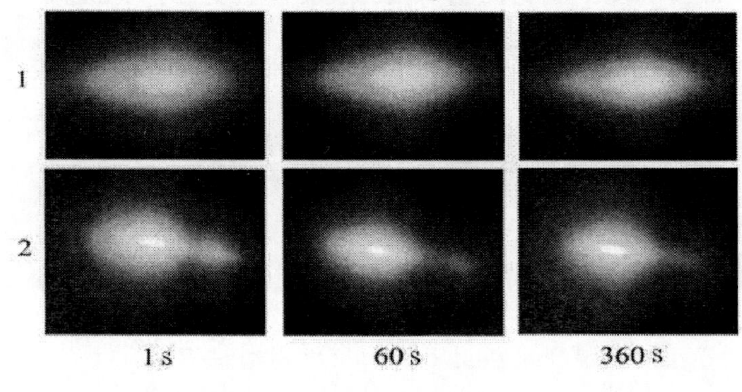

Figure 4. Temporal dynamics of Photo-induced Light Scattering (PLS) in the third group of $LiNbO_3$ crystals: 1) $LiNbO_3$:Y(0.46%), 2) $LiNbO_3$:Y(0.24%):Mg(0.63%)

We recorded Raman spectra in $LiNbO_3$ and $LiTaO_3$ monocrystals at different light scattering geometries: X(ZZ)Y; Z(XX, YY, XY)Z (forward scattering) and Z(XX, YY and XY)$\underline{Z}$ (backward scattering). Here the Z axis corresponds to the polar direction of ferroelectric crystals. According to group theory predictions [6, 7] for $LiNbO_3$ and $LiTaO_3$ crystalline structures, there exist the following types of modes in the vibrational spectra of optical branches at k=0:

$$T_{opt} = 4A_1(Z) + 9E(X,Y) + 5A_2 \quad (1)$$

Our experimental investigations led to us observing all the transversal and longitudinal modes of A_1(Z) and E(X,Y) in the Raman spectra. We found that the A_2 mode forbidden in the Raman and infrared spectra appears in the Raman spectra of doped $LiNbO_3$ and $LiTaO_3$ crystals, its intensity depending on the quantity of defects. Figure 5 illustrates a corresponding Raman spectrum at the X(ZZ)Y light scattering geometry for a $LiNbO_3$ monocrystal, for which selection rules do not forbid A_1(Z)-TO modes.

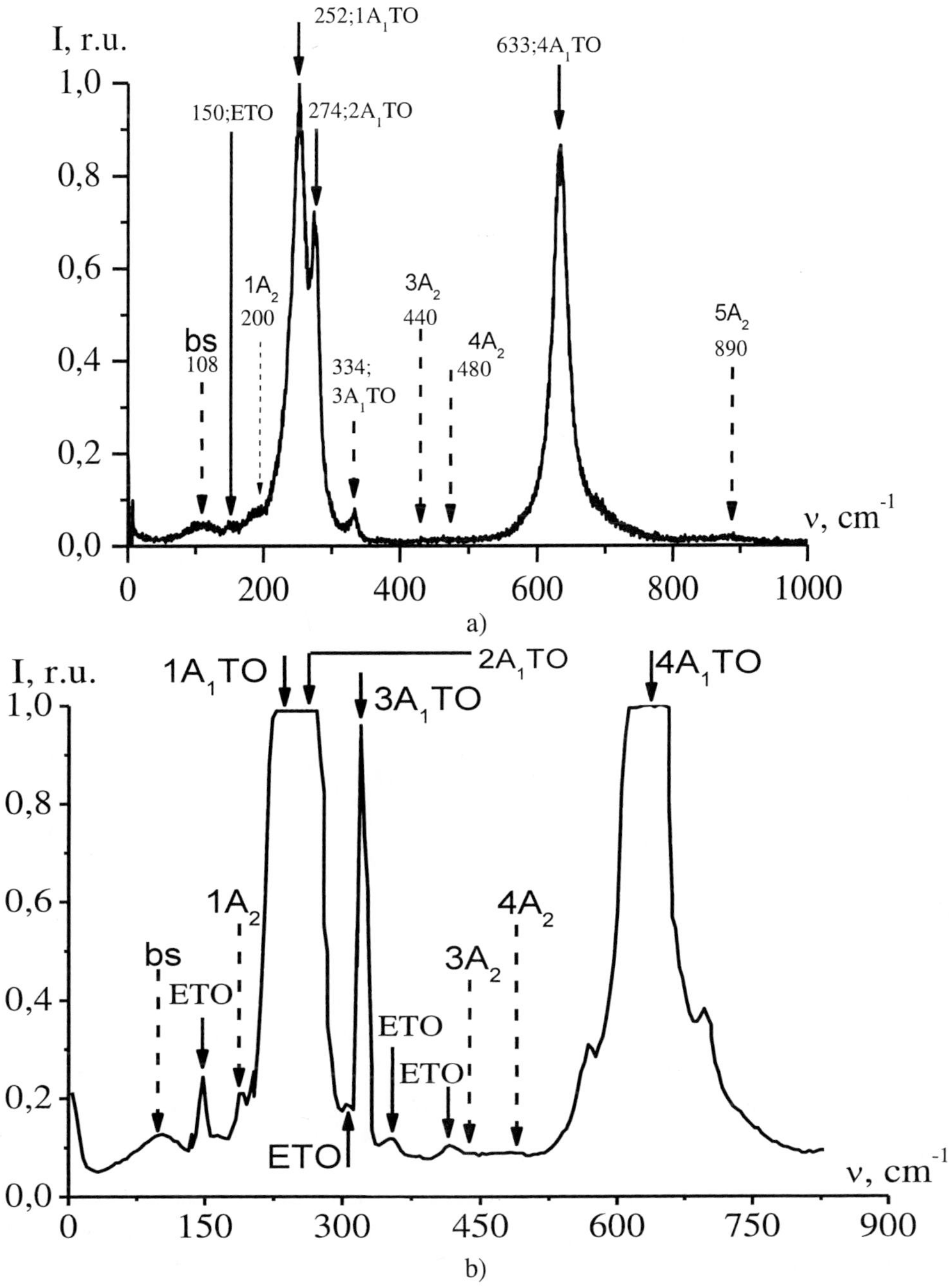

Figure 5. Raman spectra of lithium niobate monocrystals at *T*=300K(a) and 80K(b) for the X(ZZ)Y scattering geometry; solid arrows correspond to fundamental A_1 and E modes: dotted arrows specify extra modes: A_2 and two-phonon bound states (bs)

In this figure we can see peaks at 108 and 200 cm^{-1} apart from four permitted A_1(Z)-TO modes. The first one (108 cm^{-1}) corresponds to the two-photon bound state (bs), the so-called biphonon; the second (200cm^{-1}) matches the forbidden A_2-type vibration, a pseudoscalar-type mode. We estimate that the Raman intensity of the A_2 mode increases with increasing impurity concentration. So, we may explain detecting such a Raman peak as a result of violating selection rules in the presence of defects, when the point group symmetry correspondingly lowers from C_{3v} to C_3.

Figure 6 illustrates the evolution of Raman spectra when lithium tantalate crystals are heated over a wide range of temperatures, including the region near the ferroelectric phase transition (T_c=898 K), for the X(ZZ)Y scattering geometry corresponding to the excitation of a fully symmetric polar mode A_1(Z)(TO).

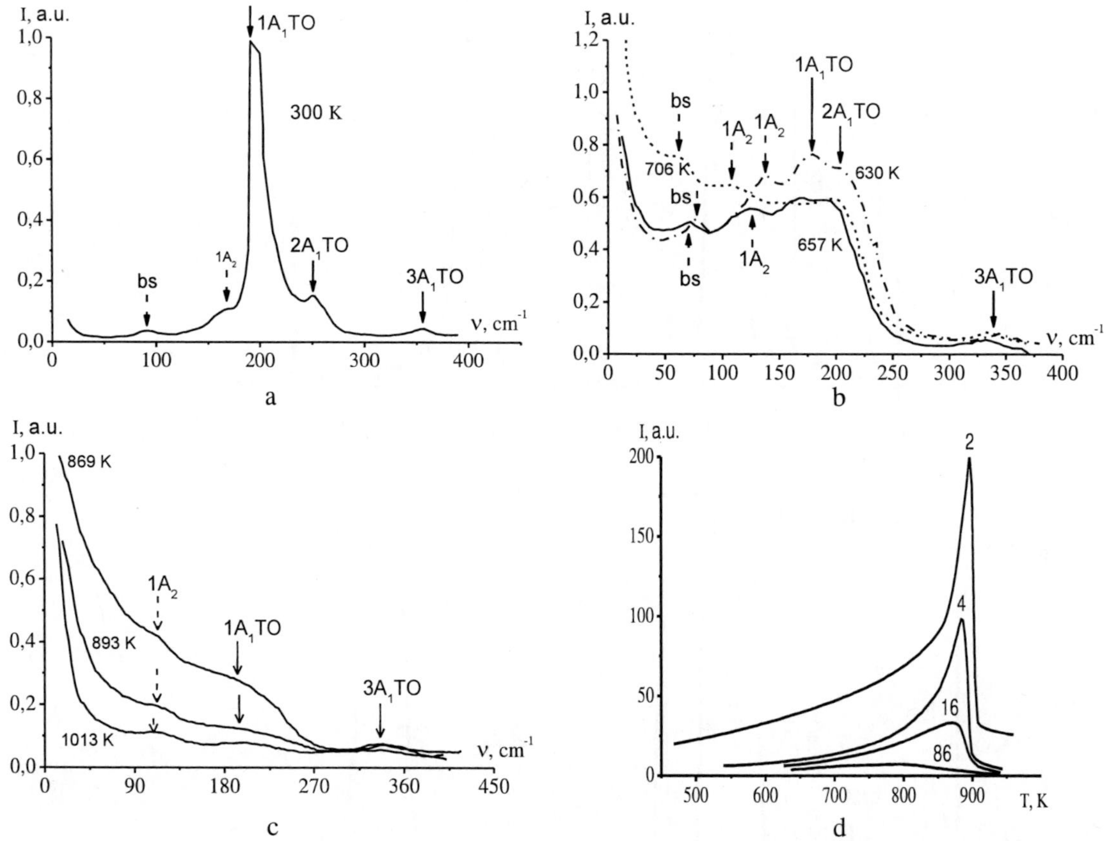

Figure 6. Temperature evolution of low-frequency Raman spectra in a lithium tantalate crystal at the following temperatures: a) T=300 K, b) T=630, 657 and 706 K, c) T=869, 893 and 1013 K(T_C=898 K); solid arrows correspond to the fundamental A_1(Z)TO-mode; d) isofrequency curves for spectral intensity of Raman scattering for the X(ZZ)Y geometry with fixed frequencies: 2, 4, 16 and 86 cm^{-1}.

According to the established theory of phase transition, a so-called soft mode is expected to exist in the Raman spectrum, whose frequency should tend to zero as the temperature of the crystal approaches the transition point. The theory predicts [8, 9] that such a mode should be a fundamental oscillation of the $1A_1$(Z)TO-type. In the observed Raman spectra (see figure 6) the frequency of the fundamental mode decreases only slightly when the crystal is heated. At the same time a significant redistribution of the intensity in the low-frequency range is observed. In this case the "softness" [10-12] is transferred from the fundamental $1A_1$(Z)TO mode to the "extra" pseudoscalar mode A_2 (converted to the scalar mode A(Z)) and then to the biphonon A_1(Z)TO (the lowest-frequency "extra" line, see figures 5 and 6a).

Figure 6d plots the isofrequency curves of the Raman intensity recorded at several fixed frequency values in the low-frequency region of the spectrum. As we can see from this figure, there is a sharp increase in the spectral intensity near the point of the ferroelectric phase transition in the lithium tantalite crystal at a 2 cm^{-1} frequency (the isofrequency opalescence effect [11, 12]), corresponding to the effective soft mode "passing" through a fixed frequency specified by the monochromator. The isofrequency opalescence characteristics observed are in satisfactory agreement with the theory of the effective soft mode developed earlier [13, 14] to describe isofrequency curves of Raman intensity near phase transition points in crystals.

4. Conclusions

Thus, we synthesized $LiNbO_3$ and $LiTaO_3$ crystals containing various dopants (Mg^{2+}, Zn^{2+}, Gd^{3+}, B^{3+} and Cu^{+}). We showed that the patterns of photoinduced light scattering observed in $LiNbO_3$ crystals significantly depend on the types of impurities introduced during crystal growth. We recorded Raman spectra of doped and defective $LiNbO_3$ and $LiTaO_3$ crystals. The Raman spectra at X(ZZ)Y geometry featured an A_2 type mode, forbidden by selection rules. We discovered a strong interaction of the forbidden A_2 mode and the soft A_1(Z)-TO mode. We observed a central peak isofrequency anomaly near the phase transition point in $LiTaO_3$ at 898 K.

Acknowledgments
This work was supported in part by RFBR grants 15-02-02882, 16-52-00026 (BRFFR 16R-063), and Ministry of Education and Science of the Russian Federation Projects 5top100 and N3.9896.2017/BY.

References

[1] Quintanilla M *et al* 2010 *Opt. Express* **18** 5449
[2] Zhang Y *et al* 2001 *J. Cryst. Growth* **233** 537
[3] Barker A S *et al* 1967 *Phys. Rev.* **158** 443
[4] Henry C H 1965 *Phys. Rev. Letters* **15** 964
[5] Gorelik V S *et al* 1983 *Sov. J. Quantum. Electron* **13** 1300
[6] Anderson A 1973 *The Raman Effect* (Marcel Dekker, New York)
[7] Landau L D and Lifshitz E M 1977 *Course of Theoretical Physics, Vol. 3: Quantum Mechanics: Non-Relativistic Theory* (Pergamon, New York)
[8] Ginzburg V L *et al* 1980 *Phys. Rep.* **57** 151
[9] Ginzburg V L 1989 *Some Remarks on Ferroelectricity, Soft Modes, and Related Questions Inelastic Light Scattering in Crystals* ed M Sushchinsky (New York: Nova Science Publishers Inc) pp 1-21
[10] Gorelik V S *et al* 1989 *Polarization and Angular Distribution of the Raman Scattered Light in Nonlinear Crystals Inelastic Light Scattering in Crystals* ed M Sushchinsky (New York: Nova Science Publishers Inc) pp 53-93
[11] Gorelik V S 1985 *Izv. Akad. Nauk SSSR, Ser. Fiz.* **49** 282
[12] Gorelik V S 1996 *Spectroscopy of Inelastic Light Scattering of the Phase State in Solids Optics and Lasers* ed G Petrash (New Tork: Nova Science Publishers Inc) pp 1-23
[13] Gorelik V S 1989 *Anomalies Spectral Intensity of Inelastic Light Scattering Near the Phase Transition Point in Crystals Inelastic Light Scattering in Crystals* ed M Sushchinsky (New York: Nova Science Publishers Inc) pp 207-231
[14] Gorelik V S 1986 *J. Mol. Struct.* **143** 83

Magnetoresistance effect in ferromagnetic metal foil

S L Timchenko[1]

[1]Assist.Professor, Bauman Moscow State Technical University, Physics Department, Russia

E-mail: svtimchenko@yandex.ru

Abstract. The paper presents the results of investigating how electric current and a permanent external magnetic field affect the magnitude of magnetoresistance effect (MRE) in polycrystalline nickel foils. For a current density of $j = 2.5 \cdot 10^8$ A/m^2 the maximum MRE we obtained for our foil samples (25x2x0.005 mm strips) was 11.6%. We studied the effect of external magnetic fields, when H_o = (0.5…6) $\cdot 10^5$A/m, on the MRE magnitude in polycrystalline nickel foils through which we simultaneously passed electric current with densities in the $j = (1...4) \cdot 10^8$ A/m^2 range. We found that the maximum MRE value (10.7 %) results from increasing the external magnetic field H_o up to $6 \cdot 10^5$ A/m when the current in the sample is fixed at $j = 3 \cdot 10^8$ A/m^2. These results may stem from the fact that the electric current may cause electroplastic softening of the foil material, which leads to additional magnetization. The current value is the threshold for starting the dislocation motion and restructuring the conductive material. Varying the external magnetic field controls the decrease in magnetization dispersion in nickel foils.

Keywords: electric current, magnetic field, ferromagnetic metal, foil, magnetization, magnetoresistance effect

1. Introduction

A high density electric current $j \geq 10^7$ A/m^2 in a metal does not only contribute to its heating, but also impacts its mechanical stress state. This stress state has anisotropic properties. Experiments [1-3] confirmed this: among the phenomena observed there were movement of dislocations along the current, oriented recrystallization and dynamic texturing of metals and alloys of arbitrary crystal structure, which may lead to changes in the electrical resistance of a conductor.

Using ferromagnetic materials as magnetoresistive transducer elements requires detailed information about the properties of such a material when it is exposed to external pressure or magnetic fields [4]. It is necessary to consider both the amplitude of the external effect and its direction. In this regard, we carried out a study of magnetoresistance of a polycrystalline ferromagnetic metal sample subjected to high current densities. We took the direction of the external factors into account in the first place.

2. Experiments and results

We conducted two series of experiments. In the first series we varied the current flowing through the sample in the I = (0.1...10) A range, and varied the orientation of the DC external magnetic field. In the second series of experiments we varied the magnitude of the DC external magnetic field in the H_o

= (0.5...6)·10^5 A/m range and its orientation relative to the current, keeping the electric current passed through the sample constant. We used polycrystalline nickel foil in the form of 25x2x0.005 mm strips as samples. We used an electromagnet to create a constant external magnetic field, which we measured both with a magnetometer and independently by measuring the electric current passing through the windings of the electromagnet. We employed a standard four-probe method to measure *R*(*j*) and *R*(*H*) magnetoresistance. Figure 1 schematically shows the factors we exposed our samples to: $\mathbf{H_0}$and **j**, while θ is the angle between the external field $\mathbf{H_0}$ and the current density vector **j** in the sample. In both series of our experiments we obtained current-voltage characteristics (CVC) as well.

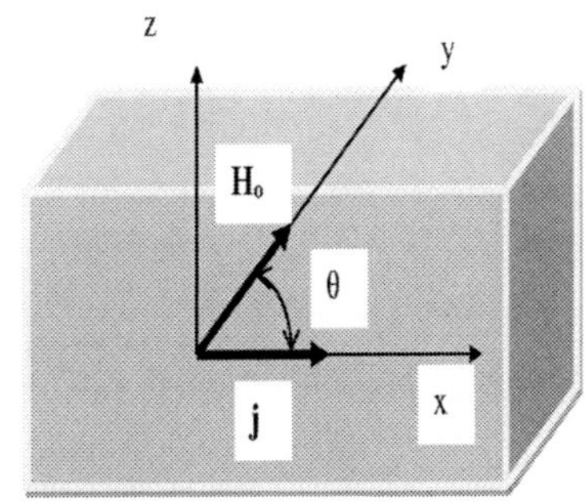

Figure 1. Diagram of external factors.

Figure 2 shows how magnetoresistance depends on the electric current density in the sample for the external magnetic field $H_0 = 3\cdot10^5$A/m, when the conducting sample is positioned differently with respect to the external magnetic field $\mathbf{H_0}$. Curve 1 represents the dependence of the sample magnetoresistance *R*(*j*) on the electric current density when θ = 0°. Curve 2 plots *R*(*j*) when θ = 90°. Let us denote $R(j)\,|_{\theta=0°} = R_{||}(j)$; $R(j)\,|_{\theta=90°} = R_{\perp}(j)$. The *R*(*j*) dependence lies between the curves 1 and 2, if the external magnetic field is missing.

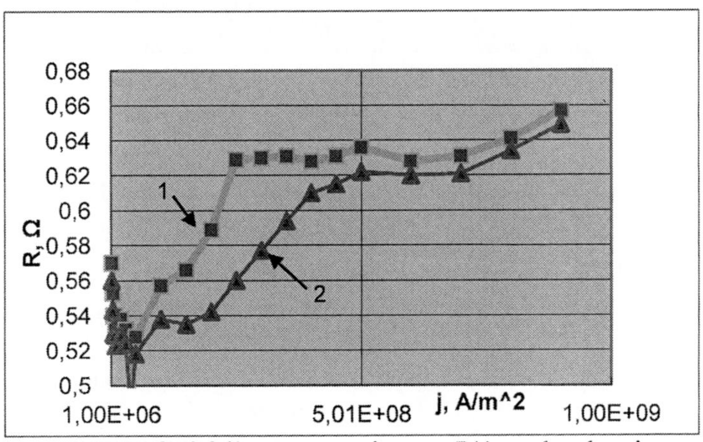

Figure 2. The dependence of Ni foil magnetoresistance *R*(*j*) on the electric current density for an external magnetic field $H_0 = 3\cdot10^5$A/m. Curve 1: θ = 0°; curve 2: θ = 90°.

We determined the dependence of the magnetoresistance effect (MRE) on the electric current density *j* in the foil sample by measuring the $R_{||}$(j) and $R_{\perp}(j)$ components:

$$\frac{\Delta R}{R_o}(j) = \frac{R_{//}(j) - R_{\perp}(j)}{R_o(\theta = 0^o) + R_o(\theta = 90^o)} \qquad (1)$$

This dependence is shown in figure 3.

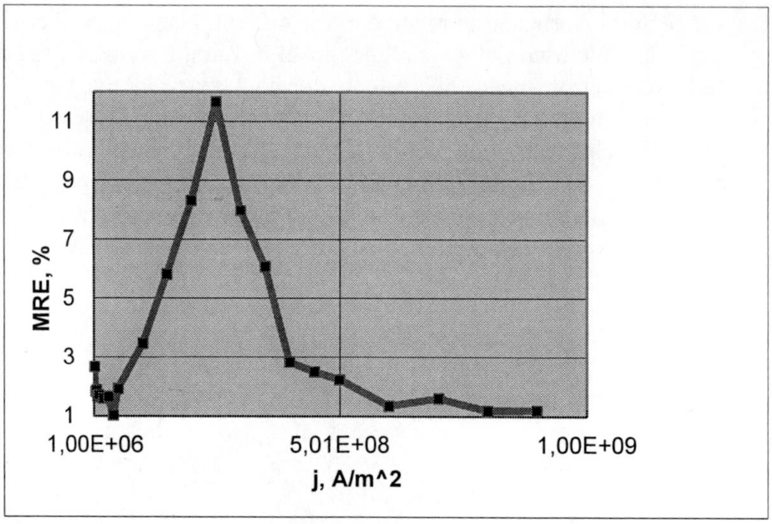

Figure 3. The dependence of MRE on current density in nickel foil for an external magnetic field $H_o = 3 \cdot 10^5$A/m.

Figure 4 shows how the MRE in nickel foils depends on the external magnetic field $H_o = (0.5...6) \cdot 10^5$A/m at various fixed current densities ($j = (1...4) \cdot 10^8$ A/m^2) passed through the foil.

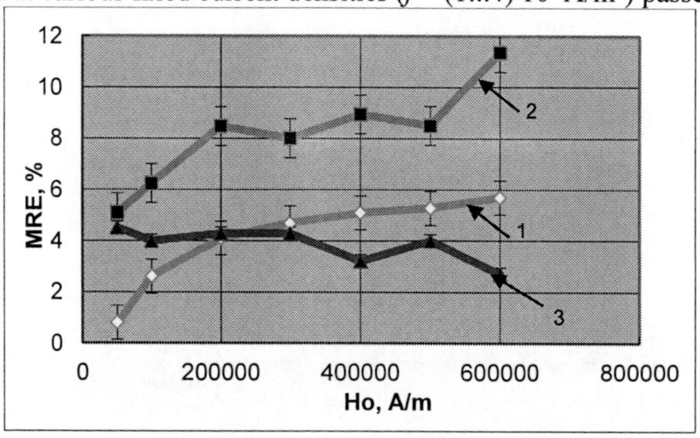

Figure 4. The dependence of the MRE on the external magnetic field at a fixed current in the sample: $j = 10^8$, $3 \cdot 10^8$, $4 \cdot 10^8$ A/m^2 in curves 1-3 respectively.

3. Discussion

It is evident that the dependence of $R_{||}$ (j), $R_{\perp}$(j) (figure 2) and the dependence $\frac{\Delta R}{R_o}(j)$ (figure 3) have three characteristic areas. In the $j = (0...0.5) \cdot 10^8$ A/m^2 range the magnetoresistance is almost independent of the current, and the MRE takes on a value of 1.8%. Then, the magnetoresistance increases in the $0.5 \cdot 10^8 < j < 2.5 \cdot 10^8$A/m^2 range of the electric current, and when the current is equal to the value of $j = 2.5 \cdot 10^8$ A/m^2, the value of the MRE reaches an abnormal value of 11.6%. The magnetoresistance $R_{||}$ does not depend on j anymore when $j > 2.5 \cdot 10^8$A/m^2, while $R_{\perp}$ continues to grow until $j = 4.5 \cdot 10^8$A/m^2. The value of $R_{\perp}$ does not depend on the electric current density j for $j > 4.5 \cdot 10^8$A/m^2. Thus, there is a dramatic decrease in the MRE value for $j > 2.5 \cdot 10^8$A/m^2. Then, when $j = (6...10) \cdot 10^8$A/m^2, the magnetoresistance effect magnitude depends weakly on the current density and drops to approximately 1.2 %.

Previous publications [5] state that the maximum MRE value in nickel is 2.5 %, which is consistent with the results of the experiment in the region of small currents $j < 0.5 \cdot 10^8$ A/m^2 (figure 1, curve 3). However, the results of the experiment indicate an increase in MRE (11.6 %) for $j = 2.5 \cdot 10^8$ A/m^2 by 6.4 times in comparison to the MRE value (1.8 %) for the case of low currents when $j = 0.5 \cdot 10^8$ A/m^2.

Considering that the MRE magnitude is proportional to magnetization squared, namely $\frac{\Delta R}{R_0} \sim M^2$ [5], it is logical to assume that the effective magnetization of the sample should increase by 2.5 times. Therefore, the appearance of additional magnetization can explain the abnormally high MRE value (11.6 %) that is achieved when the current in the foil is $j = 2.5 \cdot 10^8$ A/m^2.

On the other hand, the influence of Joule-Lenz heat cannot explain the experimental dependence of the MRE in Ni foil on the electric current density j, since we know [5] that MRE will decrease with increasing sample temperature. The decrease in MRE when $j > 2.5 \cdot 10^8$ A/m^2 is not associated with the heating of the sample either as the experiment was conducted in conditions similar to isothermal. Solving the stationary equation of thermal conductivity for nickel foil allowed us to estimate the drop in temperature across the thickness d of the sample, which is $\Delta T \cong 6$ K if $j = 10 \cdot 10^8$ A/m^2. The following sample parameters were used to calculate the temperature difference: thickness $d = 5 \cdot 10^{-6}$ m, specific conductivity $\rho = 7.0 \cdot 10^{-8}\,\Omega \cdot$m, coefficient of thermal conductivity $\lambda = 92$ W/(m·K), heat transfer coefficient $\alpha = 3 \cdot 10^4$ W/(m^2·K), initial sample temperature $T_0 = 293$ K. It means that we can neglect the sample temperature dependence on electric current density if our thin-film metallic conductor is subjected to intensive cooling.

The decrease in MRE for $j > 2.5 \cdot 10^8$ A/m^2 can be explained by an increase in the internal magnetic field $H_\sigma = \frac{3\lambda\sigma}{M}$ due to the magnetostriction effect, which leads to a change in the orientation of the magnetization vector **M** relative to the external magnetic field $\mathbf{H_0}$. However, we should bear in mind that the effects of a high-density electric current ($j = (0...12) \cdot 10^8$ A/m^2) in thin metallic magnetic films (TMF) have a threshold character as regards current, and yet they always decrease [6]. The exception is the case of a strong external magnetic field $H_0 > 4 \cdot 10^5$ A/m, when the value of MRE does not depend on the electric current density in TMF.

When studying the influence of a constant external magnetic field $H_0 = (0.5...6) \cdot 10^5$ A/m on the magnitude of magnetoresistance and MRE (1) in polycrystalline nickel foil samples, we were able to identify the value of the current that maximises the effect that the magnetic field has on the MRE magnitude (figure 4). The MRE increases from $\approx$1% to $\approx$6%, i.e. by 6 times (curve 1, figure 4) for a fixed current density in the sample $j = 1 \cdot 10^8$ A/m^2 when the external magnetic field H_0 increases from $0.5 \cdot 10^5$ A/m to $6 \cdot 10^5$ A/m. The maximum MRE value (10.7 %) is achieved by increasing the external magnetic field H_0 up to $6 \cdot 10^5$ A/m when the current in the sample is fixed at $j = 3 \cdot 10^8$ A/m^2 (curve 2, figure 4). Increasing the external magnetic field leads to a decrease in MRE from a value of $\sim$ 5 % to a common tabulated value of approximately 2.5% (curve 3, figure 4) which we can observe when the current density in the sample is $j = 4 \cdot 10^8$ A/m^2.

The results of our experiments can be explained by the state that the sample structure is in. For instance, applying an external magnetic field and increasing it further leads to a decrease in magnetization dispersion and an increase in the MRE [6], which is observed at low currents ($j = 1 \cdot 10^8$ A/m^2). Electroplastic softening process in the sample material leads to the appearance of additional magnetization; an increase in H_0 then reduces the scatter of the magnetization vector **M** in the foil. Consequently, the MRE increases (curve 2, figure 4). We observe this when $j = (2.5 - 3) \cdot 10^8$ A/m^2. As the plastic softening process approaches its final stage, the MRE is reduced to tabulated values; we observe this when $j = 4 \cdot 10^8$ A/m^2. In this range of currents, magnetic field has virtually no effect on the MRE magnitude (curve 3, figure 4).

4. Conclusion

The dependence of the MRE on the electric current density and a constant external magnetic field, as well as its anomalous value of 11.6% at $j = 2.5 \cdot 10^8$ A/m^2, can be attributed to structural changes in the sample caused by the current accompanied by plastic deformations. Namely, the electronic subsystem causes additional directional movement of atoms, affecting the crystal lattice of the conductor. This lattice possesses structural inhomogeneity, which in turn causes the movement of defects. As a result, the process affects the physical properties of the material. Increasing the external magnetic field when the currents in the sample are above the threshold causes additional magnetization, which leads to a decrease in magnetization dispersion. This anomalous dependence of magnetoresistance on current density in nickel foils when the conducting sample is positioned differently with respect to the external uniform magnetic field may be used to increase the sensitivity of electromagnetic signal sensors based on MRE in particular.

References

[1] Spitzin V I and Troitzkii O A 2000 *Electroplastic deformation of metals* (Moscow: Science).
[2] Conrad H, Karam N, Mannan S and Sprecher A F 1988 *Scr. met.* **22** 235-238
[3] Kornev Y V, Sidorenkov V V and Timchenko S L 2001 *Reports of the Academy of Sciences of the USSR* **46**(10) 690-93
[4] Kucherenko S S, Paschenko V P, Polyakov P I, Charzev S I and Shtaba V A 2001 *Letters to Journal of Technical Physics* **27**/11 24-29
[5] Vonsovsky S V 1971 *Magnetism* (Moscow: Science) p 1032
[6] Sementsov D I, Sidorenkov V V and Timchenko S L 1991 *Letters to Journal of Technical Physics* **17**/4 28-32

IOP Conf. Series: Journal of Physics: Conf. Series **918** (2017) 012019 doi:10.1088/1742-6596/918/1/012019

Measurement the Finsler transformation parameters of surround space *(when it's "the long-awaited doomsday")*

P I Ivashkin[1], G N Izmailov[2], V V Priimochenko[3], A M Tshovrebov[4] and L N Zherikhina[4]

[1]A.M.Prokhorov General Physics Institute of the Russian Academy of Sciences, Moscow Russia [2]Moscow Aviation Institute, Moscow Russia [3]National Research Nuclear University (MEPHI), Moscow Russia [4]P.N.Lebedev Physical Institute of the Russian Academy of Sciences, Moscow Russia

Email : zherikh@sci.lebedev.ru

Abstract. The present work is considered a scheme in which by means of SQUID / magnetostrictor system serves to carry out measurements attempt Finsler parameters surrounding space.

1. Introduction and background

Nowadays it's getting popular to describe the universal-galactic scale's spaces with various models [1], which includes the Finsler [2, 3] geometry representation. If the "normal" Riemann geometry of curved space allows a local reduction to Euclidean (pseudo) in a small neighborhood of each point, the Finsler geometry of curved space "locally straightens" to projective. It must have been a pole in projective geometry, with its fractional / linear transformations (like $x' = \dfrac{x}{1-\mathbf{K}x}$), i.e. the point from which all parallel (in the usual sense) to each other lines proceed, and the set of points forms a horizon line (medieval painting with perspective - figure 1) like many focus points of the lens (either of which corresponds to a paraxial beam directed at a certain angle to the optical axis) it forms a focal plane.

Now let's turn to the case of reverse perspective (which is supposedly found by some "wisenheimers" in the illustration to the Russian annals [4] - figure 2) in the sense that the observing sizes goes up from the future into the past. The "true distance" to the celestial object $r = ct$ (where t - time of arrival of the light signal "from it to us") is transformed as $r' = \dfrac{r}{1-\mathbf{K}r}$, wherein respectively inverse transform is $r = \dfrac{r'}{1+\mathbf{K}r'}$. Let's define the rate of change of the distance $\upsilon = \dfrac{\partial r'}{\partial t} = \dfrac{\partial}{\partial t}\left(\dfrac{ct}{1-\mathbf{K}ct}\right) = \dfrac{c}{1-\mathbf{K}ct} + \dfrac{\mathbf{K}c^2 t}{(1-\mathbf{K}ct)^2} = \dfrac{cr'}{r} + \dfrac{c\mathbf{K}(r')^2}{r}$, where the first term has no physical meaning, because there is differentiation of unchanged distance r itself (in contrast to the 2-nd term), not the result of the transformation distance. As a result, $\upsilon = \frac{\partial r'}{\partial t} = \frac{\kappa c(r')^2}{r} = c\kappa r'(1+\kappa r')$, wherein in turn, the first term $v_1 = c\kappa r'$ is on the same form as the Hubble law $v = Hd$, and the second $v_2 = c(\kappa r')^2$ can be compared with the acceleration of the expansion of galaxies, which is associated now with the presence of dark energy. The value of the Hubble constant $H \approx 2,2 \times 10^{-18}$ Hz enables, by analogy $v_1 = c\kappa r' \leftrightarrow v = Hd$, to evaluate the coefficient $\kappa = H/c \approx 2,2 \times 10^{-18}$ Hz / 3 × 10^8 m/s ≈ $0,73 \times 10^{-26}$ m^{-1}.

Figure 1. The Battle of San Romano (1432) by Paolo Uccello (triptych, tree, tempera). The first in the history of mankind, a lot of the objective battle scene, depicted with the explicit use of elements of the theory of perspective (ie, in fact, projective geometry).

Figure 2. Andrei Rublev "Holy Trinity". Elementary analysis shows that within the concept of perspective this image is absurd in approximately the same way as all illiterate falsified photographs allegedly made by American astronauts on the Moon.

According to the estimates made on the original independent manner by I.Ya.Mostovoy [5], k can be significantly more than order $\kappa \approx 10^{-15} m^{-1}$. Let the "receiving" rod has "unperturbed" length $\ell = L_2 - L_1$, where the coordinates of its ends are measured from the point at which is "transmitting" mass M is located (figure 3). Let us write down how the coordinates of the "transmission" mass are converting relative to the ends of the rod $L_1' = \frac{L_1}{1-\kappa L_1} = L_1(1+\kappa L_1 + (\kappa L_1)^2 + (\kappa L_1)^3 + \ldots)$

$L_2' = \frac{L_2}{1-\kappa L_2} = L_2(1+\kappa L_2 + (\kappa L_2)^2 + (\kappa L_3)^3 + \ldots)$. Let the mass M oscillates along the axis of the rod with an amplitude δL_0 and frequency ω. Then the difference of the transformed coordinates (in fact, of the rod length), or $\ell(t) \approx \ell(1 + 2\kappa(L_0 + \delta L_0 \sin \omega t))$, or $\frac{\delta\ell(t)}{\ell} \approx 2\kappa\, L_0 \sin \omega t$, wherein L_0 - distance from the stationary to the middle of the rod M, $\frac{\delta\ell(t)}{\ell}$ - its elongation. Thus, if the "transmitting" body M fluctuates with the "scale" $\delta L_0 \approx$ 1m the elongation of "receiving" rod in accordance with the first estimate ($\kappa_I \approx 0{,}73 \times 10^{-26}$m^{-1}) amounted to approximately $1{,}5 \times 10^{-26}$, and in accordance with the second ($\kappa_{II} \approx 10^{-15}$m^{-1}) of about 2×10^{-15}.

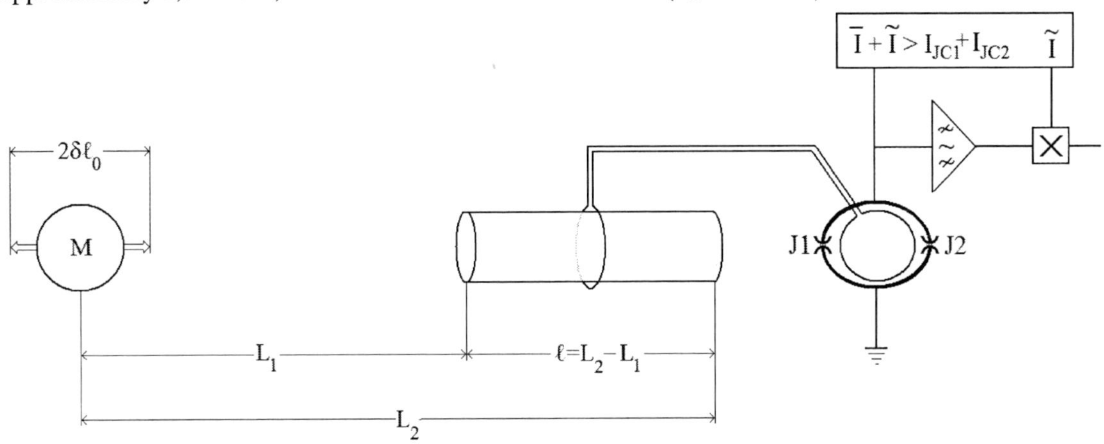

Figure 3. Scheme of using the SQUID/magnetostrictor system for measuring Finsler parameters of the surrounding space. The crosses (J1, J2) denote a pair of Josephson tunneling transitions of the DC-SQUID (ie, a superconducting quantum direct-current interferometer).

2. SQUID/magnetostrictor system

Dilatometer with a sensitivity that sufficient to detect such small extensions ($1{,}5 \times 10^{-26} \div 2 \times 10^{-15}$), can be built on the basis of SQUID/magnetostrictor system (figure 3), which was previously planned to be used in projects for the creation of a compact detector of gravitational waves [6]. High values of the inverse magnetostrictive effect constant $\Lambda^{(-1)} = \frac{\partial B}{\partial P} = 10^{-5} T/Pa$ (or to respond to the magnetic flux on the force $\Lambda^{(-1)} = \frac{\partial \Phi}{\partial F} = 10^{-5} Wb/N$) of the "good" magnetostrictive materials (alloys Pt54%Fe46%, TbFe, superparamagnetics) let you at $S \approx 10^{-3}$m^2 rod section with a Young's modulus E $\approx$ 100GPa obtain magnetic responses in the range $\delta\Phi = 1{,}5 \times 10^{-23} \div 2 \times 10^{-12}$Wb $\approx 0{,}6 \times 10^{-8} \div 1000\Phi_0$, corresponding to said interval extensions $\delta\ell/\ell = 1{,}5 \times 10^{-26} \div 2 \times 10^{-15}$. Meaning that the resolution of the modern superconducting quantum interferometer (SQUID) reaches values $10^{-6}\Phi_0/\sqrt{Hz}$ (where $\Phi_0 = \pi\hbar/e \approx 2{,}07 \times 10^{-15}$Wb - quantum of magnetic flux), the time required for the registration of extension $\delta\ell/\ell \approx 1{,}5 \times 10^{-2}$ corresponds 3×10^4s ≈ «half a day", while at $\delta\ell/\ell \approx 1{,}5 \times 10$-^{15} elongation recorded almost instantaneously. In [6] a dilatometer design with hub mechanical strain, virtually eliminating the loss of the magnetic signal transferred to a SQUID through a superconducting flux transformer.

IOP Conf. Series: Journal of Physics: Conf. Series **918** (2017) 012019 doi:10.1088/1742-6596/918/1/012019

Conducting such an experiment (transmitted vibrations "transmission" of the body M - reception by dilatometry "reception" of the rod) allows not only to measure the coefficient k, but also, apparently, within the framework of hypothesis of "reverse perspective" to estimate when a doomsday $T_U \approx \kappa/c$ come to the universe.

It is impossible to ignore just another question, "... if, although slightly (to the extent that κ), but everywhere, the movement of the random body affects the size of the others, wouldn't it lead in adding all effects to some" crazy "paradoxes, leading to the "monstrous" divergences. Of course, this question is difficult to answer without experimental data on κ, but, for example, such a "universal" and thus intelligibly-unexplained phenomenon of the 1/f-noise ratio could well be the result of cumulative effects of the movements of all the bodies.

A number of V.G.Zhotikov's studies [1] formulated somewhat different motivation of the Finsler geometry's manifestations in the surrounding space. Here, in contrast to the above "passive" version (where $\kappa \neq 0$ is a "particular" property of the surrounding space, even in conditions of complete immobility), Finsler conversions are "induced" local coordinate transformations of material points and now obviously depend on the momentum p of these points. The coefficient κ is already a monotonically increasing function of the momentum $\kappa = \kappa(p)$, in its simplest form is represented as $\kappa(p) = p/\hbar$, where $\bar{H} >> \hbar$ (and much / many orders of magnitude). To estimate $\bar{H}$, that is included in "induced" Finsler the simplest form transformations, it should be proceeded dilatometric measurements SQUID/magnetostrictor system, located near the flow of matter, in which at least briefly a huge boost is evolving. Apparently, it is possible to consider the cumulative build metal as a champion of the flow velocity, and the availability of technical realization. So the legendary German Panzerfaust the Panzerfaust-30 is higher than the 1st Space (10 km/s [7]).

3. Instead of concluding

The mention of the first cosmic velocity provides a convenient occasion finally to escape from the rather controversial features of manifestation of Finsler features in the surrounding space and refer as something more real to the Tsiolkovsky rocket equation $v(m) =_J \ln(M/n$, where $v(m)$ is a velocity of the rocket, the mass of which is the initial value of M reduced fuel consumption due to the creation of the jet to m, v_J - velocity of the exhaust. As is clear, the growth of the final rocket speed $v(m)$ due to the increase of fuel inventory M - a very inefficient way of (only a logarithmic dependence), but the increase before the logarithmic factor v_J is dramatically effective. In addition, how can we not recall the old/good Panzerfaust-30, which could be used as a high-performance pulse rocket booster...

References

[1] Zhotikov V G 2014 Introduction to Finsler geometry and its generalizations (for physicists) (Moscow: MPTI Press)

[2] Finsler P 1918 Über Kurven und Flächen in allgemeinen Räumen (dissertation, Göttingen) JFM 46.1131.02. Reprinted by Birkhäuser (1951)

[3] Rashevskii P K 1941 Polymetric geometry (Proceedings of the seminar on vector and tensor analysis, Issue 5, Moscow)

[4] Raushenbakh B V 1975 Spatial constructions in Old Russian painting (Moscow: Science)

[5] Mostovoi I Ya Privet communication; see also Mostovoi I Ya and Trufanov E Yu 2015 Narrative of the partial Moon Defense technology **7** 96-140 (Moscow)

[6] Golovashkin A I Zherikhina L N Tskhovrebov A M and Izmailov G N 2012 Supersensitive SQUID/magnetostrictor detecting system Quantum Electronics **42**(12) 1140–1146

[7] Vasilchakov A V 2008 "Faustniki" in the battle. (Moscow: YAUZA-PRESS)

IOP Conf. Series: Journal of Physics: Conf. Series **918** (2017) 012020 doi:10.1088/1742-6596/918/1/012020

Mesoporous anodic aluminum oxide photonic crystalline films and its applications

V S Gorelik[1, 2], A V Pudovkin[1], Y P Voinov[1], V V Filatov[2], Dongxue Bi[2], Guo Liang Shang[3] and Guang Tao Fei[3]

[1] Lebedev Physical Institute of the Russian Academy of Sciences, Moscow, 119991, Russia

[2] Bauman Moscow State Technical University, Moscow, 105005, Russia

[3] Key Laboratory of Materials Physics and Anhui Key Laboratory of Nanomaterials and Nanotechnology, Institute of Solid State Physics, Hefei Institutes of Physical Science, Chinese Academy of Sciences, Hefei, China

E-mail: gorelik@sci.lebedev.ru

Abstract. We have calculated the optical characteristics of mesoporous photonic crystalline film of aluminum oxide. We obtained the dispersion relation for electromagnetic waves in the film, calculated an effective refractive index of photonic crystal, worked out group velocities and effective masses of photons and polaritons in mesoporous medium. The results of theoretical analysis are in the good agreement with the measured reflection spectrum. It is shown that the mesoporous photonic crystalline films are good as the radiation amplifiers and highly efficient selective mirrors and filters.

1. Introduction

New gain medium, high efficient selective mirrors, narrow-band filters developments are actual issues in laser science now [1]. Photonic crystals and photonic crystalline films may be used to resolve such issues [2]. Applications of mesoporous globular photonic crystals (based on amorphous SiO_2) as optical resonator mirrors for different laser types were analyzed [3]. Mesoporous photonic crystalline films (based on Al_2O_3) manufacturing technology was announced [4].

Reflection coefficient of photonic crystal film is very high in photonic stop band region. Our tasks were: to analize optical properties of anodic mesoporous photonic crystalline films based on Al_2O_3, to perform experimental measurement of these properties and compare our calculations with experimental data. The possibilities of such films usage in laser science and spectroscopy were analyzed.

2. Experimental part

Photonic crystalline films of anodic aluminum oxide with thickness about 20 μm were the object of research. These films were produced from aluminum (see figure 1) in electrolyte with acid [5].

In the process of electrolysis the electric current was changing periodically in electrolyte (see figure 2). As a result the periodical crystalline structure was produced. This crystalline structure is formed from two kinds of dielectric layers with refractive indexes n_1 and n_2.

Photographs of the film produced out of aluminum substrate are shown in figure 3 (a, b, c). Film color depends on observation angle: this property is consistent with Bragg's law. The film is transparent in wide frequency region: in ultraviolet, visible and infrared.

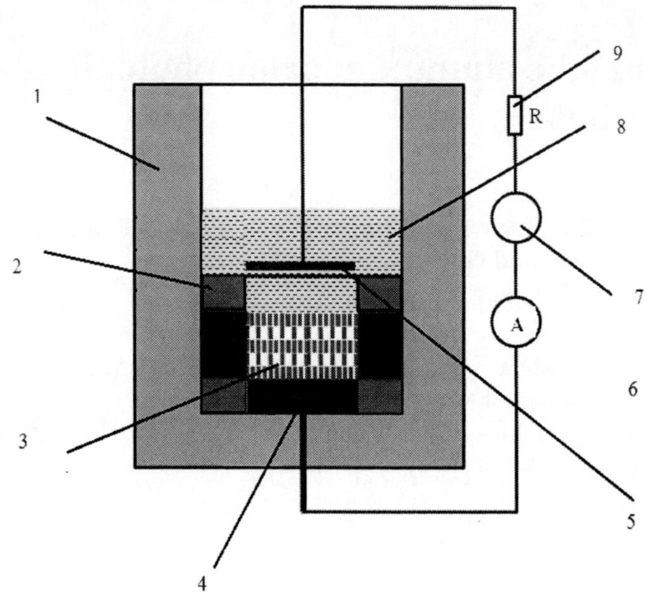

Figure 1. Photonic crystalline film production plant layout: 1 – dielectric cuvette, 2 – dielectric holder, 3 – Al_2O_3 film, 4 – cathode, 5 – anode, 6 – ammeter, 7 – current source, 8 – electrolyte with acid, 9 – resistance.

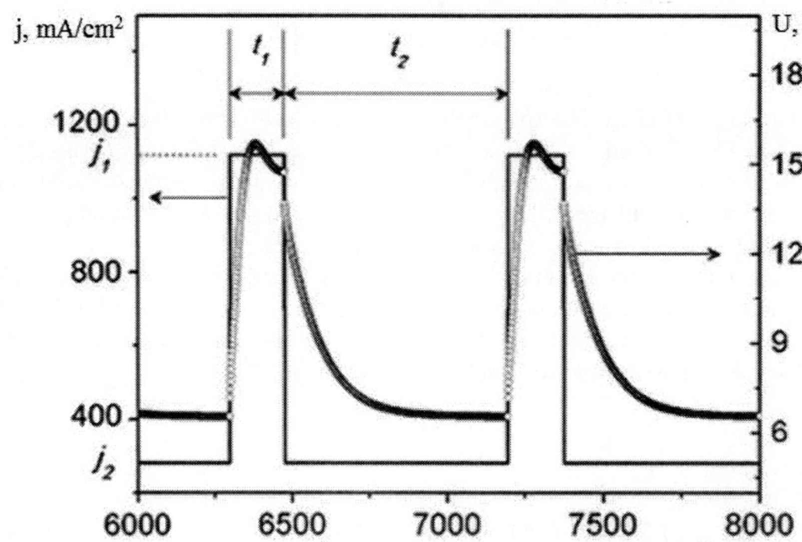

Figure 2. Current-time dependence is periodical: electric current with higher density j_1 exposure film with shorter duration t_1 and then electric current with lower density j_2 exposure film with longer duration t_2.

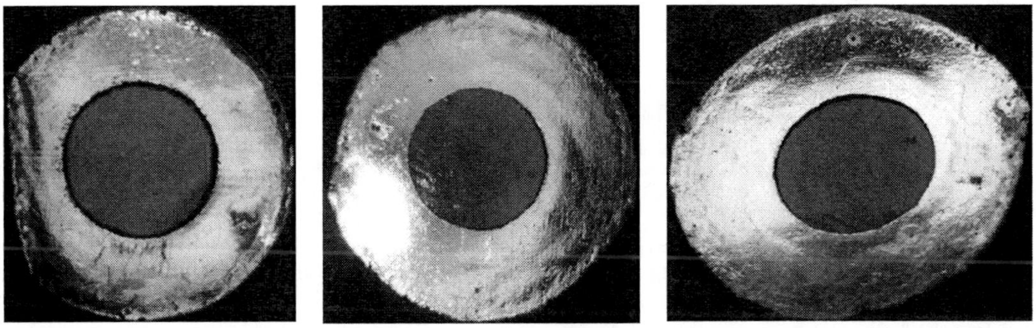

(a) (b) (c)

Figure 3. Photonic crystalline film surface photo taken from different observation angles.

Schematic view of aluminum oxide photonic crystalline film structure is shown in figure 4. We can see, that Al_2O_3 photonic crystalline film (1) consists of two kinds of layers with different refractive indexes: this film is surrounded by aluminum frame (2) which was not affected by acid. The lattice constant of fabricated anodic aluminum oxide mesoporous crystalline films may be changed at 200-500 nm range.

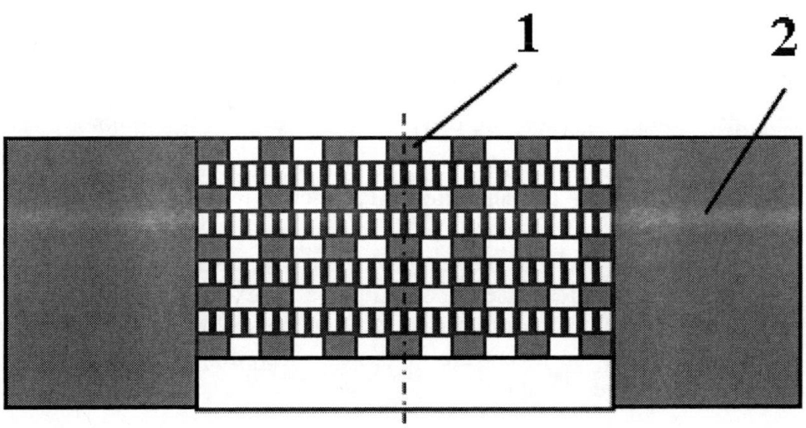

Figure 4. Photonic anodic aluminum oxide mesoporous crystalline film nanostructure: 1 – Al_2O_3 photonic crystalline film, 2 – aluminum substrate.

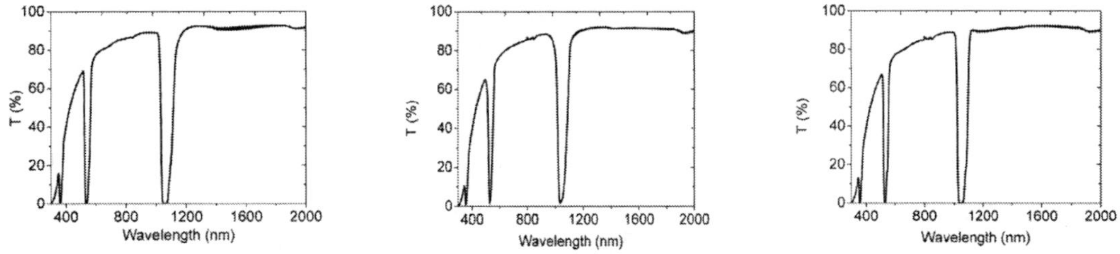

(a) (b) (c)

Figure 5. Photonic crystalline film transmission spectra for zero angle of registration: (a), (b), (c) – three types of anodic aluminum oxide crystalline films.

Vertical light transmission spectra of incandescent light through film surface for three samples are presented on figure 5 (a, b, c). There are three stop bands in the spectrum of these films. Spectral position of the first stop band is in infrared area, the second stop band is located in visible region and the third one – in near UV.

3. Theoretical analysis

One-dimensional plane-layered medium model [6] was used to render optical characteristics of the photonic crystalline film. According to this approximation, the propagation of electromagnetic waves in such structure satisfies to dispersion relation [7]:

$$\cos ka = \cos k_1 a_1 \cdot \cos k_2 a_2 - \frac{1}{2}\left(\frac{n_1}{n_2} + \frac{n_2}{n_1}\right)\sin k_1 a_1 \cdot \sin k_2 a_2 \,. \tag{1}$$

Here n_1 and n_2 – refractive indexes of layers with thicknesses a_1 and a_2 correspondingly; structure period $a=a_1+a_2$; $k_1=(\omega/c)\cdot n_1$, $k_2=(\omega/c)\cdot n_2$; $c=3\cdot 10^8$m/sec – speed of light in vacuum. Film transmittance spectrum is inverse to its reflectance spectrum $T(\omega)=1-R(\omega)$. Film reflectance spectrum is determined by Fresnel condition [8]. For normal incidence we have:

$$R(\omega) = \left|\left(\frac{ck(\omega)}{\omega} - 1\right)\Big/\left(\frac{ck(\omega)}{\omega} + 1\right)\right|^2 . \tag{2}$$

In the case of thin film (which contains N layers), it is more accurate to use the following relation [7]:

$$R_N(\omega) = \left[1 + \left|\left(\frac{\sin ka}{\sin Nka}\right)\Big/\left(\frac{1}{2}\left(\frac{k_1}{k_2} - \frac{k_2}{k_1}\right)\sin k_2 a_2\right)\right|^2\right]^{-1} . \tag{3}$$

On the basis of dispersion relation (1) we can calculate electromagnetic wave parameters in photonic crystalline film: group velocity $V(\omega) = \frac{\partial \omega}{\partial k}$, (4); photon effective mass $m(\omega) = \left(\frac{\partial^2 E}{\partial p^2}\right)^{-1} = \hbar \cdot \left(V(\omega)\frac{dV(\omega)}{d\omega}\right)^{-1}$, (5); and effective refractive index $n(\omega) = \pm\frac{c\cdot k(\omega)}{\omega}$, (6), here plus or minus signs match with signs of electromagnetic wave group velocity (4).

So, it is necessary to know thicknesses and refractive indexes of layers (a_1, a_2, n_1 and n_2) to calculate optical properties of the film. In our case the thicknesses of the layers are equal $a_1=a_2$, and period $a=a_1+a_2$ and porosity of the layers are unknown. According to Bragg relation, the spectral positions of stop bands satisfy the condition

$$2an_{eff} = m\lambda_m , \tag{7a}$$

Here m – peak sequential number, λ_m – reflectance peak spectral position, $n_{eff} = \sqrt{\frac{1}{2}n_1^2 + \frac{1}{2}n_2^2}$ – film effective refractive index. At different incident angles instead of relation (7a) we have:

$$2a\sqrt{n_{eff}^2 - \sin^2\theta} = m\lambda_m , \tag{7b}$$

On the other hand, spectral line width depends on refractive indexes contrast of layers:

$$\frac{\Delta\lambda}{\lambda_m} \approx \frac{2}{\pi}\cdot\frac{\Delta n}{n_{eff}} . \tag{8}$$

Reflectance peak spectral position (λ_m = 575 nm) and full width at half maximum ($\Delta\lambda$ = 25 nm) are known from experiment, see figure 4(c). So, it is possible to calculate film parameters: a_1 = 95 nm, a_2 = 95 nm, n_1 = 1.43, n_2 = 1.58 (and layers porosity is 52% and 31% respectively).

Dispersion relation (1) for given parameters is shown in figure 6. Wave vector tends to zero with frequency decreasing. There are several branches; breaks between them on frequency axis are stop bands. Odd (2m–1) numbers of stop bands are wider than even (2m) ones. Light group velocity in the area of the 2-nd stop band (and other even stop bands) is close to zero according to (4).

Comparison of calculated transmission spectra (based on equations (2) and (3)) with experimental data (see figure 5c) is shown in figure 7. We can see quite accurate match of stop band spectral positions, their widths and decreasing of transmittance with wavelength decrease. Experimental transmittance is lower than theoretically rendered one especially in UV area, supposedly because of unaccounted absorption.

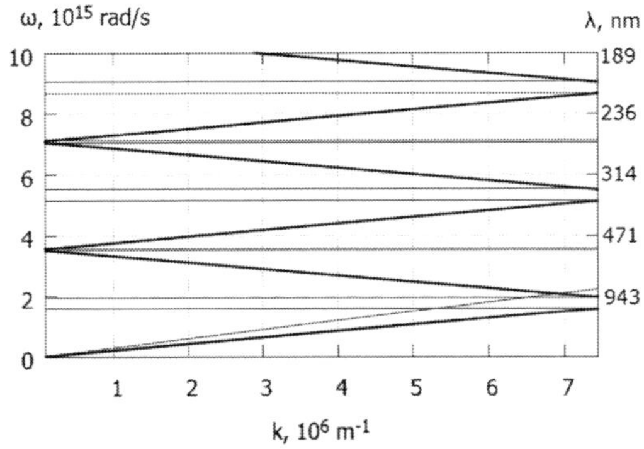

Figure 6. Dispersion relation ω = ω(k) for photonic crystalline film (thin line is the dispersion relation for light in vacuum ω = c·k).

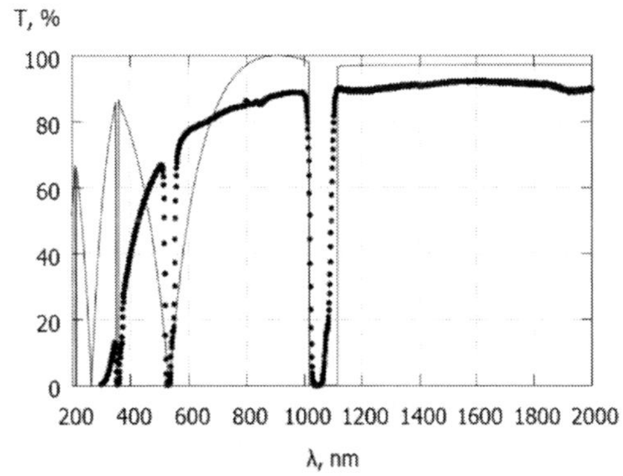

Figure 7. Comparison of theory (grey curves) with experiment (black dots).

Group velocity of light (see figure 8) may be negative or positive in some ranges and may be close to zero near to the 2nd stop band. This property may be used for coherent light generation when photons turn into the state of Bose–Einstein condensate. Convenient conditions for such effect observation are the using of laser emission, with frequency value close to the Γ- point (k=0) of the second stop band (see figure 6: higher branch). The temperature of Bose-Einstein condensation is given by the known relation:

$$T_{B-E} = \frac{3{,}31\hbar^2}{k_B m_r} \left(\frac{N}{V}\right)^{\frac{2}{3}}. \tag{9}$$

The effective film refractive index of investigated photonic crystalline films is shown in figure 9. Refractive index can be positive or negative. In the recent case the photonic crystalline film is left-handed material in some wavelengths ranges.

Areas corresponding to the stop bands of photonic crystalline film are confined by vertical lines (see figures 8, 9).

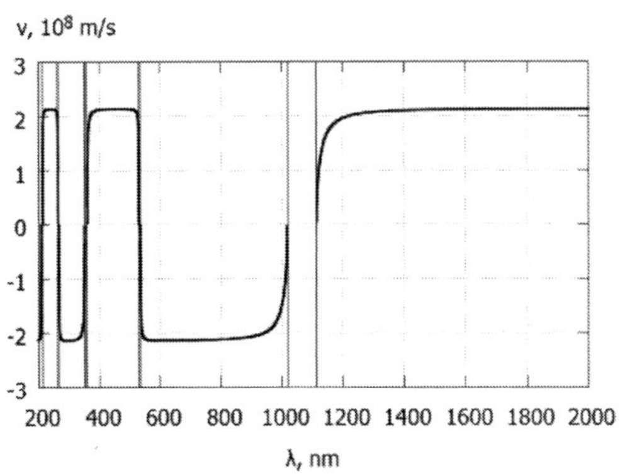

Figure 8. Group light velocity in the film calculated using (4).

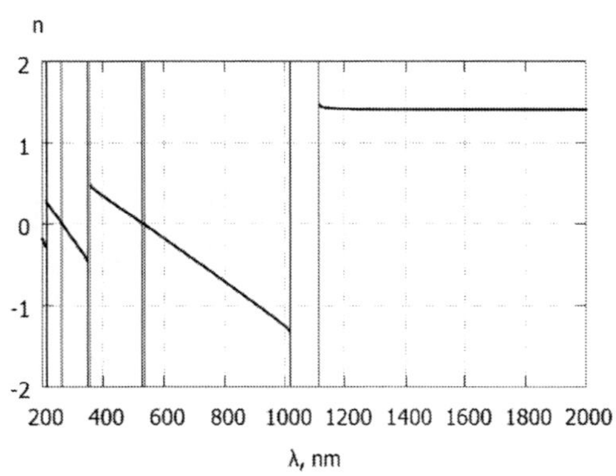

Figure 9. Effective refractive index of the film calculated using (6).

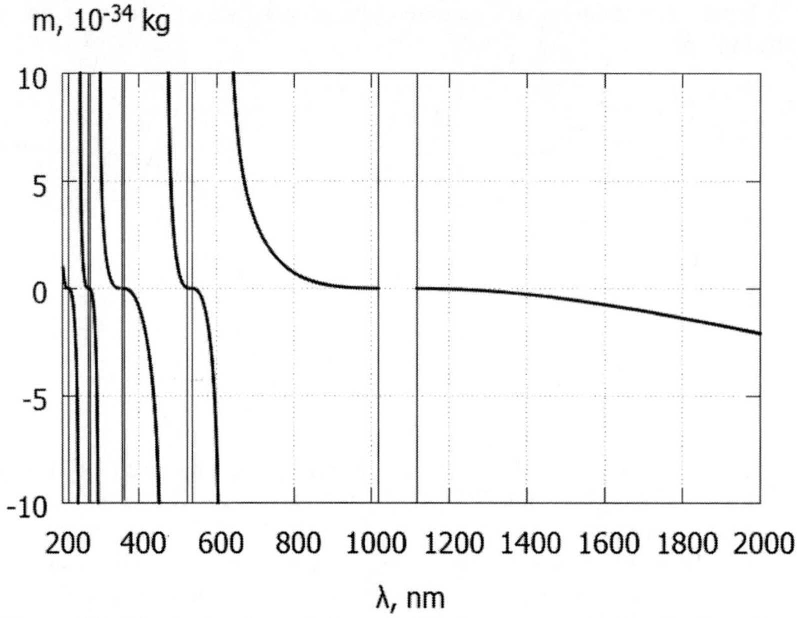

Figure 10. Dispersion law of photon effective mass calculated using (5).

Effective photon mass m(λ) depends on corresponding wavelength and this corresponding dependence is illustrated in figure 10. The mass meaning may be positive or negative and its value changes in wide frequency region. The rest mass m_r, corresponding to the edge of the second stop band, is about 2 eV, i.e. essentially lower than the electron rest mass.

4. Applications of mesoporous anodic aluminum oxide photonic crystalline films

Mesoporous anodic aluminum oxide photonic crystalline films have specific reflection spectrum with distinct stop band. Reflection coefficient in stop band area is very high (up to 99.9%). Such properties are usable in selective optical filters and for splitting of multimode laser emission lines (see figure 11).

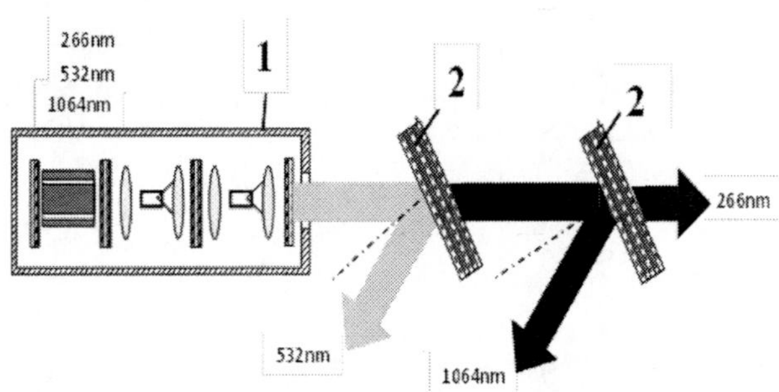

Figure 11. Scheme of laser lines division: 1 – multimode laser, 2 – photonic crystalline films, which reflect wavelengths of 532 and 1064 nm.

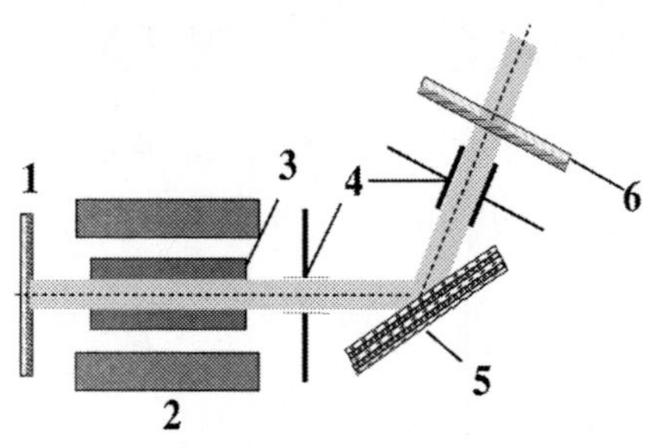

Figure 12. Scheme of multimode to single-mode modification of laser: 1 – high reflection mirror, 2 – pump source, 3 – gain medium, 4 – diaphragms, 5 – photonic crystalline film, 6 – output mirror.

Total reflection of chosen emission line can be realized in three different methods: it's possible to specify crystalline period while anodization (by changing current-time periods); stop band area may be moved by adjusting of angle of incidence/reflection; stop band area may be transformed [9] by injection of dielectric material inside photonic crystalline film pores.

Another probable application is to improve conditions of Raman scattering spectrum registration. See scheme in figure 12: incident photons are filtered by photonic crystalline film but scattered ones pass through it. Sensor sensitivity may be increased because exciting line doesn't interrupt to measure Stokes and anti-Stokes components.

Photonic crystalline films may be used as detectors of molecular structures. Molecular structure may be injected in photonic crystalline film pores using different techniques: in form of wetting solution, in form of alloy, etc. Raman scattering cross section increases anomalously when spontaneous Raman scattering spectra are registered under conditions of an anomalous increase in the density of the corresponding photon states [10]. When pores of photonic crystal film are filled with Stimulated Raman Scattering active media, it is possible to generate low-threshold Stimulated Raman Scattering due to an anomalous increase in the density of photonic states [10] in the case when the frequencies of the exciting or generated radiation fall on the edge of the corresponding stop bands. The value of the angle θ at which the Wulff-Bragg's condition is satisfied is given by the well-known relation:

$$2a\sqrt{n^2 - \sin^2\theta} = m\lambda_B, m = 1,2,3... \quad (10)$$

Here $\lambda_B = \lambda_0$ - the values of exciting or Stimulated Raman Scattering radiation wavelengths falling into the edge of the corresponding stop band.

Group light velocity is significantly reduced in this case (see figure 8). This leads to increase of photon density. Similar effect can be used to analyze thin dielectric molecular layers deposited on a photonic-crystalline film.

Photonic crystalline films are usable for filtering of laser generation. In figure 11 radiation from multimode laser (1) is filtered by photonic crystalline films (2). One film reflects 532 nm part and transmits 266 and 1064 nm parts, another film reflects 1064 nm part and only 266 nm band is in output. It is possible to modify optical cavity construction from figure 11 to allow only one frequency mode generation. Such scheme is illustrated by figure 12. In this case one of the cavity mirrors is replaced by photonic crystalline film (5). This film reflects only one frequency radiation (based on incident angle). So the single frequency emission will be in output although there are some other allowed modes of the cavity.

Compact analyzer for Raman spectroscopy may be constructed by using photonic crystalline films (figure 13). Laser radiation (6) incidents at an angle θ on the sample (1). Sample is placed in the pores of photonic crystalline film (2). Lens (4) converge scattered radiation which goes to refocusator (7) via optical fiber (5). Another photonic crystalline film is placed inside the refocusator at an angle θ to the beam. Laser component is reflected by this film while Stokes and anti-Stokes components pass through.

Another Raman spectroscopy analyzer is depicted in figure 14. For this case laser radiation is reflected by film (1), filled by substance for Raman spectra recording. Raman emission goes to tip (4) and then to spectrometer (7) via optical fiber (5). Halogen lamp (6) is needed for spectrometer calibration.

Figure 15 illustrates the setup for generation of the second optical harmonics emission ($\lambda' = \lambda_0/2$) by using of mesoporous photonic crystalline film (3), filled by nonlinear optical substances ($LiIO_3$, $NaBrO_3$, $BaTiO_3$) and rotated at an angle θ to the beam (see figure15) for coinciding of exciting line (λ_0) or the second harmonic emission ($\lambda' = \lambda_0/2$) wavelengths with the edges of the photonic stop band wavelength. There is possible also to observe the opposite phenomenon: the subharmonic transformations of exciting laser emission (see figure 16), when frequency doubling ($\omega' = \omega_0/2$; $\lambda' = 2\lambda_0$) by photonic crystalline film takes place.

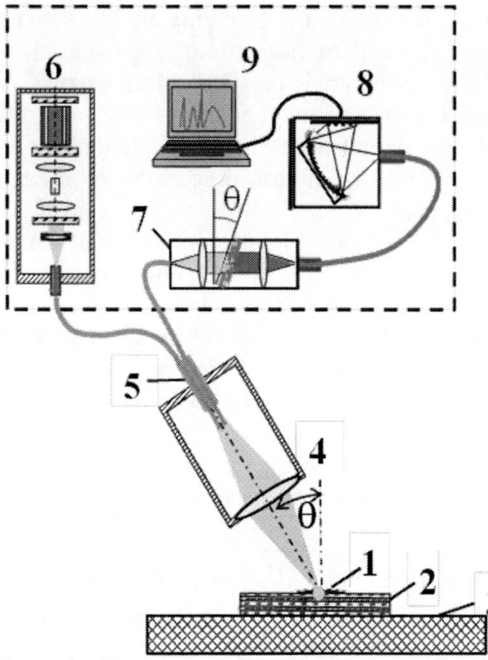

Figure 13. Compact analyzer for Raman spectroscopy: 1 – sample, 2 – photonic crystalline film, 3 – substrate, 4 –lens, 5 – optical fibers, 6 – laser, 7 – refocusator, 8 – mini-spectrometer with multielement radiation receiver, 9 – computer.

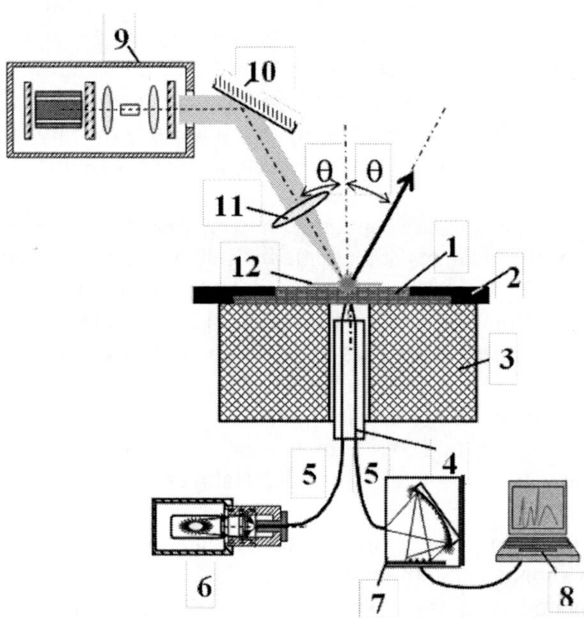

Figure 14. Raman spectroscopy analyzer: 1 – photonic crystalline film, 2 – Al_2O_3 base, 3 – fluoropolymeric bracket, 4 – optical fiber probe, 5 – optical fibers, 6 – tungsten halogen, 7 – spectrometer, 8 – computer, 9 – laser, 10 – mirror, 11 –lens , 12 – sample.

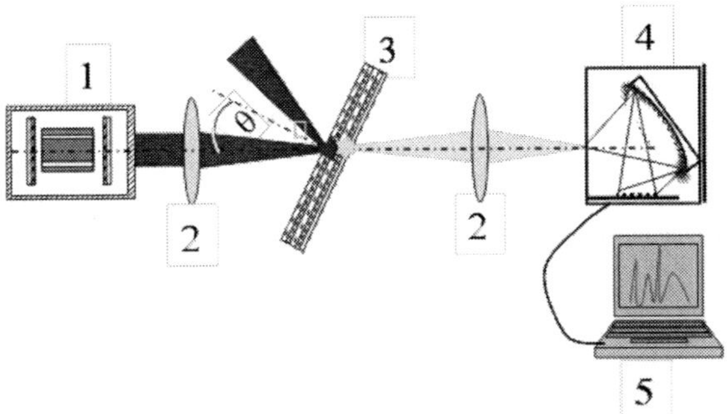

Figure 15. Setup for exciting laser wavelength doubling (the second optical generation) by photonic crystalline film: 1 – laser, 2–lens, 3 – photonic crystalline film, embedded by nonlinear substance, 4 – spectrometer, 5 – computer.

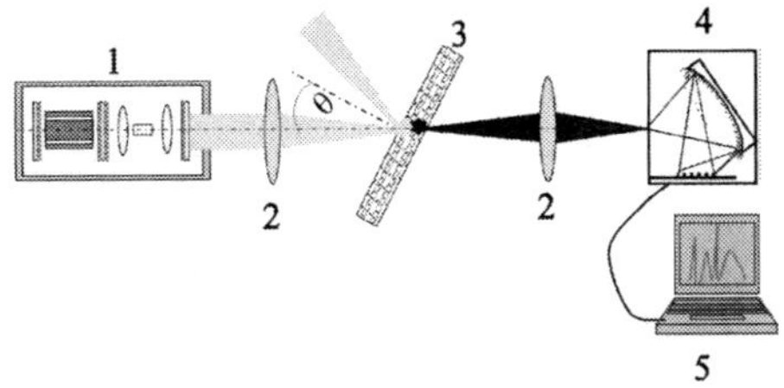

Figure 16. Setup for exciting laser frequency doubling (subharmonic generation) by photonic crystalline film: 1 – laser, 2 – converging lens, 3 – photonic crystalline film, embedded by nonlinear substance, 4 – spectrometer, 5 – computer.

Stimulated Raman Scattering analyzers scheme is shown in figures 17. The same photonic crystalline mesoporous film 3, filled by Raman active media, is used both for Stokes light amplification and for laser beam filtration. Laser beam incidents at an angle θ to the film surface under the condition (10), providing an density increase of the photonic states. The film 3 in figures 17 reflects laser beam while Stimulated Raman Scattering is generated inside of this film.

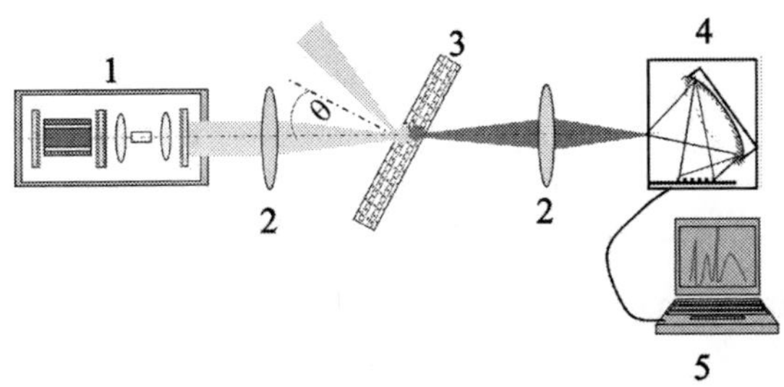

Figure 17. Stimulated Raman Scattering with mesoporous anodic aluminum oxide photonic crystalline films, filled by Raman active medium: 1 – laser, 2 – converging lens, 3 – photonic crystalline film, 4 – spectrometer, 5 – computer.

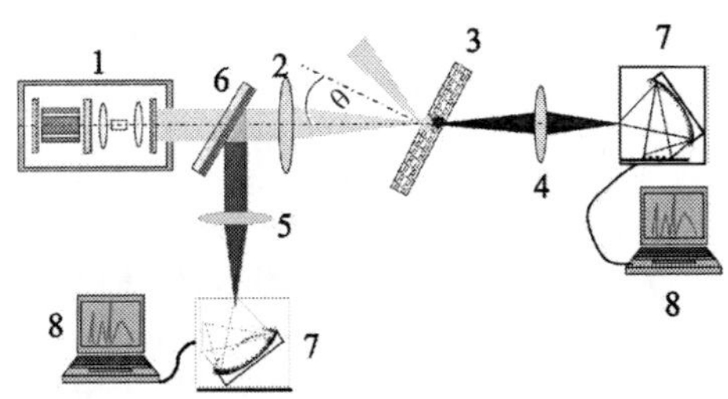

Figure 18. Experimental setup for generation of entangled photonic states during two-photon subharmonic emission excitation: 1 – laser, 2, 4, 5 –lenses, 3, 6– photonic crystalline films, embedded by nonlinear substance, 7 – spectrometers, 8– computer.

In figure 18 the experimental scheme for entangled photonic states generation and its recording with the help of mesoporous anodic aluminum oxide photonic crystalline film 3, embedded by nonlinear optics media, is presented. Photonic crystalline film surface is turned at such angle θ so the wavelength of exciting light emission from laser 1 is close to the edge the second stop band of photonic crystal at wave vector q=0. In this case the efficiency of subharmonic generation increases and subharmonic waves with opposite wave vectors ($k_1 = -k_2$) may be detected with the help of two spectrometers 4.

5. Conclusion

Thus we have shown that the high reflection coefficient in narrow stop band range may be realized in mesoporous anodic aluminum oxide photonic crystalline films for wide frequency region: in UV, visible and infrared. Stop band spectral position and its width may be optimized with the help of lattice constant and effective refractive index changing. Such properties open the wide field for applications [11-17]. One way for it is the filling of mesoporous film void by different substances, including liquids, ferroelectrics, luminophores etc. Mesoporous anodic aluminum oxide photonic crystalline films may be explored as selective filters in laser resonators and in devices for registration of spontaneous Raman scattering (see figures 10-13). Mesoporous anodic aluminum oxide photonic crystalline films, embedded by nonlinear optics substances ($NaNO_2$, $LiIO_3$, $NaBrO_3$, $BaTiO_3$), may be used as nonlinear optic elements for the second harmonics and subharmonics generation and for observation of Stimulated Raman Scattering.

Acknowledgments
This work was supported by the Russian Foundation for Basic Research (RFBR) (grants 15-02-02882, 16-08-00618, 16-02-00488, 16-52-00026) and China Scholarship Council.

References

[1] Kaiwei Li, Ting Zhang, Nan Zhang, Mengying Zhang, Jing Zhang, Tingting Wu, Shaoyang Ma, Junying Wu, Ming Chen, Yi He and Lei Wei 2016 Integrated liquid crystal photonic bandgap fiber devices *Frontiers of Optoelectronics* **9** 466–481

[2] Svyakhovskiy S E, Maydykovskiy A I, Novikov V B, Kompanets V O, Skorynin A A, Bushuev V A, Chekalin S V, Murzina T V and Mantsyzov B I 2015 Dynamical Bragg Diffraction in the Laue Geometry in 1D Porous Silicon-Based Photonic Crystals *Journal of Russian Laser Research* **36** 588–601

[3] Gorelik V S and Filatov V V 2012 Spectroscopy of stop bands in artificial opals filled with an alcohol solution of potassium iodide *Optics and Spectroscopy* **113** 301–305

[3] Gorelik V S, Klimonsky S O, Napolskii K S and Filatov V V 2016 Optical properties of one-dimensional photonic crystals based on porous films of anodic aluminum oxide *Optics and Spectroscopy* **120** 32–38

[4] Yisen Liu, Yi Chang, Zhiyuan Ling, Xing Hu and Yi Li 2011 Structural coloring of aluminum *Electrochemistry Communications* **13** 1336–39

[5] Rytov S M 1955 Electromagnetic properties of a finely layered medium *JETP* **2** 466–475

[6] Yariv A and Yeh P 1984 *Optical waves in crystals. Propagation and control of laser radiation.* (New York: Wiley) p 589

[7] Born M and Wolf E 1999 *Principles of Optics: Electromagnetic Theory of Propagation, Interference and Diffraction of Light* (Cambridge: Cambridge Univ. Press)

[8] Gorelik V S and Pudovkin A V 2013 Resonance Globular Photonic Crystals Filled with Al_2O_3:(Cr^{3+}) Nanoparticles *Herald of the Bauman Moscow State Technical university. Natural sciences* **49** 43–49

[9] Almohamed Y, Barille R, Vodchits A I, Voynov Yu P, Gorelik V S, Kudryavtseva A D, Orlovich V A and Tcherniega N V 2015 Reduction of the threshold of stimulated Raman scattering in Raman-active media introduced into pores of a globular photonic crystal *JETP Letters* **101** 399–404

[10] Burdanova M G and Gorelik V S 2015 Polariton waves in capillary fibers doped with rare-earth ions *Herald of the Bauman Moscow State Technical university. Natural sciences* **63** 46–51

[11] Gorelik V S 2015 Photon-boson conversion in ferroelectrics and amino acids *Herald of the Bauman Moscow State Technical university. Natural sciences* **61** 23–36

[12] Gorelik V S 2014 Possibilities of dark matter elementary particles registration *Herald of the Bauman Moscow State Technical university. Natural sciences* **57** 3–23

[13] Tappura K, Luomahaara J, Haatainen T, Hassel J and Vehmas T 2016 Influence of Substrate on Plasmon-Induced Absorption Enhancements *Plasmonics* **11** 627–635

[14] Kaniber M, Flassig F, Reithmaier G, Gross R and Finley J J May 2016 Integrated superconducting detectors on semiconductors for quantum optics applications *Applied Physics* B **122** 115:1–10
[15] Gorelik V S, Dovbeshko G I and Pyatyshev A Yu 2016 Photoluminescence spectra of DNA and ADP in photon traps under ultraviolet excitation *Herald of the Bauman Moscow State Technical university. Natural sciences* **65** 25–33
[16] Gorelik V S, Pudovkin A V and Filatov V V 2016 Mesoporous Photonic-Crystal Films for Amplification and Filtering of Electromagnetic Radiation *Journal of Russian Laser Research* **37** 604–610

Modeling sound propagation in a waveguide with a gas-saturated sedimentary layer

M V Yarina[1,2]

[1]Bauman Moscow State Technical University, Moscow, Russia

[2]A. M. Prokhorov General Physics Institute Russian Academy of Sciences, Moscow, Russia

E-mail: marina-malysheva@autorambler.ru

Abstract. There was developed an acoustic wave propagation model in a waveguide, where the bottom is represented as a gas-saturated layer. This study uses the ray theory because the investigation of shallow reservoirs with a gas-saturated bottom requires modeling the sound field on short distances. The theory takes into account the rays passing through a gas-saturated layer. The obtained model was used in order to define the distance and the depth of the receiving array (in a horizontal position) elements. The experiment was carried out in the Klyazma reservoir in 2014. In accordance with the peculiarities of the experiment (short distance between receiving array and radiator; irregular array of the radiated signal) there was designed an algorithm agreed with the processing environment in the time domain.

1. Introduction

Many problems in underwater acoustics require mathematical modeling of sound propagation under various conditions. For example, localization of underwater receiving or radiating array elements in an acoustic waveguide belongs to this class of problems. It can be solved with the use of Matched Field Processing (MFP) [1]. The essence of MFP lies in comparison of experimental signals with modeled ones, calculated for different conditions of propagation. The objectives of this work are obtaining the analytical expressions for the calculation of sound field complex amplitude in a waveguide, where the bottom is represented as a gas-saturated layer overlaying a liquid half-space and localization of a receiving array in this waveguide. Note that the sound speed in a gas-saturated sedimentary layer is less than the sound speed in the water [2,3]. Ray approach of sound propagation is one of the methods of simulating acoustic signals [4,5].

2. Ray structure of the sound field in a waveguide with a gas-saturated sedimentary layer

Within the framework of ray approach an analytical expression for amplitude of sound field at the frequency *f* in a waveguide with a gas-saturated sedimentary layer (Fig. 1) is obtained.

$$P(f;r,z)=\frac{\exp(ikR_0)}{R_0}-\frac{\exp(ikR_1)}{R_1}+\sum_{U,L}\exp(ikR_{u1}+ik_1R_{d1})S_1(U,L)\sum_i C_{i1}(U,L)-$$
$$-\sum_{U,L}\exp(ikR_{u2}+ik_1R_{d2})S_2(U,L)\sum_i C_{i2}(U,L)-\sum_{U,L}\exp(ikR_{u3}+ik_1R_{d3})S_3(U,L)\sum_i C_{i3}(U,L)+$$
$$+\sum_{U,L}\exp(ikR_{u4}+ik_1R_{d4})S_4(U,L)\sum_i C_{i4}(U,L); \quad (1)$$

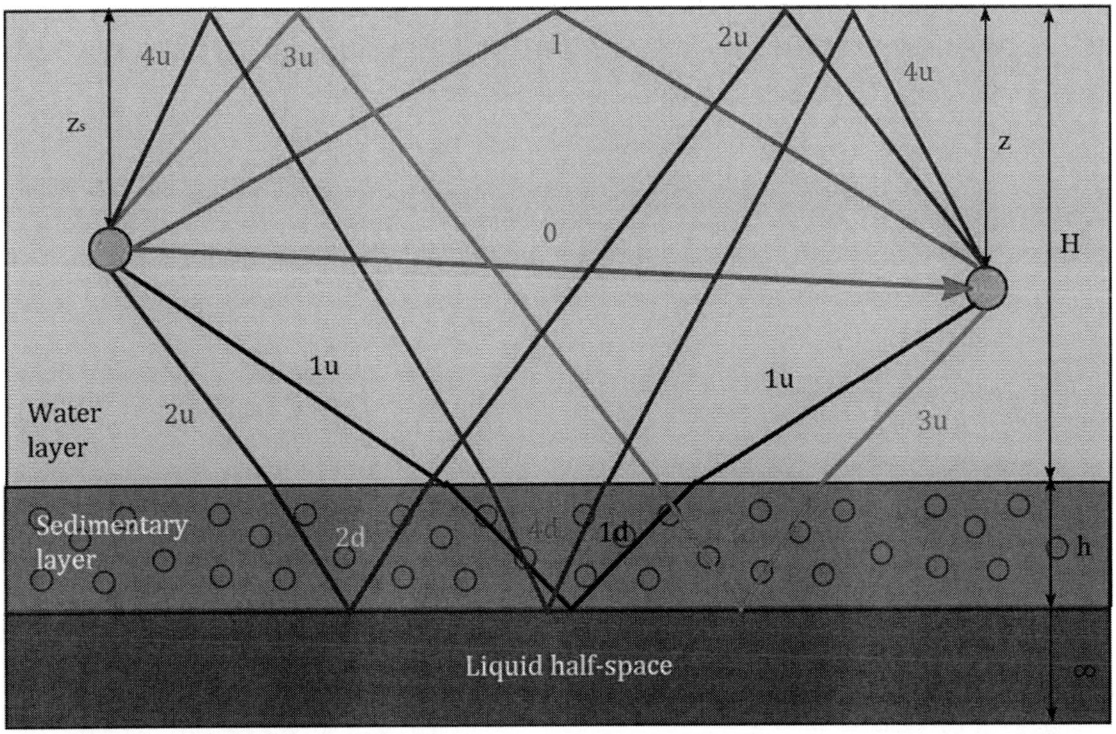

Fig. 1. Paths of the first two rays (0, 1) and rays with a single transmission in the sedimentary layer. The indices u and d denote the parts of the rays passing through the water and sedimentary layers, respectively. The reflections from the water-sediment boundary are not shown in the figure.

where the lengths of ray parts, passing through water (*Ru1*, *Ru2*, *Ru3*, *Ru4*) and gas-saturated layer (*Rd1*, *Rd2*, *Rd3*, *Rd4*), are calculated with various combinations of signs "+" and "-" in the formulas:

$$R_u = \frac{2UH}{\cos(\theta)} + \frac{H \pm z_s}{\cos(\theta)} + \frac{H \pm z}{\cos(\theta)};\ R_d = \frac{2Lh}{\cos\left[\arcsin\left(\frac{c_1}{c}\sin(\theta)\right)\right]};$$

$R_0 = \sqrt{r^2 + (z_s - z)^2}$ is the length of a straight ray; zs is the source depth, z is the receiver depth, H is the water layer depth, h is the thickness of a gas-saturated layer, r is the distance between source and receiver (on the horizontal line), $S_j(U,L) = \left[r\tan(\theta)\left(\frac{\partial r}{\partial \theta}\right)\right]^{-1/2}_{\theta=\theta_j(U,L)}$ is the coefficient of geometrical spreading, *U* is the number of reflections from the upper boundary, *L* is the number of reflections from the boundary between sedimentary layer and bottom half-space, *c* and *c1* are the sound speeds in water and in sedimentary layer, respectively, $k = 2\pi f / c$ and $k_1 = 2\pi f(1+\alpha_1/2)/c_1$ are the wavenumbers in water and in gas-saturated layer. Incidence angles for each ray are the solution of a transcendental equation:

$$2UH\tan(\theta)+2Lh\tan\left[\arcsin\left(\frac{c_1}{c}\sin(\theta)\right)\right]+(H\pm z_s)\tan(\theta)+(H\pm z)\tan(\theta)=r.$$

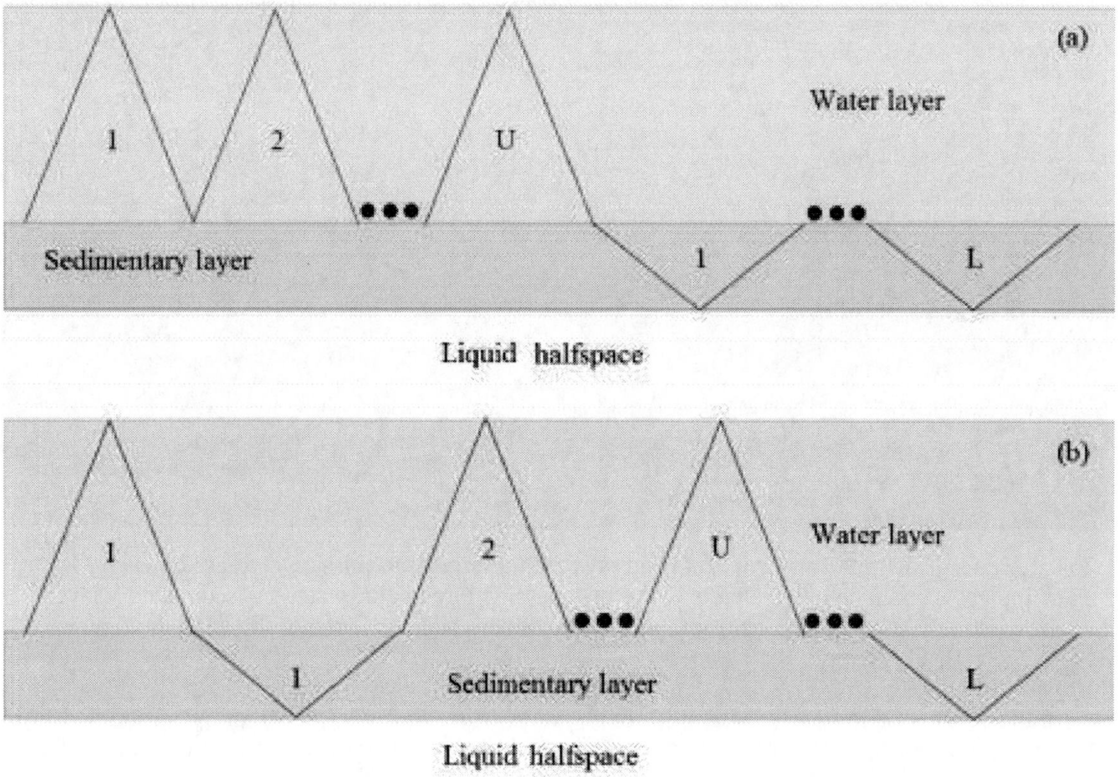

Fig. 2. Representation of the rays in order to find the combinatorial coefficients of transmission and reflection (a) - grouped "triangles" (b) - after a reshuffle.

For the fixed U and L combinatorial coefficients can be represented as follows (Fig. 2):

$$C_i(U,L)=\frac{(U+1)!}{(U+1-i)!i!}\frac{(L-1)!}{(L-i)!(i-1)!}(W_{12}W_{21})^i V_{21}{}^{L-i} V_{12}{}^{U-i+1} V_{23}{}^{L}(-1)^U; i=\overline{\min(1,L),\min(U+1,L)};$$

where $V_{12}=\frac{Z_1-Z}{Z_1+Z}$; $V_{21}=-V_{12}$; $W_{12}=1+V_{12}$; $W_{21}=1+V_{21}$; $Z=\frac{\rho c}{\cos(\theta)}$; V_{12} and V_{21} are the reflection coefficients from the boundary between water and sedimentary layer (↓ and ↑), V23 is the reflection coefficient between sedimentary layer and liquid halfspace, $W12$ and $W21$ are the transmission coefficients, Z is the impedance. In the case when the rays are not passing into sedimentary layer the coefficient has the form $C_0=(-1)^U V_{12}{}^{U+1}$.

3. Localization of underwater receiving array elements using MFP

Using the above-described model (Eq. 1) and MFP the problem of localization of a receiving array is solved for the experimental data, obtained in the Klyazma reservoir in 2014 [3]. The experiment was conducted in an area with a constant depth (H), which, according to the measurement

IOP Conf. Series: Journal of Physics: Conf. Series **918** (2017) 012021 doi:10.1088/1742-6596/918/1/012021

made with the use of echo sounder, was $H = 8.5$ m. The gas-saturated layer had the thickness of $h = 1$ m and the sound speed of $c1 = 100$ m/s. An autonomous chain of 10 hydrophones fixed along a rigid slat was set horizontally near the bottom. A piezoceramic sound source (with a shape of sphere) of 8 cm in diameter, was set at the depth of $zs = 7.1$ m. Chirp signals in the band of 100 to 1400 Hz with a duration of 5 s were radiated. The received signals were cross-correlated with radiated ones in to distinguish the signals coming in separate rays. To match the modeling results, calculated for various points of the waveguide with fixed coordinates of the sound source with experimental data, the following two-step matched field processing algorithm proved to be effecient:

1. Reduction to absolute time by use of correlation function. Let us assume that the signal $S(t)$ is obtained in the time interval [$t1$, $t2$]. $P(t;r,z) = \int P(f;r,z)e^{-i2\pi ft}df$ is the model of sound pressure field in the waveguide.

   $$B(\tau;r,z) = \int_{t_1}^{t_2} P(t;r,z)S(t+\tau)dt;$$

   τ_m is the time point, at which the function $B($; r, $z)$ has a maximum. Then $P(t$; r, $z)$ was shifted in time by τ_m .
2. Finding the most probable position of the receiver by means of the least squares method for envelopes:

$$K(r,z) = \left(\sum_k \left(\hat{S}(t_k) - \hat{P}(t_k;r,z) \right)^2 \right)^{-1} \qquad (2)$$

$\hat{S}$ and $\hat{P}$ are the envelopes of experimental and simulated signals respectively.

Modeled signals were calculated for a grid Δr=0.2 m, Δz=0.1 m, maximum range is 35 m. Example of the algorithm application is shown on Fig.3, where the distribution of $K(r, z)$ is presented for three different hydrophones. The lightest spots on the pictures correspond to hydrophone positions. The coordinates of hydrophones, obtained with the MFP are shown on Fig.4. The comparison of envelopes of the simulated and experimental time series demonstrates a good agreement between real and modeled data (Fig. 5).

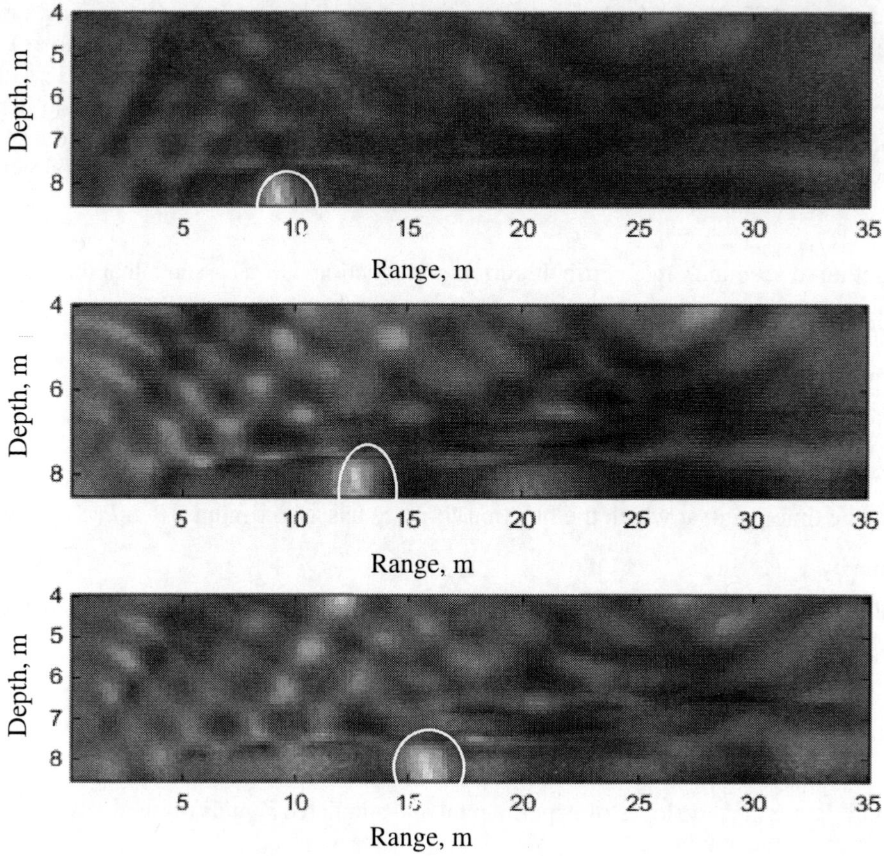

Fig.3. Distribution of $K\,(r, z)$ for three different receivers.

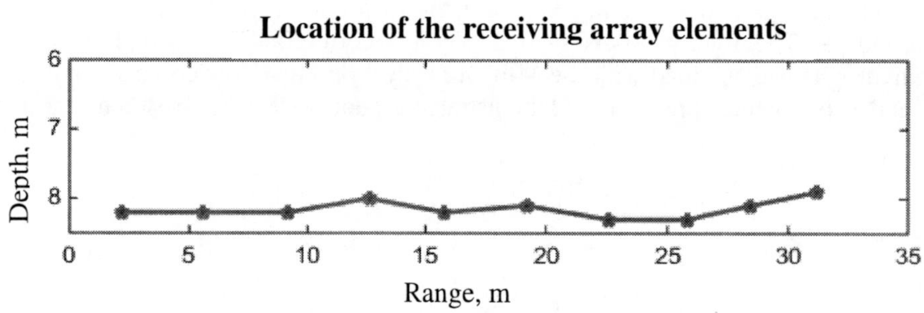

Fig.4 Location of the receiving array elements.

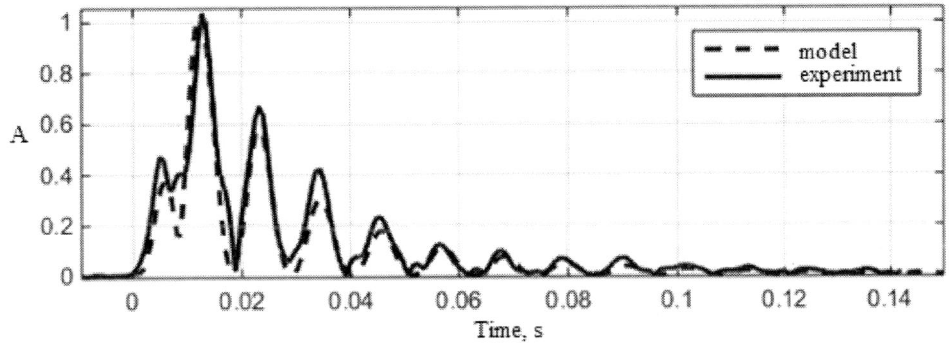

Fig.5. Comparison of the experimental and simulated envelopes of the signal.

4. Conclusion

As a result, expressions for the sound pressure field in a waveguide with a gas-saturated sedimentary layer were obtained. Determination of depth and range to the receiving array elements was performed using time domain matched field processing for experimental data, obtained in 2014 in the Klyazma reservoir.

Acknowledgements

The work was supported by the RFBR, projects Nos 16-32-60194, 16-29-02036.

References

[1] Katsnelson B, Petnikov V, Lynch J 2012 Fundamentals of shallow water acoustics. *Springer Science & Business Media*, p 540

[2] Grigorev V A, Lunkov A A, Petnikov V G 2015 Attenuation of sound in shallow-water areas with gas-saturated bottoms. *Acoustical Physics* **61** (1) pp 85-95

[3] Lunkov A A, Petnikov V G, Chernousov A D 2015 Estimating of the effective sound speed in the bottom in shallow water areas. *Acoustical Physics* **61**(5) pp 707-714

[4] Brekhovskikh, L M 2003 Fundamentals of ocean acoustics. *Springer Science & Business Media.*

[5] Tindle C T, Bold G E J 1981 Improved ray calculations in shallow water. *J. Acoust. Soc. Am.* **70**(3).

IOP Conf. Series: Journal of Physics: Conf. Series **918** (2017) 012022 doi:10.1088/1742-6596/918/1/012022

Modes of nanosatellite aerodynamic oscillations in atmosphere

Yu V Gerasimov[1], E A Ivanov[2], G K Karetnikov[1], I A Konstantinova[2] and A B Selivanov[1]

[1] Associated professor, Bauman Moscow State Technical University, Moscow, RF
[2] Master Degree, Bauman Moscow State Technical University, Moscow, RF

E-mail: drgerasimov.y@gmail.com

Abstract. The paper is devoted to the results of investigating the dependencies of nanosatellite aerodynamic oscillations frequency on attack angle at different altitudes up to 70 km are defined. The oscillations bandwidths are determined with respect to the geometric parameters for a nanosatellite with 10 kg mass and 6000 kg/m3 average density. The model allows estimating the bandwidth aerodynamic oscillations in the suborbital nanosatellite trajectory based on the given geometry and mass-dimensional parameters.

1. Introduction

The paper investigates the aerodynamic oscillations bandwidth of an axisymmetric high-fineness-ratio configuration body with an aerodynamic skirt in a hypersonic flow. This objective is relevant in the context of the nanosatellite electromagnetic launch [1]. This research is concerned with suborbital motion at altitudes up to 70 km. The nanosatellite aerodynamic configuration is shown in Fig. 1.

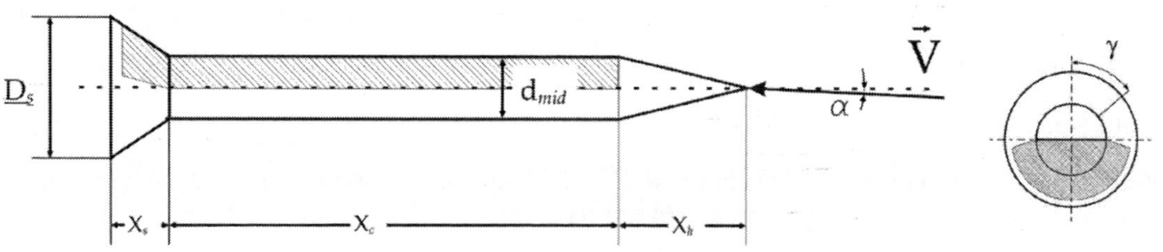

Figure 1. Nanosatellite aerodynamic configuration.

The nanosatellite aerodynamic coefficients in hypersonic flow in the body-fixed coordinate system are estimated. The aerodynamic coefficients are calculated using the Newtonian flow theory [2]. The corresponding formulas for the coefficients of the axial force, normal force and pitching moment are:

$$\begin{aligned} C_R &= \frac{2}{S_{mid}} \int_0^l dx \cdot r \cdot tg\beta \int_0^{\gamma_{Shadow}} d\gamma \cdot \overline{p}; \\ C_N &= \frac{2}{S_{mid}} \int_0^l dx \cdot r \int_0^{\gamma_{Shadow}} d\gamma \cdot \overline{p} \cdot \cos\gamma; \\ m_Z &= \frac{2}{l \cdot S_{mid}} \int_0^l dx \cdot r \cdot x \int_0^{\gamma_{Shadow}} d\gamma \cdot \overline{p} \cdot \cos\gamma; \end{aligned} \qquad (1)$$

IOP Conf. Series: Journal of Physics: Conf. Series **918** (2017) 012022 doi:10.1088/1742-6596/918/1/012022

where
r is the local radius of a rotation body;
l is the nanosatellite length;

$$\overline{p} = 2(\sin\beta \cdot \cos\alpha - \sin\alpha \cdot \cos\beta \cdot \cos\gamma)^2 \quad (2)$$

is the surface pressure coefficient;
β is a local angle of the streamlined surface to the nanosatellite axis;
α is an angle of attack;
γ is an integrating parameter equal to the angle between the attack angle plane and the meridian plane. The range $\gamma > \gamma_{Shadow}$ corresponds to the aerodynamic shadow area, it is marked by shading in Fig. 1a.

The external moment in relation to the center of mass is calculated using the formula below [2]:

$$M_z = \left(m_z - C_N \cdot \frac{X_{CM}}{l} \right) \cdot l \cdot S_{mid} \cdot q, \quad (3)$$

where X_{CM} is the center of mass position relative to nanosatellite forebody;

$q = \frac{\rho \cdot V^2}{2}$ is dynamic pressure.

We assume the speed of nanosatellite to be V = 8000 m/s, and the atmosphere parameters correspond to those at 30 km altitude. In accordance with [3], $\rho = 1.84 \cdot 10^{-2}$ kg/m^3, so the dynamic pressure is q = 588 kPa.

The following nanosatellite parameters are used in the work: head cone aspect ratio is assumed to be constant $X_l = 4\ d_{mid}$; the length of cylindrical part is $X_{cil} = 20\ d_{mid}$; the skirt length is $X_s = 2.2\ d_{mid}$. The skirt diameter D_s varies from 1.3 d_{mid} to 2.2 d_{mid}; The midsection diameter d_{mid} must enable 10 kg satellite mass, with the uniform distribution of material of 6000 kg/m^3 density,.

The dependencies of the aerodynamic moment on the attack angle M (α) for varying skirt diameters are shown in Fig. 2. The corresponding behavior patterns are numbered:

- $D_s = 1.85\ d_{mid}$ – curve I;
- $D_s = 1.5\ d_{mid}$ – curve II;
- $D_s = 1.3\ d_{mid}$ – unnumbered curve;
- $D_s = 2.2\ d_{mid}$ – curve III.

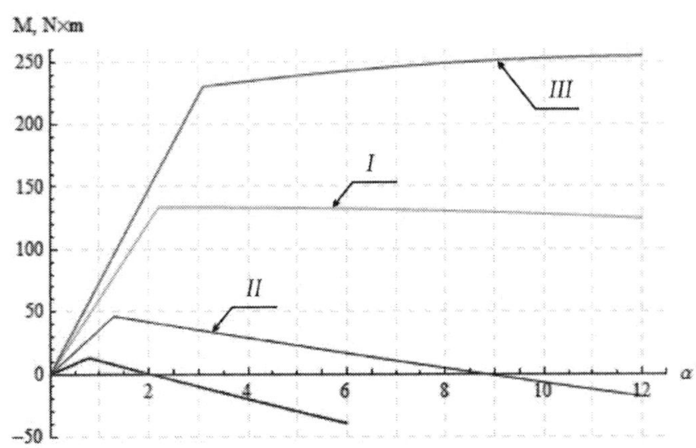

Figure 2. Aerodynamic moment as a function of the attack angle for varying skirt diameters

The curves clearly show the influence of the aerodynamic shadow on the nanosatellite flight dynamics: the break in the curves corresponds to the critical angle of attack α_*, where the aerodynamic shadow from the cylindrical part reaches the edge of the skirt.

It is evident that the dependence of the aerodynamic moment on the angle of attack can assume three characteristic patterns. Worth noting is the geometry benefit that enables Pattern I of the M (α) dependence: When the moment is constant, in order to apply the control pulse, only the direction of the deviation from the zero angle of attack must be known – consequently, the control system problem is simplified.

Let us investigate the possible bandwidths of nanosatellite aerodynamic oscillations for different geometric configurations.

2. Behavior pattern I

The dependence may be approximated by a piecewise linear function:

$$M_z = \begin{cases} \lambda \cdot \alpha, & \alpha < \alpha_*; \\ \lambda \cdot \alpha_*, & \alpha > \alpha_*. \end{cases} \tag{4}$$

where $\lambda = \frac{\partial M_z}{\partial \alpha}$ is a derivative of the aerodynamic moment at small angles of attack.

The attack angle dynamics is described by the equation:

$$J \cdot \ddot{\alpha} + M_z = 0. \tag{5}$$

where J is the nanosatellite's moment of inertia relative to the transverse axis through the center of mass. Taking into account (4), the dynamics equation is written down as:

$$\ddot{\alpha} + \tfrac{\lambda}{2J}\left(|\alpha_* - \alpha| - |\alpha_* + \alpha|\right) = 0. \tag{6}$$

The oscillations period depends on the amplitude α_0. When the amplitude is less than critical, harmonic oscillations occur; at large angles nonlinear oscillations take place:

$$T = \begin{cases} 2\pi\tau_0, & \alpha_0 < \alpha_*; \\ 4\tau_0\left(\frac{\sqrt{1-\bar{\mu}^2}}{\bar{\mu}} + \arcsin(\bar{\mu})\right), & \alpha_0 > \alpha_*. \end{cases} \tag{7}$$

where two parameters with the temporal dimension and their ratio are introduced:

$$\tau_0 = \sqrt{\frac{J}{\lambda}}; \quad \tau_1 = \frac{\alpha_*}{\dot{\alpha}(0)}; \quad \bar{\mu} = {\tau_1}/{\tau_0} \tag{8}$$

Here $\dot{\alpha}(0)$ is amplitude angular velocity, α_* is critical angle.

For further analysis the results are conveniently presented in form of the reduced oscillation frequency $\bar{\nu} = 2\pi\frac{\tau_0}{T}$ as a function of the reduced amplitude obtained by the integration of dynamic equations (4, 5):

$$\bar{\alpha}_0 = \frac{\alpha_0}{\alpha_*} = \frac{1+\bar{\mu}^4}{2\bar{\mu}^2},$$

The corresponding graph is shown in Fig. 3a. The graphs in Fig. 3b allow estimating the bandwidth of nanosatellite aerodynamic oscillations in hypersonic flow with dynamic pressure q = 588 kPa, D_s = 1.85 d_{mid}. With the amplitude angles up to 12° the nanosatellite undergoes oscillations in the bandwidth ν = 8 - 15 Hz.

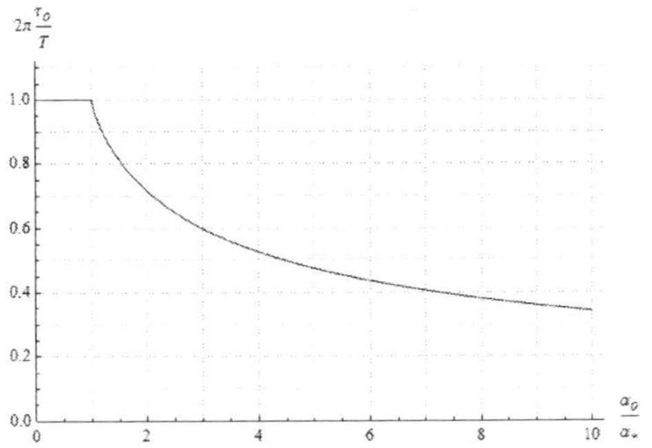

Figure 3a. Reduced oscillation frequency as a function of reduced amplitude

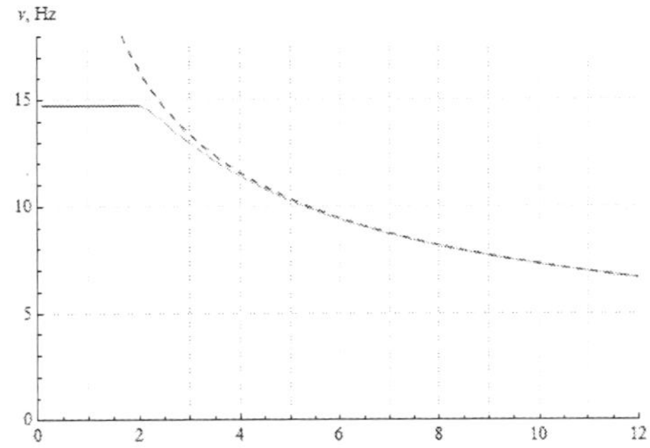

Figure 3b. Frequency as a functionof amplitude

The limiting form of (4) is of interest:

$$M_z = M_* \cdot Sign(\alpha) \quad (9)$$

The formal solution (5, 9) is a periodical temporal function, with each half-period having a parabola form. The frequency dependence on the amplitude is plotted as a dashed line in Fig. 3b and takes the following form:

$$\nu(\alpha_0) = \frac{1}{4}\sqrt{\frac{M_*}{2J \cdot \alpha_0}} \quad (10)$$

Thus, determining the bandwidth boundaries does not involve the motion equations (5, 4). The upper limit is determined by a derivative of the aerodynamic moment at small angles of attack. The lower limit is provided by the asymptotics (10), and is supported by the assumption of the amplitude angle not exceeding 5°:

$$\bar{\nu} \in \left(\frac{1}{2},\ 1\right) \quad (11)$$

3. Behavior patterns II and III

M(α) dependence is represented by a piecewise-linear approximation with angular coefficients in the following form:

$$M_z = \begin{Bmatrix} \lambda \cdot \alpha, & \alpha_0 < \alpha_*; \\ \lambda_r \cdot \alpha + (\lambda - \lambda_r) \cdot \alpha_*, & \alpha_0 > \alpha_*. \end{Bmatrix} \quad (12)$$

$\lambda_r > 0$ corresponds to Pattern III of angular momentum dependence on the angle; $\lambda_\gamma < 0$ corresponds to Pattern II. The lower limit of the dynamic systems bandwidth (5, 12) is determined by the angular coefficient λ_r, the upper limit is determined by λ. Fig. 4 a, b show the dependencies for $\lambda_\gamma > 0$ and the formal representation of the dependence for $\lambda_\gamma < 0$, respectively. Asymptotics is plotted as a dashed line.

Formal asymptotics:

1. when $\lambda_r > 0$ is $\bar{\nu} \to \sqrt{\frac{\lambda_r}{\lambda}}$ for $\alpha_0 \to \infty$;
2. when $\lambda_r < 0$ is $\bar{\nu} \to -\sqrt{\frac{|\lambda_r|}{\lambda}}$ for $\alpha_0 \to \infty$.

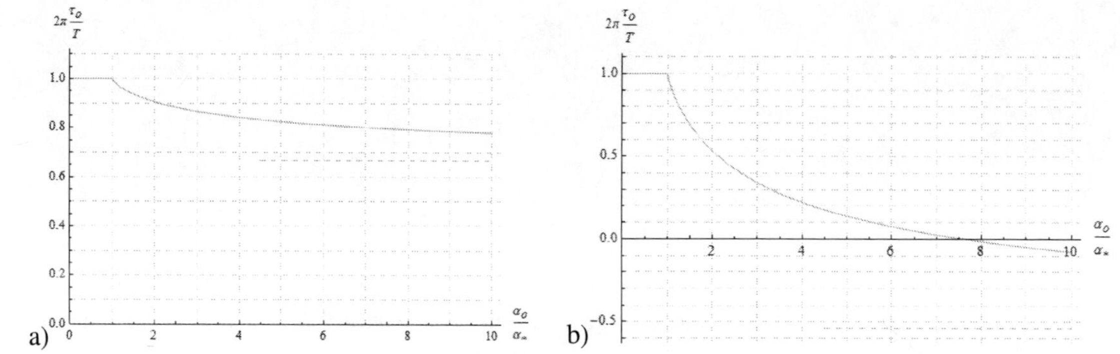

Figure 4. Oscillation frequency as a function of amplitude for behavior patterns II (a) and III (b)

We propose, by analogy with (11), an estimate of the frequency range:

1. $\bar{\nu} \in \left(\frac{1}{2}(1+\sqrt{\frac{\lambda_r}{\lambda}}),\ 1 \right)$ for $\lambda_r > 0$;

2. $\bar{\nu} \in \left(\frac{1}{2}(1-\sqrt{\frac{|\lambda_r|}{\lambda}}),\ 1 \right)$ for $\lambda_r < 0$.

The bandwidths for the given altitude and geometry parameters are obtained by scaling by a factor of $\sqrt{\frac{J}{\lambda}}$ in y direction, and by a factor of α in x direction.

4. Conclusions.

The dependencies of nanosatellite aerodynamic oscillations frequency on attack angle at different altitudes up to 70 km are defined. The oscillations bandwidths are determined with respect to the geometric parameters for a nanosatellite with 10 kg mass and 6000 kg/m^3 average density. The model allows estimating the bandwidth aerodynamic oscillations in the suborbital nanosatellite trajectory based on the given geometry and mass-dimensional parameters [1].

References

[1] Gerasimov Yu V., Karetnikov G K., Selivanov A B and Fionov A S 2013 Assessment of the Relative Payload Mass of a Nano-satellite Launched into near-Earth Space with a Gun Launch and Corrective Engines. *Herald of the Bauman Moscow State Technical University. Series Mechanical Engineering* **3**

[2] Krasnov N F 1970 *Aerodynamics of bodies of revolution* (New York)

[3] *Standard Atmosphere* 1975 (International Organization for Standardization) ISO 2533:1975

Nonlinear localized elastoplastic waves in a rod

V.I. Erofeev[1], A.V. Leonteva[1], A.O. Malkhanov[1]
[1]Mechanical Engineering Research Institute of RAS, 603024, Belinskogo str., 85, Nizhny Novgorod, Russia

Abstract. It has been shown that longitudinal nonlinear stationary elastic-plastic waves can be formed in a rod. Waves can be both periodic and solitary (solitons), which is determined by the index of strain hardening of the material.The velocity of propagation of such waves is less than the speed of the elastic shear wave. Along with classical solitons, so-called "embedded solitons" can be formed, that is, double solitons, one of which with a smaller amplitude and widthis located inside the other with a larger amplitude and width.

The main problems of the propagation of elastoplastic waves in media with a nonlinear stress-strain dependence for longitudinal and transverse shockswere considered in [1, 2]. An analytical solution of such problems is possible only when considering the simplest relationships that determine the behavior of a material. The consideration of real rheological models leads to the need to use numerical methods [3].

The propagation of a longitudinal wave in a rod with allowance for the high-frequency dispersion of its phase velocity (the Bishop model) is described by the equation [4]:

$$\rho F\frac{\partial^2 u}{\partial t^2}-\rho\nu^2 I_0\frac{\partial^4 u}{\partial x^2\partial t^2}+\nu^2\mu I_0\frac{\partial^4 u}{\partial x^4}=F\frac{\partial\sigma}{\partial x} \quad (1)$$

here $u(x,t)$ – longitudinal displacement of the points of the median line of the rod, F – the area of the cross section of the rod, ρ – the density of the material, ν – Poisson's ratio, I_0 – polar moment of inertia of the rod section, μ – shear modulus, σ – axial stress.

In the case of a uniaxial stress state $\sigma_i=\sigma$, the relation, linking the deformation $\varepsilon_i=\varepsilon$ and displacement u, has the following form:

$$\varepsilon=\left(1+\frac{1}{2}\frac{\partial u}{\partial x}\right)\frac{\partial u}{\partial x}. \quad (2)$$

The dependence of stresses on deformations is given by the Hollomon equation [5]:

$$\sigma=K\varepsilon^n, \quad (3)$$

where K – hardening module. Through n we denote the value of the strain hardening index. The values of this constant lie in the interval from zero to one. The value $n=0$ corresponds to the ideal plastic material, $n=1$ – ideally elastic material. Most metals and their alloys have the value n from 0,1 up to 0,5 [6].

Taking into account the relations (2), (3) (under the condition of the presence of plastic deformations) leads (1) to the form:

$$\frac{\partial^2 u}{\partial t^2} - v^2 r^2 \frac{\partial^2}{\partial x^2}\left(\frac{\partial^2 u}{\partial t^2} - c_\tau^2 \frac{\partial^2 u}{\partial x^2}\right) = \frac{K}{\rho}\frac{\partial}{\partial x}\left(\left[\frac{\partial u}{\partial x} + \frac{1}{2}\left(\frac{\partial u}{\partial x}\right)^2\right]^n\right) \quad (4)$$

where $r = \sqrt{\frac{I_0}{F}}$ – polar radius of inertia, $c_\tau = \sqrt{\frac{\mu}{\rho}}$ – the velocity of shear wave propagation in the material.

The solution to (4) we look in the form of stationary waves. In the equation we introduce the running variable $z = x - Vt$, where V – velocity of a linear wave. After a one-time integration and the denotation $U = du/dz$, the partial differential equation (4) reduces to an ordinary differential equation of the second order with a fractional nonlinearity

$$\frac{d^2U}{dz^2} + m_1 U - m_2\left(U + \frac{1}{2}U^2\right)^n + C_1 = 0, \quad (5)$$

where C_1 – the integration constant, and the coefficients of the equation are positive and have the form

$$m_1 = \frac{V^2}{v^2 r^2 \left(c_\tau^2 - V^2\right)},\ m_2 = \frac{K}{\rho r^2 v^2 \left(c_\tau^2 - V^2\right)}.$$

The presence in the equation (5) of the fractional index n makes it possible to obtain the approximate equation

$$\frac{d^2U}{dz^2} + m_1 U - m_2\left(U^n + \frac{1}{2}nU^{n+1}\right) + C_1 = 0, \quad (6)$$

the solution to which can be found by the method of truncated expansions [7] at specific values n.

Equation (6) has a pole of fractional order $p = \frac{2}{n}$. When $n = 1/2$ the equation has the following form:

$$U'' + m_1 U - m_2\sqrt{U} - \frac{1}{4}m_2\sqrt{U^3} + C_1 = 0. \quad (7)$$

The function replacement $U = y^2$ allows us to get rid of square root in the equation (7)

$$\left(y^2\right)'' + m_1 y^2 - m_2 y - \frac{1}{4}m_2 y^3 + C_1 = 0. \quad (8)$$

The solution of the equation has a second-order pole. The solution is sought in the form $y(z) = b_0 Y^2(z) + b_1 Y(z) + b_2$, where $Y(z)$ satisfies the Riccati equation $Y' = -Y^2 + B_0$, whose solution is known and expressed through a hyperbolic tangent. After substitution into equation (8), we obtain the values of the coefficients

$$b_0 = \frac{80}{m_2},\ b_1 = 0,\ b_2 = \frac{5}{3}\frac{m_1}{m_2} - \frac{2}{3}\frac{\sqrt{25m_1^2 - 20m_2^2}}{m_2},\ B_0 = \frac{\sqrt{25m_1^2 - 20m_2^2}}{80}$$

and the expression for the integration constant

$$C_1 = -\frac{1}{27}\frac{25m_1^3 - 30m_1 m_2^2 + \left(5m_1^2 - 4m_2^2\right)\sqrt{25m_1^2 - 20m_2^2}}{m_2^2}.$$

If the integration constant is set equal to zero, then the coefficients of equation m_1 and m_2 have the following relation

$$15m_1^2 - 16m_2^2 = 0. \qquad (9)$$

The velocity of a nonlinear wave with allowance for (9) assumes a fixed value

$$V = \frac{2}{\sqrt[4]{15}}\sqrt{\frac{K}{\rho}}.$$

The solution to equation (8) under condition (9) has the form

$$y = \frac{2\sqrt{15}}{3}\operatorname{th}^2(kz),\ \text{где}\ k = \sqrt{\frac{m_2}{8\sqrt{15}}}$$

Or, expressing in terms of the exponent,

$$y = \frac{2\sqrt{15}}{3} - \frac{8\sqrt{15}}{3}\frac{\exp(2kz)}{(\exp(2kz)+1)^2}.$$

Returning to the original function, we obtain a solution of the approximate equation (6) for $n = 1/2$ and $C_1 = 0$

$$U = \frac{20}{3}\operatorname{th}^4(kz).$$

The graphs of the dependence $y(z)$ and $U(z)$ are shown in the Figure 1 and Figure 2 respectively. The solutions have the form of a classical Figure 1 and non-classical Figure 2 symmetric solitons of negative polarity. As the parameter K increases or the parameter ρ decreases, the soliton became narrower and its velocity increases, however, the amplitude remains constant $A = 20/3$. The width of the soliton is the following $\Delta = \sqrt{8\sqrt{15}/m_2}$.

The phase portraits Figure 3 and Figure 4 constructed for equations (8) and (7), respectively, show that in systems along with the separatrix solutions (in the figures are marked in red), periodic solutions exist near the equilibrium position.

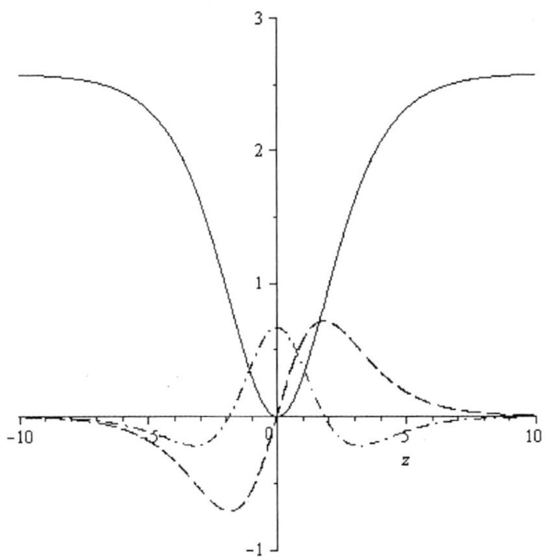

Figure 1. Dependencies $y(z)$ (solid curve), $y'(z)$ (dashed curve), $y''(z)$ (dot-dash curve)

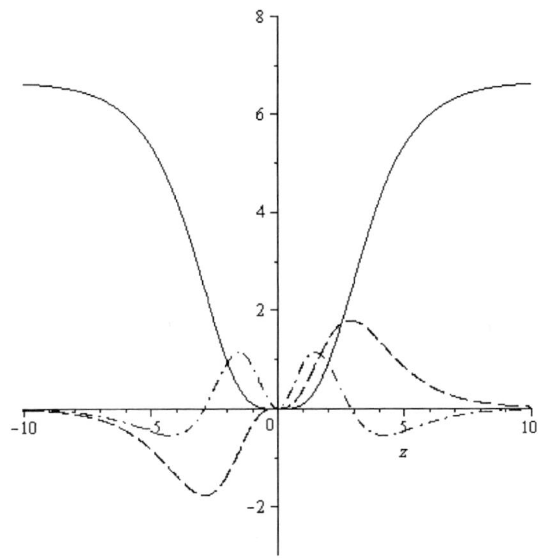

Figure 2. Dependencies $U(z)$ (solid curve), $U'(z)$ (dashed curve), $U''(z)$ (dot-dash curve)

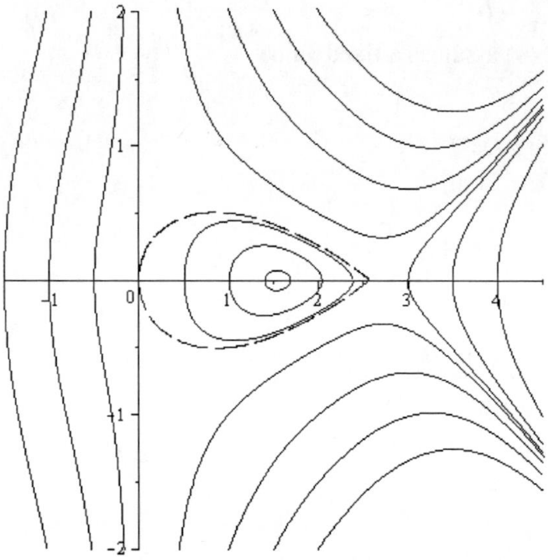

Figure 3. Phase portrait (y, y')

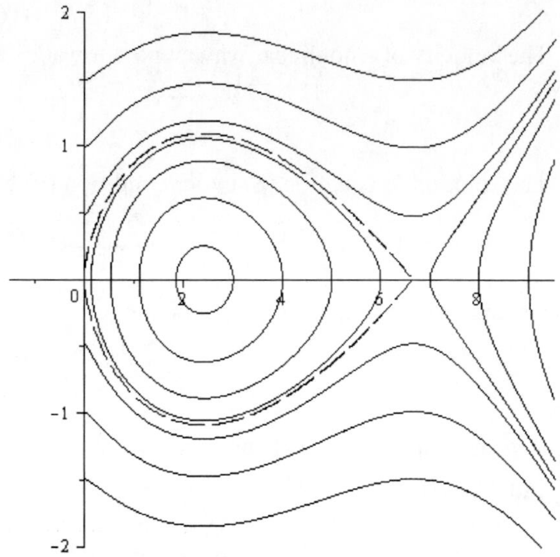

Figure 4. Phase portrait (U, U')

Consider the approximate equation (6) for $n = 1/3$

$$U'' + m_1 U - m_2 \sqrt[3]{U} - \frac{1}{6} m_2 \sqrt[3]{U^4} + C_1 = 0. \quad (10)$$

Taking into account the substitution $U = y^3$, we get the following equation

$$\left(y^3\right)'' + m_1 y^3 - m_2 y - \frac{1}{6} m_2 y^4 + C_1 = 0, \quad (11)$$

Solutiontowhichhasthepoleof the second order.Analogously to the previous case, we seek the solution in the form $y(z) = b_0 Y^2(z) + b_1 Y(z) + b_2$, where $Y(z)$ satisfies the equation $Y' = -Y^2 + B_0$. After the substitution into the equation (11), we find the values of unknown coefficients

$$b_0 = \frac{252}{m_2},\ b_1 = 0,\ b_2 = -\frac{7}{3}\frac{m_1}{m_2},\ B_0 = \frac{1}{36} m_1,$$

where the velocity of the nonlinear wave and the integration constant have certain values $V = \sqrt[3]{\frac{9}{7\sqrt{8}}} \sqrt{\frac{K}{\rho}}$, $C_1 = 0$.

The solution to equation (11), expressed in terms of the hyperbolic tangent and the exponent, has the form

$$y = \frac{\sqrt[3]{21}}{2}\left(3\,\mathrm{th}^2(kz) - 1\right),\ y = \sqrt[3]{21} - 6\sqrt[3]{21}\frac{\exp(2kz)}{(\exp(2kz)+1)^2},\ \text{где}\ k = \sqrt{\frac{\sqrt[3]{21} m_2}{168}}.$$

The solution to the equation (10):

$$U = \frac{21}{8}\left(3\,\mathrm{th}^2(kz) - 1\right)^3.$$

The solution $y(z)$ Figure 5 has the form of a classical soliton. The graph of dependency $U(z)$ Figure 6 has the form of a double soliton: one soliton having a smaller amplitude and width is located inside another soliton having large amplitude and a larger width ("embedded solitons"). The

IOP Conf. Series: Journal of Physics: Conf. Series **918** (2017) 012023 doi:10.1088/1742-6596/918/1/012023

amplitudes of the larger and smaller solitons are equal, respectively $A = 21$ and $A = 21/8$. The width of the common soliton is $\Delta = \sqrt{\dfrac{168}{\sqrt[3]{21} m_2}}$, when K decreases or ρ increases the soliton expands, its velocity decreases, the amplitude remains constant.

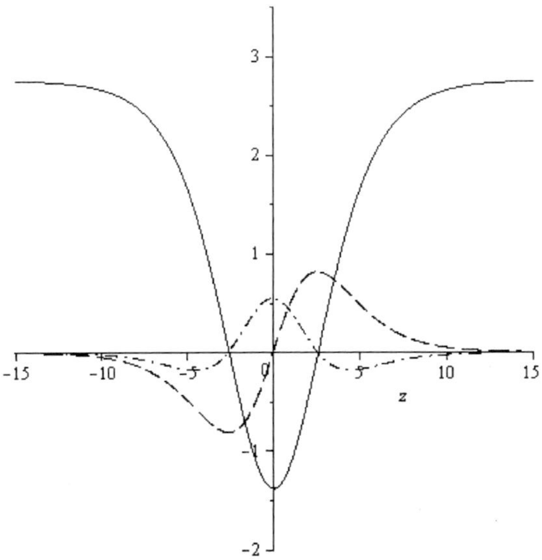

Figure 5. Dependencies $y(z)$ (solid curve), $y'(z)$ (dashed curve), $y''(z)$ (dot-dash curve)

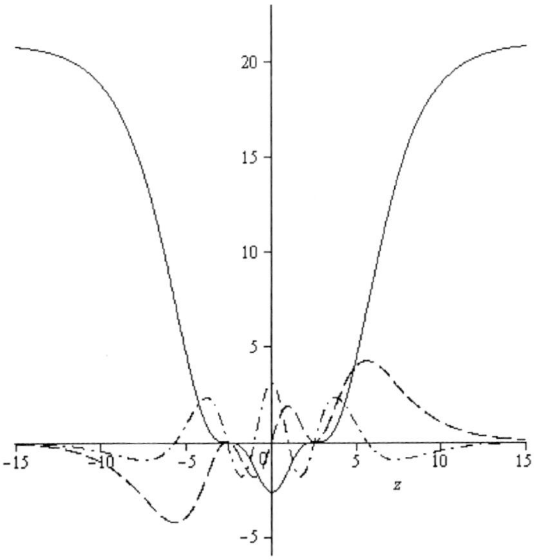

Figure 6. Dependencies $U(z)$ (solid curve), $U'(z)$ (dashed curve), $U''(z)$ (dot-dash curve)

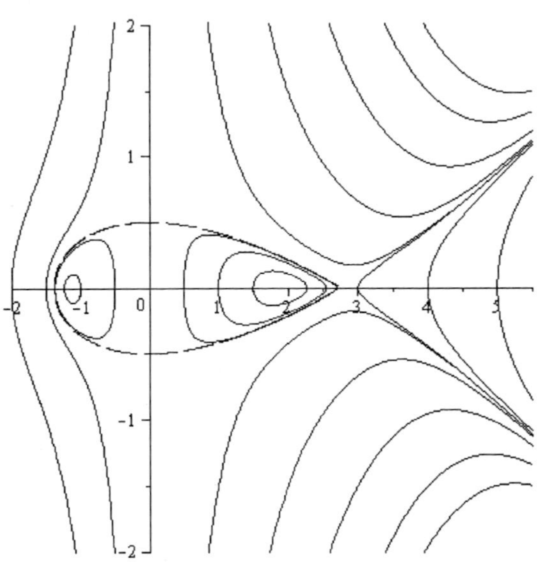

Figure 7. Phase portrait (y, y')

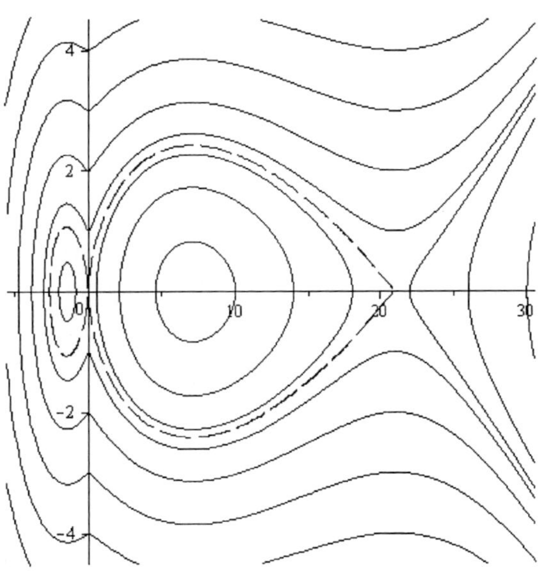

Figure 8. Phase portrait (U, U')

IOP Conf. Series: Journal of Physics: Conf. Series **918** (2017) 012023 doi:10.1088/1742-6596/918/1/012023

From the phase portraits of equations (11) Figure 7 and (10) Figure 8, it can be seen that in the system periodic motion near two stable equilibrium positions is possible.
Exact solutions to the original equation (5) we could not find yet. However, the approximate equation (6) has analytic solutions for $n = 1/2$ and $n = 1/3$.

This work was supported by the Russian Scientific Foundation (project no. 14-19-01637).

References

[1] Nowacki W.K. 1978 Stress Waves in Non-Elastic Solids (Oxford, New York, Pergamon Press).
[2] Goldsmith W. 2001 Impact. The Theory and Physical Behavior of Colliding Solids (Mineola, New York, Dover Publications).
[3] Molinari A. and Ravichandrana G. 2004 J. Appl. Phys.95 1718.
[4] Doronin A. M. et al 2016 J. Mach. Manuf. Reliab.45 6.
[5] Kleemola H.J. and Nieminen M.A. 1974 Metallurg. Trans. 5 1863.
[6] Callister Jr. and William D. 2005 Fundamentals of Materials Science and Engineering (New York, John Wiley & Sons).
[7] Kudryashov N. A. 2004 Analytical Theory of Nonlinear Differential Equations (Moscow, Izhevsk, Institute for Computer Research) (in Russian).

Nonreciprocal optical element of ring laser gyroscope based on the effect of light entrainment with moving medium

V O Gladyshev[1], D D Goryushkina[2], V N Kuryatov[3]

[1] Professor of Physics Department, N. Bauman MSTU, Russia
[2] Bachelor of Physics Department, N. Bauman MSTU, Russia
[3] Doctor of Science, Scientific Research Institute "Polius", Russia
E-mail: vgladyshev@mail.ru

Abstract. This article describes the use of Fresnel-Fizeau effect in order to reduce the entrapment effect in a ring laser gyroscope. The study gives an analysis of the nonreciprocal optical element impact on the gyroscope output characteristic. Moreover, it numerically estimates the method's effectiveness.

Due to a number of indisputable advantages, laser gyroscopes have received widely practical application in strapdown navigation systems. A promising area of research in this field is the development of technical methods for improving the accuracy and angular displacement sensitivity of a laser gyroscope.

In general terms, a laser gyro is an optical ring resonator, in which two independent oppositely directed traveling waves are generated. The turn of the contour having an area S and perimeter L in inertial space with angular velocity Ω, whose vector is perpendicular to the plane of the contour, leads to a difference frequency of ingoing waves, expressed by the formula:

$$\Delta\nu = \frac{4S\Omega}{L\lambda} = k\Omega, \tag{1}$$

where $k = \frac{4S}{L\lambda}$ is a scaling coefficient of the laser gyro. Any effect that leads to a deviation of the characteristics from this dependence, is seen as an error source. Among the errors which are critical in the laser gyroscope design, there is frequency synchronization of oncoming waves (the entrapment), which is caused by microscopic defects on the optical surfaces. They serve as centers for backscattering of the energy of one beam in the direction of propagation of the other [1]. The backscattering leads to the fact that at low speeds of angular rotation Ω, which are lower than a certain critical value (the entrapment threshold), traveling waves are synchronized, and the device becomes insensitive to the angular displacement.

One of the most common ways to eliminate mutual synchronization is based on the use of alternating dithering, providing torsional vibration of the gyro body around an equilibrium position with an angular velocity, which is greater than the entrapment threshold. Thus, the operating point of the device is displaced from the entrapment zone into the area where the output characteristic varies linearly (1). The analysis of the potential of current technical means shows that it is possible to increase the gyro sensitivity by reducing the oncoming waves relationship, if the refractive index nonreciprocity is provided. This method is based on introduction of the optically transparent solid medium into the resonator, the medium moving at known speed. As this takes place, in the moving medium, which is isotropic in the fixed coordinate system, there occurs "anisotropy" induced by the motion, which leads to the entrainment of light wave polarization plane. That is, there occurs an effect similar to the Fizeau-Fresnel entrainment effect [2]. A good example of practical implementation of this method can be found in the article [3]. As a moving medium we chose a quartz disk 20 mm in thickness with a mirror coating on flat surfaces, rotating around its axis with a frequency of 100-380 Hz. To reduce the losses inserted in the resonator, the disk was tilted at Brewster's angle with respect to the incident beam.

Similar in technical implementation to a rotating disk is the device in which the dithering oscillations are simulated by oscillations of two plane-parallel transparent prisms having a refractive index greater than a unity element, and being installed on the path of the laser beam, so that the oscillating movement is parallel to the direction of light propagation. [4]

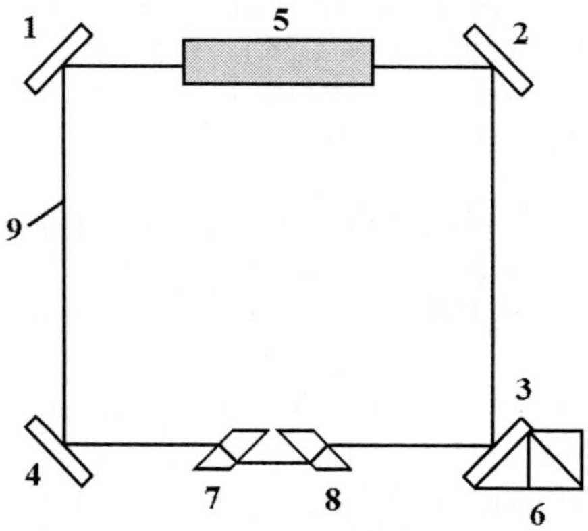

Figure 1. Scheme of laser gyro with two movable prisms

Figure 1 shows the schematic plan of the laser gyro comprising such a device. The gyroscope resonator comprises four prisms 1, 2, 3 and 4 which guide the laser beam about a closed path 9, and the active medium 5. All together form a ring laser, in which there can exist waves propagating along the optical path in clockwise and counterclockwise directions. The prism 3 passes a small portion of the light waves in the photomixer 6 which combines the light beams into one and sends it to the photodetector (not shown) which records the interference pattern. On the laser optical path there are two plane-parallel prisms 7 and 8, installed in such a way, that the laser beam intersects their surfaces at Brewster's angle. The prisms 7 and 8 are connected to the generator (not shown), which causes them to vibrate at a frequency of several hundred Hertz and an amplitude of 2 mm, and gives them an oscillating translational motion along the direction of light propagation.

For the scheme described the wave frequency shift caused by the motion of the optical medium, will be determined by the following expression:

$$\Delta\nu' = 4 \cdot \frac{\left(1 - \frac{1}{n^2}\right) v}{\lambda}, \quad (2)$$

where v is the velocity of the plates, λ is laser wavelength, and $\left(1 - \frac{1}{n^2}\right)$ is the Fresnel dragging coefficient, in which n is the refractive index of the medium.

The time integration of the difference frequency (1) with regard to the additional frequency shift (2) for the gyroscope with oscillating prisms gives the following result:

$$N = \int_0^t \Delta\nu + \Delta\nu' dt = k\gamma + 4 \cdot \frac{\left(1 - \frac{1}{n^2}\right)}{\lambda} l, \quad (3)$$

where N is the number of interference fringes that have passed in front of a photodetector in time t, γ is the total angle of the device rotation, l is the distance at which the prisms shifted over a specified time interval.

The estimate of the value $N' = 4 \cdot \frac{\left(1-\frac{1}{n^2}\right)}{\lambda} l$ is of essential interest. It characterizes an additional frequency difference between the oncoming waves, which results from the movement of the optical medium.

Since the prisms perform harmonic oscillations at a predetermined frequency f, then the distance l changes sinusoidally:

$$l = A\sin(\omega t), \tag{4}$$

where $\omega = 2\pi f$ is the cyclic oscillation frequency, A is the amplitude.

The speed of the prisms movement is incomparably small in relation to the speed of light, allowing light waves propagating in a closed circuit, to make the effective number of passes i through the optical medium in the plates semioscillation. With each passing the waves gain an elementary phase shift caused by the displacement of an optical element. Then the actual distance l, passed by the prisms in semioscillation can be represented as the sum of small displacements $\tilde{l}$,occurring in time $\tilde{t}$, during which the light propagates in the medium:

$$l = \sum_{k=0}^{i} \tilde{l}_k = i \cdot \tilde{l} = i \cdot A\sin(\omega \tilde{t}), \tag{5}$$

where $i = \frac{T/2}{L/c} = \frac{Tc}{2L}$ can be interpreted as the number of the laser beam passes of the resonator perimeter in the prisms semioscillation (c is the speed of light in vacuum), and $\tilde{t} = \frac{2nd}{c}$ (d is the thickness of a single prism).

In due consideration of (4) and (5), the expression for the value N'' will be as follows:

$$N' = (n^2 - 1)\frac{8A\pi d}{nL\lambda}, \tag{6}$$

The experimental studies showed that the efficiency of light entrainment with the moving medium in the main approximation depends on the product $\chi = (n^2 - 1)IV$, where $I = l$ is the optical length of the path run by the laser beam in the medium, and $V = 2A\pi f$ is the linear velocity of movement (in the projection on the beam) - maximum to modulo prisms speed. [3]

Figure 2 shows the dependency graph of the interference shift N' x' on the parameter χ, calculated at different values of the parameters A, d, and n optical element according to the formula (6) (λ = 633 nm, and L = 1,5 m).

As Figure 2 shows, dependency of the additional number of interference fringes on the efficiency parameter of light entrainment with the moving medium is close to direct proportionality. The graph deviation from a straight line can be explained by the fact that the change in the estimated value n was not equal-interval.

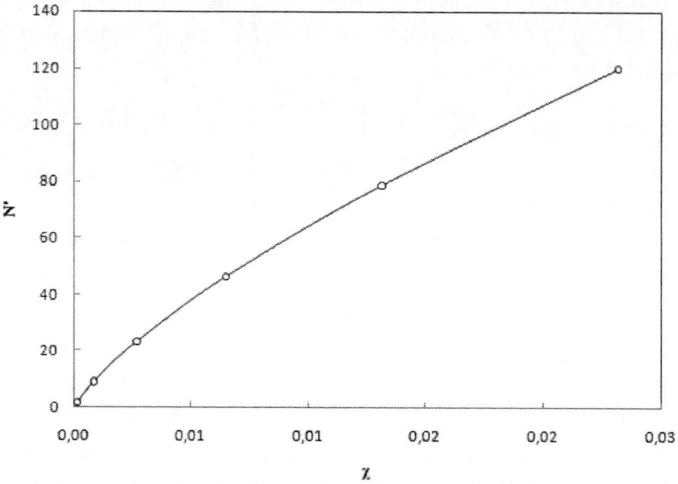

Figure 2. Dependency of the interference shift on the efficiency parameter of light entrainment.

This article describes the method of removing a laser gyro from the entrapment zone. The method is based on the effect of light entrainment with the moving medium and can be used as an alternative to replace the dithering. We carried out an analysis of the scheme comprising a phase shifter in the form of a nonreciprocal optical element. The analysis done leads to the firm conclusion that the proposed method is effective.

Moreover, the cyclic control of the electromagnetic wave phase can be used for registration of quasi-harmonic gravity waves. The phase modulation on the frequency of the expected gravitational wave burst can increase the signal / noise ratio when using a low-frequency optical resonance occurring in the Fabry-Perot interferometer [5].

The work was supported by Russian Foundation for Basic Research grants №16-02-00488 A and №16-08-00618 A.

References

[1] Bogdanov A D 1973 Giroskopy na lazerakh [Laser gyroscopes]. Moscow, *Voenizdat Publ.*, pp 42-44 (in Russ).

[2] Frankfurt U I, Frenk A M 1972 Optika dvizhuschikhsya tel [Optics of moving bodies]. Moscow, *Nauka Publ.*, pp 31-34 (in Russ).

[3] Gladyshev V O, Gladysheva T M, Dashko M, Podguzov G V 2006 The anisotropy of the space of electromagnetic radiation velocity in moving media. *Hypercomplex numbers in geometry and physics.* vol. **3**, no. 2 (6) pp 173-187.

[4] Gladyshev V, Gladysheva T, Zubarev V 2006 Propagation of electromagnetic waves in media undergoing complex motions. *Journal of Engineering Mathematics.* vol. **55**, no. 1-4 pp 235-250.

[5] Anthony J. Lindop. *Ring Lasers*, U.S. Patent 4 247 831, January 27, 1981.

[6] Gladyshev V O, Morozov A N 1999 The theory of a Fabry-Perot interferometer in a gravitational wave experiment. *Journal of the Moscow Physical Society.* vol. **6**, no. 3 pp 209-221.

IOP Conf. Series: Journal of Physics: Conf. Series **918** (2017) 012025 doi:10.1088/1742-6596/918/1/012025

On a certain laminar dynamo model

V V Novikov[1,2] and L N Fevralskikh[1,2]

[1]Department of Theoretical, Computer and Experimental Mechanics, Institute of Information Technology, Mathematics and Mechanics, Lobachevsky State University, Nizhny Novgorod, Russia
[2]Scientific Research Institute of Mechanics, Lobachevsky State University, Nizhny Novgorod, Russia

Email: vvnovikov2007@yandex.ru, grigorieva_ln@mail.ru

Abstract. We consider the motion of a conducting incompressible viscous fluid in the space between two rotating spheres. We assume the centres of symmetry of spheres to lie a short distance from each other in a plane perpendicular to the axes of rotation. If we ignore the effect of the magnetic field on the motion of the liquid and regard the flow as laminar, then we are able to solve the hydrodynamics system of equations analytically. We use this solution to assess the possibility of a magnetic dynamo.

According to Cowling's theorem [1], two-dimensional or axisymmetric flows of a conducting liquid are not able to maintain a long-term magnetic field. For a dynamo to occur, the geometric configuration of the flow must be more complex and contain a radial component, which is necessary for the formation of convective fluid flows. Determining the steady-state mode of magnetic field generation should be based on solving the magnetohydrodynamics equations together, taking into account the mutual influence of fluid velocity and magnetic field strength. Since solving the complete system of equations is difficult, it makes sense to use a kinematic approach, which assumes that the fluid flow is given, and consider the problem of the evolution of the magnetic field intensity. Such a formulation reduces the problem to finding eigenvalues with a positive real part that ensure the excitation of magnetic modes.

Many works based on the kinematic approach deal with the issue of magnetic field generation depending on the characteristics of the magnetic properties the liquid has. Works [2-3] investigate conditions for a helical fluid flow exciting a magnetic field in a cylinder and a torus. Publication [4] presents the results of investigations conducted by a number of authors who use the laminar dynamo model, created by fluid flows in the sphere. Studying the latter is of great interest for understanding the generation mechanism of the Earth's magnetic field.

This paper considers the problem of laminar motion of a conducting viscous fluid enclosed between two rotating spherical surfaces. The inner sphere constantly shifts in reference to the outer one in a direction perpendicular to the axes of rotation, which disturbs the axial symmetry of the flow. We consider this model because it reflects the modern concept of the internal structure of the Earth. It is known [5] that the solid core of the Earth is offset relative to the geocenter by a distance of 5…15 km. Moreover, according to the Reynolds number estimates given in [6], the mode of fluid flow in the outer liquid core of the Earth can be considered intermediate between laminar and turbulent. Article [7] presents a solution to the hydrodynamic problem of laminar fluid motion between rotating nonconcentric spheres. It shows that the axial asymmetry of the flow leads to a radial flow appearing, which is necessary to create convection. Our paper investigates the possibility of generating a magnetic field via the flow described in [7].

Let us consider the motion of a homogeneous incompressible conducting viscous fluid that fills the space between two uniformly rotating nonconcentric spheres. The inner sphere of radius r_1 with the center at the point O_1 rotates with the angular velocity ω_1 around the axis $O_1 z_1$ of the fixed system $O_1 x_1 y_1 z_1$, and the outer sphere of radius r_2 with the center at the point O_2 rotates with the angular velocity ω_2 around the axis $O_2 z_2$ of the fixed system $O_2 x_2 y_2 z_2$. The axes of rotation $O_1 z_1$ and $O_2 z_2$ of the spheres are parallel; a small distance δ $(\delta << 1)$ separates them. Without loss of generality, we assume that the center O_1 in the system $O_2 x_2 y_2 z_2$ has the coordinates $O_1(\delta,0,0)$ (figure 1).

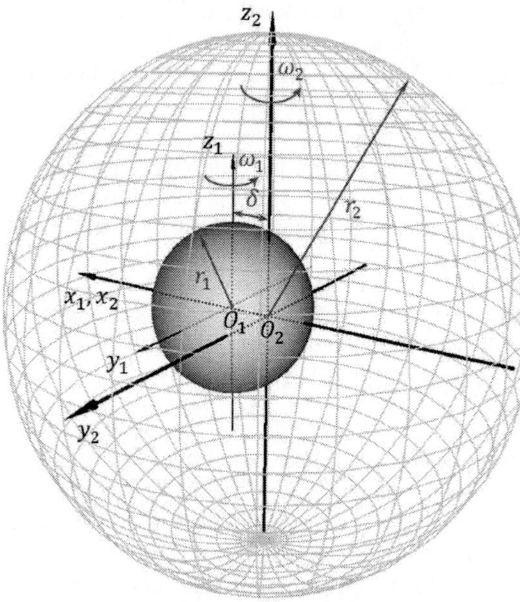

Figure 1. Model of fluid flow between rotating spherical surfaces.

We then switch to a spherical coordinate system consistent with the system $O_2 x_2 y_2 z_2$. This is the the inner surface equation in the spherical system, accurate to terms of the first order of smallness and expressed in terms of the offset parameter δ:

$$r = r_1 + \delta \sin\theta \cos\varphi.$$

We apply the curl operator to the Navier-Stokes equation, eliminating the fluid pressure:

$$rot\left(\frac{\partial \vec{v}}{\partial t} + (\vec{v},\nabla)\vec{v}\right) = \frac{1}{4\pi\rho} rot(\vec{H},\nabla)\vec{H} + \nu\, rot\, \Delta\vec{v}.$$

The following equation describes the evolution of the magnetic field strength in a moving conducting medium:

$$\frac{\partial \vec{H}}{\partial t} = rot\left[\vec{v},\vec{H}\right] + \frac{c^2}{4\pi\sigma}\Delta\vec{H}.$$

We write the continuity equation and the solenoidal condition for the magnetic field as follows:

$$div\,\vec{v} = 0, \qquad div\,\vec{H} = 0.$$

In the equations of magnetic hydrodynamics we adopt the following dimensionless variables and parameters:

$$\varepsilon = \frac{c^2}{4\pi\sigma\nu}, \quad s = \frac{I^2\sigma}{\rho\nu c^2}, \quad R_m = \frac{4\pi\sigma v_* R}{c^2}, \quad t_* = \frac{4\pi\sigma R^2}{c^2},$$

where R is the length scale, t_* is the time scale, I is the current intensity scale, v_* is the characteristic flow velocity, R_m is the magnetic Reynolds number.

The complete system of magnetohydrodynamic equations in dimensionless notation is as follows:

$$\varepsilon\, rot\left(\frac{\partial \vec{v}}{\partial t} + (\vec{v}, \nabla)\vec{v}\right) = s\, rot(\vec{H}, \nabla)\vec{H} + rot\, \Delta\vec{v}, \tag{1}$$

$$\frac{\partial \vec{H}}{\partial t} = R_m\, rot\left[\vec{v}, \vec{H}\right] + \Delta\vec{H}, \tag{2}$$

$$div\, \vec{v} = 0, \qquad div\, \vec{H} = 0. \tag{3}$$

Assuming a laminar fluid flow, the parameter ε multiplying the nonlinear inertia force component in the equation of motion can be considered small. Selecting the right scale factor I ensures that the parameter s is also small.

The smallness conditions:

$$\varepsilon << 1, \qquad s << 1$$

allow us to seek a solution to the system (1) - (3) in the form of an expansion in powers of ε, s:

$$\vec{v} = \vec{v}_0 + \varepsilon\vec{v}_1 + s\vec{v}_2 + \ldots,$$

$$\vec{H} = \vec{H}_0 + \varepsilon\vec{H}_1 + s\vec{H}_2 + \ldots.$$

This reduces the problem of determining the velocity field $\vec{v}$ and the magnetic field strength vector $\vec{H}$ to sequentially finding the approximations $\vec{v}_0$, $\vec{H}_0$, $\vec{v}_1$, $\vec{H}_1$, ... This allows us to separate equations (1), (2) and to validate employing the kinematic approach at the initial stage of magnetic field excitation.

The initial approximation of the velocity field $\vec{v}_0$ is determined from solving the following hydrodynamic equations:

$$rot\, \Delta\vec{v}_0 = 0, \tag{4}$$

$$div\, \vec{v}_0 = 0. \tag{5}$$

The equations below define the magnetic field strength $\vec{H}_0$ created by the flow $\vec{v}_0$ in the space between the spheres $r_1 + \delta\sin\theta\cos\varphi \le r \le r_2$:

$$\frac{\partial \vec{H}_0}{\partial t} = R_m\, rot\left[\vec{v}_0, \vec{H}_0\right] + \Delta\vec{H}_0, \tag{6}$$

$$div\, \vec{H}_0 = 0. \tag{7}$$

To solve the systems of equations (4) - (5) and (6) - (9) we used vector spherical harmonics [8] that form a complete system of linearly independent vector functions:

$$\vec{Y}^{+}_{lm}(\theta,\varphi) = (l+1)Y_{lm}(\theta,\varphi)\vec{e}_r - grad\, Y_{lm}(\theta,\varphi), \quad l = 0, 1, 2, \ldots,$$

$$\vec{Y}^{-}_{lm}(\theta,\varphi) = l\, Y_{lm}(\theta,\varphi)\vec{e}_r + grad\, Y_{lm}(\theta,\varphi), \quad l = 1, 2, \ldots,$$

$$\vec{Y}^{0}_{lm}(\theta,\varphi) = \left[\vec{e}_r, grad\, Y_{lm}(\theta,\varphi)\right], \quad l = 1, 2, \ldots.$$

The work [7] describes in detail how to solve the equations of hydrodynamics (4) - (5) together with the conditions for adhesion of the liquid to the surfaces bounding it. We write the equations accurate to terms of the first order of smallness with respect to the offset parameter δ:

$$\vec{v}_0(r,\theta,\varphi) = -\frac{2}{3}\frac{\sqrt{6\pi}}{r_2^3 - r_1^3}\left(\left(\omega_2 r_2^3 - \omega_1 r_1^3\right)r + \frac{r_1^3 r_2^3(\omega_1 - \omega_2)}{r^2}\right)\vec{Y}_{10}^0 +$$
$$+\frac{1}{3}i\sqrt{\pi}\delta Q_1\left[\left(\eta_1 r^2 - \frac{\eta_2}{r} + \frac{\eta_3}{r^3}\right)\left(\vec{Y}_{11}^+ + \vec{Y}_{1,-1}^+\right) + \left(\eta_4 - 5\eta_1 r^2 - \frac{4\eta_2}{r}\right)\left(\vec{Y}_{11}^- + \vec{Y}_{1,-1}^-\right)\right] +$$
$$+\sqrt{\frac{\pi}{5}}\xi_2\delta Q_2\left(r^2 - \frac{r_2^5}{r^3}\right)\left(\vec{Y}_{21}^0 - \vec{Y}_{2,-1}^0\right),$$

where

$$Q_1 = \frac{1}{(r_2 - r_1)^4(4r_1^2 + 7r_1r_2 + 4r_2^2)}, \quad Q_2 = \frac{r_1^3}{r_2^5 - r_1^5},$$
$$\eta_1 = r_1\left[4\xi_2 r_1^3 - (\xi_1 + 4\xi_2)r_1^2 r_2 + \xi_1 r_2^3\right]$$
$$\eta_2 = r_1 r_2\left[(-\xi_1 + 5\xi_2)r_1^5 - 5\xi_2 r_1^3 r_2^2 + \xi_1 r_2^5\right]$$
$$\eta_3 = r_1^3 r_2^3\left[(-\xi_1 + 5\xi_2)r_1^3 - 9\xi_2 r_1 r_2^2 + (\xi_1 + 4\xi_2)r_2^3\right]$$
$$\eta_4 = r_1\left[4(-\xi_1 + 5\xi_2)r_1^5 - 5(\xi_1 + 4\xi_2)r_1^2 r_2^3 + 9\xi_1 r_2^5\right]$$
$$\xi_1 = \sqrt{2}(\xi_0 + 2\omega_1), \quad \xi_2 = \xi_0 - \omega_1,$$
$$\xi_0 = \frac{1}{r_2^3 - r_1^3}\left[\left(\omega_2 r_2^3 - \omega_1 r_1^3\right) - 2r_2^3(\omega_1 - \omega_2)\right]$$

The term that does not depend on the offset δ corresponds to the solution to the classical problem of a viscous fluid moving between two rotating concentric spheres. The offset of the inner sphere relative to the external one in the direction perpendicular to the rotation axes leads to the emergence of additional harmonics with a weight $l = 1, 2$. In this case, the radial component of the velocity vector $\vec{v}_0$ becomes nonzero.

We now proceed to the problem of determining the strength of a magnetic field generated by the flow $\vec{v}_0$. We seek a solution to the nonstationary problem (6) - (7) in the space between spheres in the form of a series of vector spherical harmonics:

$$\vec{H}_0(t,r,\theta,\varphi) = e^{\lambda t}\sum_{l,|m|\le l}\left(\Psi_{lm}^+(r)\vec{Y}_{lm}^+(\theta,\varphi) + \Psi_{lm}^0(r)\vec{Y}_{lm}^0(\theta,\varphi) + \Psi_{lm}^-(r)\vec{Y}_{lm}^-(\theta,\varphi)\right).$$

Using a previously known expression for the field $\vec{v}_0$, we represent the vector product $\left[\vec{v}_0, \vec{H}_0\right]$ in the form of an expansion:

$$\left[\vec{v}_0, \vec{H}_0\right] = e^{\lambda t}\sum_{l,|m|\le l}\left(\Phi_{lm}^+(r)\vec{Y}_{lm}^+(\theta,\varphi) + \Phi_{lm}^0(r)\vec{Y}_{lm}^0(\theta,\varphi) + \Phi_{lm}^-(r)\vec{Y}_{lm}^-(\theta,\varphi)\right),$$

in which the functions $\Phi_{lm}^\nu(r)$ are linear combinations of unknown radial functions $\Psi_{lm}^\nu(r)$.

Substituting the expansions $\vec{H}_0$ and $\left[\vec{v}_0, \vec{H}_0\right]$ into the equation (6) and grouping the coefficients before the vector spherical harmonics with the same indices, we obtain a linear system of ordinary differential equations with respect to the unknown radial functions $\Psi_{lm}^\nu(r)$:

$$\frac{\partial^2\Psi_{lm}^+}{\partial r^2} + \frac{2}{r}\frac{\partial\Psi_{lm}^+}{\partial r} - \left[\frac{(l+1)(l+2)}{r^2} + \lambda\right]\Psi_{lm}^+ + R_m\frac{l}{2l+1}\left(\frac{\partial\Phi_{lm}^0}{\partial r} - \frac{l}{r}\Phi_{lm}^0\right) = 0,$$

$$\frac{\partial^2\Psi_{lm}^0}{\partial r^2} + \frac{2}{r}\frac{\partial\Psi_{lm}^0}{\partial r} - \left[\frac{l(l+1)}{r^2} + \lambda\right]\Psi_{lm}^0 + R_m\left(\frac{\partial\Phi_{lm}^-}{\partial r} - \frac{l-1}{r}\Phi_{lm}^- - \frac{\partial\Phi_{lm}^+}{\partial r} - \frac{l+2}{r}\Phi_{lm}^+\right) = 0,$$

$$\frac{\partial^2 \Psi_{lm}^-}{\partial r^2} + \frac{2}{r}\frac{\partial \Psi_{lm}^-}{\partial r} - \left[\frac{(l-1)l}{r^2} + \lambda\right]\Psi_{lm}^- - R_m \frac{l+1}{2l+1}\left(\frac{\partial \Phi_{lm}^0}{\partial r} + \frac{l+1}{r}\Phi_{lm}^0\right) = 0,$$

$$l = 1, 2, \ldots, \quad |m| \le l.$$

In a general case, the system has an infinite number of equations and unknowns. We then solve a truncated system, restricting ourselves to the expansion of the vector $\vec{H}_0$ by a finite number of terms of the $(l = 1, \ldots, L, |m| \le l)$series. We discard equations that do not contain derivatives of functions occurring in $\vec{H}_0$. Replacing the derivatives by difference analogues reduces the problem to finding the eigenvalues λ_k of the truncated system matrix. The λ_k values will depend on the magnitude of the magnetic Reynolds number R_m, which is a parameter of this model. If the real parts of all eigenvalues λ_k are negative, then the field $\vec{H}_0$ created by the flow $\vec{v}_0$ will damp with time. If the real part of at least one λ_k for a certain value of R_m becomes positive, the corresponding magnetic field mode begins to grow exponentially, which means that a magnetic dynamo emerges. The eigenvalue of interest has the largest real part $\mathrm{Re}\,\lambda^* = \max \mathrm{Re}\,\lambda_k$.

To estimate the value of $\mathrm{Re}\,\lambda^*$ we took into account the averaged characteristics of the Earth and the fact that the solid core rotates faster relative to the mantle [6]. We used the QR method to calculate the λ_k eigenvalues. We plotted $\mathrm{Re}\,\lambda^*$ as a function of the magnetic Reynolds number R_m for the different values of $L = 1, 2, \ldots, 9$, which determine the highest order of harmonics in the expansion of the vector $\vec{H}_0$ (figure 2). $\mathrm{Re}\,\lambda^*$ becomes positive for R_m greater than a certain critical value $R_m^{cr.}$. The critical value $R_m^{cr.}$ decreases with increasing the number of harmonics comprising $\vec{H}_0$. The $\mathrm{Re}\,\lambda^*$ dependence on the parameter R_m varies insignificantly beginning with $L = 7$. For $L = 7$ we obtained the critical value of the magnetic Reynolds number, $R_m^{cr.} = 4 \cdot 10^{-11}$.

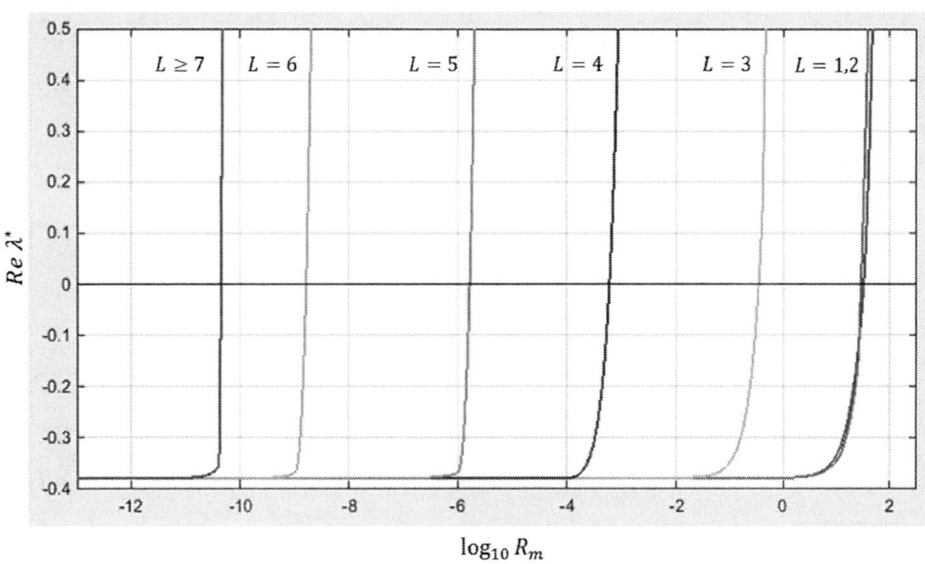

Figure 2. $\mathrm{Re}\,\lambda^*$ as a function of the magnetic Reynolds number R_m.

The resulting number $R_m^{cr.}$ is much smaller than the estimates of R_m for the liquid core of the Earth. This result does not allow us to draw a conclusion about the mechanism of the hydromagnetic dynamo. To explain the self-generation of the magnetic field in terms of this model, it is necessary to determine the following approximations: $\vec{v}_1, \vec{H}_1, \ldots$. However, at the initial stage a possible cause of magnetic field excitation may be the effect of the flow $\vec{v}_0$ that takes into account the axial asymmetry.

Acknowlegments
The work was supported by the Russian Foundation for Basic Research (project 15-01-08326).

References

[1] Cowling T 1976 *Magnetohydrodynamics* (Bristol: Adam Hilger) p 145
[2] Stepanov R A and Frik P G 1999 *Sb. nauchnykh trudov PGU: Gidrodinamika [Hydrodynamics: Proc. of the PSU]* vol 12 (Perm: PSU) pp 240-51
[3] Stepanov R A, Chupin A V and Frik P G 2010 *Materialy Mezhdunarodnoy konferentsii Vysokoproizvoditelnye parallelnye vychisleniya na klasternykh sistemakh (NRS-2010) [Proc. of the X International High-performance parallel computing on cluster systems (LDC-2010) Conference]* (Perm: PSTU) vol 2 pp 248-54
[4] Moffatt H K 1978 *Magnetic Field Generation in Electrically Conducting Fluids* (Cambridge: Cambridge University Press) p 342
[5] Barkin Yu V 1996 *Trudy Gos. Astronomicheskogo in-ta im. P.K. Shternberga - Proc. of the Sternberg Astronomical Institute* **LXV** 136-145
[6] Denisov G G, Novikov V V and Fedorov A E 2013 *Priroda - Nature* **5** 3-10
[7] Novikov V V and Fevralskikh L N 2016 *Uchen. zap. Kazan. un-ta. Ser. Fiz.-matem. nauki - Proceedings of Kazan University. Physics and Mathematics Series* **158**(1) 129–40
[8] Petrashen G I 1949 *Uchenye zapiski LGU. Seriya matematicheskikh nauk – Proc. of the Leningrad State University. Mathematics series* **17**(114) 2-27

Optimal control problem for linear fractional-order systems, described by equations with Hadamard-type derivative

Sergey Postnov
V.A. Trapeznikov Institute of Control Sciences, Profsoyuznaya str., 65, Moscow, Russia
E-mail: postnov.sergey@inbox.ru

Abstract. Two kinds of optimal control problem are investigated for linear time-invariant fractional-order systems with lumped parameters which dynamics described by equations with Hadamard-type derivative: the problem of control with minimal norm and the problem of control with minimal time at given restriction on control norm. The problem setting with nonlocal initial conditions studied. Admissible controls allowed to be the p-integrable functions ($p > 1$) at half-interval. The optimal control problem studied by moment method. The correctness and solvability conditions for the corresponding moment problem are derived. For several special cases the optimal control problems stated are solved analytically. Some analogies pointed for results obtained with the results which are known for integer-order systems and fractional-order systems describing by equations with Caputo- and Riemann-Liouville-type derivatives.

1. Introduction
Dynamics and optimal control are the actively developing topics for fractional-order systems now. There are many interesting results obtained already in this field. Note that character of these results depend on, in addition, from the type of fractional derivative definition chosen for system description [1]. So, the maximum principle, which is analogous to L.S. Pontryagin maximum principle, were proved recently for general type system described by equations with Riemann-Liouville derivative [2]. In case of systems described by equations with Caputo derivative such principle were proved for linear systems only [3]. The other effective and powerful approach for optimal control search is the moment method [4]. This method was successfully applied to investigation of optimal control problem for linear dynamic systems described by equations with Caputo derivative [5, 6, 7] and Riemann-Liouville derivative [8].

Caputo and Riemann-Liouville derivatives are most popular types of fractional derivative. In accordance with their definitions the fractional differentiation operator specified as convolution with fractional-power kernel. When the differentiation index is equal to 1, both types of the derivative reduce to ordinary first derivative. The principal difference between these two derivatives is the kind of initial conditions [9, 10, 11]. In case of equations with Caputo derivative the initial conditions are local and define the values of function and its integer-order derivatives at initial time. In case of equations with Riemann-Liouville derivative the nonlocal initial conditions used which define the value of some integral functional at initial time. Let us illustrate it by the following example from [11] (fractional derivative and integral implied here in the sense of Riemann-Liouville).

In the fractional Voight model describing the behaviour of some viscoelastic bodies under loading, the stress $\sigma(t)$ and the strain $\varepsilon(t)$ coupled by the following constitutive equation:

$$\sigma(t) = E\varepsilon(t) + K_0 D_t^\alpha \varepsilon(t),$$

where K and E — real coefficients. The relaxation after stress pulse of type $B\delta(t)$ impact will be described by the following equation:

$$\sigma(t) = E\varepsilon(t) + K_0 D_t^\alpha \varepsilon(t), t > 0.$$

For this equation the initial condition needed which define the value ${}_0D_t^{\alpha-1}\varepsilon(t) = {}_0I_t^{1-\alpha}\varepsilon(t)$ at $t \to 0+$. If we integrate the constitutive equation then we obtain:

$${}_0D_t^{-1}\sigma(t) - E_0 D_t^{-1}\varepsilon(t) = K_0 D_t^{\alpha-1}\varepsilon(t).$$

The first term here is equal to B when $t \to 0+$ [11]. Taking into account physical considerations (i.e., impossibility of instantaneous deformation) or the requirement of boundedness of function ${}_0D_t^{\alpha-1}\varepsilon(t)$, which define the initial condition, one can demonstrate that the second term vanishes when $t \to 0+$ [11]. So, we obtain:

$${}_0D_t^{\alpha-1}\varepsilon(t) = {}_0I_t^{1-\alpha}\varepsilon(t) = B/K.$$

The example considered above shown that nonlocal initial conditions for the strain $\varepsilon(t)$ can be re-written in the form of local initial conditions for stress $\sigma(t)$, which have more clear physical sense.

One of the first articles concerning the optimal control problem for fractional-order systems with nonlocal initial conditions is [10]. Later such problem statement studied by many researchers, particularly, in [2].

In this paper the optimal control problem investigated for the systems with lumped parameters, which described in terms of Hadamard fractional derivative, defined as convolution with kernel containing the logarithmic function in fractional power. This derivative type, analogously to the Riemann-Liouville derivative, need to set the nonlocal initial conditions but didn't reduce already to the first derivative when the differentiation index is equal to 1. Optimal control problems investigated by the moment method. The optimal control problem for one-dimensional linear time-invariant system studied in general setting. It's demonstrated that this problem can be reduced to the moment problem. The multi-dimensional systems of special type (the chain of integrators) considered analogously. The comparison evaluated with integer-order systems and analogous fractional-order systems, which described by equations with Riemann-Liouville derivative.

2. Problem setting

We will consider the following dynamical systems:

$${}_0D_t^{\alpha_i} q_i(t) = a_{ij}q_j(t) + b_{ij}u_j(t) + f_i(t), t \in (t_0, T], i, j = 1, ..., N, \tag{1}$$

where functions $q_i(t)$, $u_i(t)$ and $f_i(t)$ are the phase coordinates, controls and perturbations correspondingly, $T > t_0 > 0$; a_{ij} and b_{ij} are known real coefficients. The repeated indices suppose summation. Here ${}_0D_t^{\alpha_i}$ is operator of the left-side fractional Hadamard derivative (see [9, §2.7]), $\alpha_i \in (0, 1]$:

$${}_0D_t^{\alpha_i} q_i(t) = \frac{1}{\Gamma(1-\alpha_i)} t \frac{d}{dt} \int_{t_0}^{t} \left(\ln \frac{t}{\tau}\right)^{-\alpha_i} \frac{q_i(\tau) d\tau}{\tau}.$$

Functions $q_i(t)$, $u_i(t)$ and $f_i(t)$ can be considered as components of vector functions $\vec{q}(t) = (q_1(t), \ \dots, \ q_N(t))$, $\vec{u}(t) = (u_1(t), \ \dots, \ u_N(t))$ and $\vec{f}(t) = (f_1(t), \ \dots, \ f_N(t))$.

The nonlocal initial conditions are determined [9, §4.1]:

$$\lim_{t \to t_0+} \left[{}_{t_0}I_t^{1-\alpha_i} q_i(t) \right] = s_i^0, i = 1, ..., N, \tag{2}$$

where ${}_{t_0}I_t^{\alpha_i} q_i(t) = \frac{1}{\Gamma(\alpha_i)} \int\limits_{t_0}^{t} (\ln \frac{t}{\tau})^{\alpha_i - 1} \frac{q_i(\tau) d\tau}{\tau}$ is the left-side Hadamard integral of order α_i [9, §2.7].

The final conditions are determined in ordinary manner:

$$q_i(T) = q_i^T, i = 1, ..., N. \tag{3}$$

Let us consider control $\vec{u}(t) \in L_p(t_0, T]$, $p > 1$. The control norm will be defined by the following expressions [4, 13]:

$$\|u(t)\| = \left(\int\limits_{t_0}^{T} \sum_{i=1}^{N} |u_i(t)|^p \, dt \right)^{1/p}, 1 < p < \infty,$$

$$\|u(t)\| = \text{vrai} \max_{t \in (t_0, T]} \max_i |u_i(t)|, p \to \infty.$$

Let us take that functions $q_i(t)$ and $f_i(t)$, $i = 1, ..., N$ (or $\vec{q}(t)$ and $\vec{f}(t)$) possess all properties required for existence of solutions of equations studied further on, are, in particular, differentiable at least once.

Let us set the following optimal control problem: determine such control $\vec{u}(t)$, $t \in (t_0, T]$ at which the system (1) will pass from the given initial state (2) to the assigned final state (3) and herewith: 1) norm of control will be minimal with the assigned control time T (**OCP A**), control time T will be minimal provided $\|\vec{u}\| \le l$, $l > 0$, where l is the assigned constant (**OCP B**).

3. The moment problem

3.1. Preliminaries

The classical l-problem of moments can be formulated in a following way [4, 14, 15].

Let we have a system of functions $g_i(t) \in L^{p'}(t_0, T]$, $i = 1, ..., N$, $p' > 1$. Let we also have the assigned numbers c_i (calling by moments), $i = 1, ..., N$ and $l > 0$. We should find function $u(t) \in L^p(t_0, T]$, $p > 1$ that satisfies the following conditions:

$$\int\limits_{t_0}^{T} g_i(\tau) u(\tau) d\tau = c_i, \tag{4}$$

$$\|u(t)\| \le l. \tag{5}$$

and The spaces $L^p(t_0, T]$ and $L^{p'}(t_0, T]$ are adjoint:

$$\frac{1}{p} + \frac{1}{p'} = 1. \tag{6}$$

In order to the moment problem (4)-(5) will be solvable it's necessary and sufficient the existence of number $\lambda_N > 0$ and numbers $\xi_1^*, \dots, \xi_N^*$ that give a solution to the following equivalent conditional minimum problem [4, 14, 15].

To find

$$\min_{\xi_1,\ldots,\xi_N}\left(\int_0^T\left|\sum_{i=1}^N \xi_i g_i(t)\right|^{p'} dt\right)^{1/p'} = \left(\int_0^T\left|\sum_{i=1}^N \xi_i^* g_i(t)\right|^{p'} dt\right)^{1/p'} = \frac{1}{\lambda_N} \tag{7}$$

with additional condition

$$\sum_{i=1}^N \xi_i c_i = \sum_{i=1}^N \xi_i^* c_i = 1. \tag{8}$$

In case of problem (7-8) is solvable the optimal control for OCP A will be given by the following expression [4]:

$$u(t) = \lambda_N^{p'}\left|\sum_{i=1}^N \xi_i^* g_i(t)\right|^{p'-1} \operatorname{sign}\left[\sum_{i=1}^N \xi_i^* g_i(t)\right],\ t\in(t_0,T]. \tag{9}$$

OCP B may be resolved by the following formula:

$$u(t) = l^{p'}\left|\sum_{i=1}^N \xi_i^* g_i(t)\right|^{p'-1} \operatorname{sign}\left[\sum_{i=1}^N \xi_i^* g_i(t)\right],\ t\in(t_0,T^*], \tag{10}$$

where T^* is the minimal non-negative real root of equation

$$\lambda_N(T^*) = l. \tag{11}$$

For solvability of momemt problem (7-8) it is necessary and sufficient to satisfy one of two equivalent conditions [4]: (1) $\lambda_N > 0$; (2) functions $g_i(t)$ are linearly independent. The single question is the possibility of correct posing of the moment problem, which depend on the existence of norm of functions $g_i(t)$ in space $L^{p'}(t_0,T]$ and the existence of at least one nonzero component in number set c_i.

Definition 1. The moment problem (4)-(5) is calling correct if the norm of functions $g_i(t)$ is determined in space $L^{p'}(t_0,T]$, $p'>1$ and there exist at least one nonzero component in number set c_i.

System (1) with fractional derivative operator in sense of Hadamard (Riemann-Liouville, Caputo) we will name below as Hadamard (Riemann-Liouville, Caputo) system or H-system (RL-system, C-system).

It's known that optimal control problems in form of OCP A and OCP B can be reduced to the moment problem (4)-(5) in case of integer-order systems with lumped parameters [4]. It was shown also that the same is valid for Caputo systems [5, 6, 7] and for Riemann-Liouville systems [8]. Below we will demonstrate that for Hadamard systems the optimal control problem in form of OCP A and OCP B also can be reduced to the moment problem (4)-(5). We will consider the one-dimensional system of general view and special multi-dimensional system — N-fold integrator, corresponding to the system (1) at $a_{ij} = \delta_{i,j-1}$, $b_{ij} = \delta_{i,N}$, $f_i(t) = 0$.

3.2. The moment problem for one-dimensional system

The general solution of (1) in case of H-system at $N = 1$ can be represented by the following expression [9, p. 234] (subscripts are omitted):

$$q(t) = s^0\left(\ln\frac{t}{t_0}\right)^{\alpha-1} E_{\alpha,\alpha}\left[a\left(\ln\frac{t}{t_0}\right)^{\alpha}\right] + \tag{12}$$

$$+\int_{t_0}^{t}\left(\ln\frac{t}{\tau}\right)^{\alpha-1}E_{\alpha,\alpha}\left[a\left(\ln\frac{t}{\tau}\right)^{\alpha}\right]\frac{bu(\tau)+f(\tau)}{\tau}\,d\tau,$$

where $E_{\alpha,\alpha}(t)$ is two-parameter Mittag-Leffler function [9, §1.8]. Solution (12) at $t=T$ subject to the final condition (3) can be written in the form of expression (4),where

$$g(\tau)=\frac{b}{\tau}\left(\ln\frac{T}{\tau}\right)^{\alpha-1}E_{\alpha,\alpha}\left[a\left(\ln\frac{T}{\tau}\right)^{\alpha}\right], \tag{13}$$

$$c(T)=q^{T}-s^{0}\left(\ln\frac{T}{t_0}\right)^{\alpha-1}E_{\alpha,\alpha}\left[a\left(\ln\frac{T}{t_0}\right)^{\alpha}\right]- \tag{14}$$
$$-\int_{t_0}^{T}\left(\ln\frac{T}{\tau}\right)^{\alpha-1}E_{\alpha,\alpha}\left[a\left(\ln\frac{T}{\tau}\right)^{\alpha}\right]\frac{f(\tau)}{\tau}\,d\tau.$$

Theorem 1. Let us suppose $b\neq 0$ and $c\neq 0$. The moment problem (4) subject to (13) and (14) will be correct and solvable if and only if the following condition is valid:

$$\alpha>\frac{p'-1}{p'}. \tag{15}$$

Proof. Let us prove the correctness of moment problem correspondingly to the Def. 1. The moment c is non-zero by the data. For the norm of function $g(\tau)\in L_{p'}(t_0,T]$ subject to the (13) the following estimation will be valid [13]:

$$\|g(\tau)\|\leq|b|\cdot\left\|\frac{1}{\tau}\right\|\cdot\left\|\left(\ln\frac{T}{\tau}\right)^{\alpha-1}\right\|\cdot\left\|E_{\alpha,\alpha}\left[a\left(\ln\frac{T}{\tau}\right)^{\alpha}\right]\right\|. \tag{16}$$

We have $b\neq 0$ by the data. The norm of Mittag-Leffler function defined on half-interval $t\in(t_0,T]$ at $\alpha>0$ [9, p. 42]. The second multiplier in (16) can be represented by the following formulas:

$$\left\|\frac{1}{\tau}\right\|=\ln\frac{T}{t_0}$$

at $p'=1$,

$$\left\|\frac{1}{\tau}\right\|=\left[\frac{T^{1-p'}}{1-p'}\left(1-\left(\frac{t_0}{T}\right)^{1-p'}\right)\right]^{1/p'}$$

at $p'>1$. Consequently, the norm of function $\frac{1}{\tau}$ at $0<t_0<T$ also is fully defined (bounded and positive). Finally, the norm of third multiplier in (16) will be represented by the following expression:

$$\left\|\left(\ln\frac{T}{\tau}\right)^{\alpha-1}\right\|=\left(\int_{t_0}^{T}\left(\ln\frac{T}{\tau}\right)^{p'(\alpha-1)}d\tau\right)^{1/p'}.$$

Integral in the right side of the expression by substitution $\ln\frac{T}{\tau}=x$ reduces to the following view:

$$\int_{t_0}^{T}\left(\ln\frac{T}{\tau}\right)^{p'(\alpha-1)}d\tau=T\int_{0}^{\sigma}x^{p'(\alpha-1)}e^{-x}dx, \tag{17}$$

where $\sigma = \ln\left(\frac{T}{t_0}\right) > 0$. Convergence condition for integral in right side of (17) can be obtained from comparison criterion [20, Ch. 9]. Integrand in right side of (17) majorized, obviously, by function $x^{p'(\alpha-1)}$, which order of greatness equal to $p'(1-\alpha)$. Correspondingly to comparison criterion integral in right side of (17) will be converge if this order is less than 1. Consequently, we obtain the condition (15). The correctness in sense of Def. 1 is proven.

Now we will prove the solvability of the moment problem (4) subject to (13) and (14). Let us consider interpolation problem (7)-(8) for $N = 1$. One can express ξ from (8) and obtain the following unconditional minimization problem:

$$\left(\int_{t_0}^{T}\left|\frac{g(\tau)}{c}\right|^{p'}d\tau\right)^{1/p'} = \frac{1}{\lambda_N}. \tag{18}$$

Function $g(\tau)$ (in accordance with (13)) at $b \neq 0$ is positive definite function on half-interval $t \in (t_0, T]$, $0 < t_0 < T$. And $c \neq 0$ by the data. Consequently, left side in this expression is positively defined and $\lambda_N > 0$. So, the solvability condition is satisfied. Thus, theorem 1 is proven.

Remark 1. Condition (15) match with analogous conditions obtained for Caputo [5, 6, 7] and Riemann-Liouville [8] systems.

Let us consider now the special one-dimensional system — single integrator (it correspond to (1) at $a = 0$, $b = 1$, $f(t) = 0$). In this case solution of equation (1) can be obtained explicitly by its fractional integration:

$$q(t) = \frac{s^0}{\Gamma(\alpha)}\left(\ln\frac{t}{t_0}\right)^{\alpha-1} + \frac{1}{\Gamma(\alpha)}\int_{t_0}^{t}\left(\ln\frac{t}{\tau}\right)^{\alpha-1}\frac{u(\tau)}{\tau}d\tau. \tag{19}$$

When $t = T$ the expression (19) subject to final condition (3) represents the moment problem (4), where:

$$g(\tau) = \frac{1}{\Gamma(\alpha)}\left(\ln\frac{T}{\tau}\right)^{\alpha-1}\frac{1}{\tau}, \tag{20}$$

$$c = q^T - \frac{s^0}{\Gamma(\alpha)}\left(\ln\frac{T}{t_0}\right)^{\alpha-1}. \tag{21}$$

Note that expressions (20) and (21) can be derived from (13) and (14) correspondingly by passage to the limit at $a \to 0$.

3.3. The moment problem for N-fold integrator of fractional order

Let us consider system (1) at $a_{ij} = \delta_{i,j-1}$, $b_{ij} = \delta_{i,N}$, $f_i(t) = 0$:

$$\begin{cases} {}_{t_0}D_t^{\alpha_1} q_1(t) = q_2(t), \\ {}_{t_0}D_t^{\alpha_2} q_2(t) = q_3(t), \\ \dots \\ {}_{t_0}D_t^{\alpha_N} q_N(t) = u(t). \end{cases} \tag{22}$$

If we integrate each of equations in (22) in sense of Hadamard with corresponding index (α_i for i-th equation) then we will obtain the following expressions for phase coordinates:

$$\begin{cases} q_1(t) - \frac{s_1^0}{\Gamma(\alpha_1)}\left(\ln\frac{T}{t_0}\right)^{\alpha_1-1} = {}_{t_0}I_t^{\alpha_1} q_2(t), \\ q_2(t) - \frac{s_2^0}{\Gamma(\alpha_2)}\left(\ln\frac{T}{t_0}\right)^{\alpha_2-1} = {}_{t_0}I_t^{\alpha_2} q_3(t), \\ \dots \\ q_N(t) - \frac{s_N^0}{\Gamma(\alpha_N)}\left(\ln\frac{T}{t_0}\right)^{\alpha_N-1} = {}_{t_0}I_t^{\alpha_N} u(t). \end{cases}$$

Substituting sequentially phase coordinates from the last equation to penultimate one etc. and using the semigroup property of Hadamard fractional integration operators we will obtain:

$$\begin{cases} q_1(t) - \sum_{k=1}^{N} \frac{s_k^0}{\Gamma(\alpha_1+\ldots+\alpha_k)}\left(\ln\frac{T}{t_0}\right)^{\alpha_1+\ldots+\alpha_k-1} = {}_{t_0}I_t^{\alpha_1+\ldots+\alpha_N} u(t), \\ q_2(t) - \sum_{k=2}^{N} \frac{s_k^0}{\Gamma(\alpha_2+\ldots+\alpha_k)}\left(\ln\frac{T}{t_0}\right)^{\alpha_2+\ldots+\alpha_k-1} = {}_{t_0}I_t^{\alpha_2+\ldots+\alpha_N} u(t), \\ \dots \\ q_N(t) - \frac{s_N^0}{\Gamma(\alpha_N)}\left(\ln\frac{T}{t_0}\right)^{\alpha_N-1} = {}_{t_0}I_t^{\alpha_N} u(t) \end{cases} \tag{23}$$

System (23) at $t = T$ subject to final conditions (3) represent the N-dimensional moment problem (4), where the following expressions are valid:

$$g_k(\tau) = \frac{1}{\Gamma(\alpha_k+\ldots+\alpha_N)}\left(\ln\frac{T}{\tau}\right)^{\alpha_k+\ldots+\alpha_N-1}\frac{1}{\tau}, \tag{24}$$

$$c_k = q_k^T - \sum_{l=k}^{N}\frac{s_l^0}{\Gamma(\alpha_l+\ldots+\alpha_N)}\left(\ln\frac{T}{t_0}\right)^{\alpha_l+\ldots+\alpha_N-1}. \tag{25}$$

Theorem 2. Let us suppose that at least one of moments c_i is non-zero. Then the moment problem (4) subject to (24), (25) will be correct and solvable if and only if the following condition is valid:

$$\alpha_N > \frac{p'-1}{p'}. \tag{26}$$

Proof. Correctness of moment problem (4) for system (23) can be proven analogously to the Theorem 1. And condition (26) will be caused by existence of norm of function $g_N(\tau)$. Existence of norm for other functions $g_i(\tau)$, $i = 1, \ldots, N-1$ caused by the following conditions:

$$\alpha_k + \ldots + \alpha_N > \frac{p'-1}{p'}.$$

These conditions will be satisfied if condition (26) is valid since $\alpha_i > 0$ by the data.

Solvability of the moment problem (4) subject to (24), (25) caused by linear independency of functions $g_i(\tau)$. Thus, the theorem is proven.

Remark 2. Condition (26) match with analogous conditions obtained later for Caputo [5, 7] and Riemann-Liouville [8] systems.

4. Moment problem solution and optimal control properties

4.1. One-dimensional system

In this case we can solve the problem (7)-(8) explicitly and construct the optimal control for OCP A and OCP B.

Theorem 3. Let the one-dimensional system (1) given with initial and final conditions (2) and (3). Let $b \neq 0$, $c \neq 0$ (where moment c defined by (14)) and condition (15) satisfied (i.e., moment problem (4)-(5) subject to (13)-(14) is correct and solvable). Then:

IOP Conf. Series: Journal of Physics: Conf. Series **918** (2017) 012026 doi:10.1088/1742-6596/918/1/012026

(i) OCP A solution at $p \to \infty$ is given by the following expression:

$$u(t) = \frac{c}{b} \frac{a}{E_\alpha \left[a \left(\ln \frac{T}{t_0} \right)^\alpha \right] - 1}, \tag{27}$$

where $\alpha > 0$, $t \in (t_0, T]$, $E_\alpha(t) = E_{\alpha,1}(t)$ — one-parameter Mittag-Leffler function;

(ii) OCP A solution at $1 < p < \infty$ is given by the following expression:

$$u(t) = \frac{c}{bK_1^{p'}(t_0, T)} \left[\frac{1}{t} \left(\ln \frac{T}{t} \right)^{\alpha - 1} E_{\alpha,\alpha} \left[a \left(\ln \frac{T}{t} \right)^\alpha \right] \right]^{p'-1}, \tag{28}$$

where $\alpha > 1/p$, $t \in (t_0, T]$, $K_1(t_0, T) = \left(\int\limits_{t_0}^{T} \left| \frac{E_{\alpha,\alpha}\left[a\left(\ln \frac{T}{\tau}\right)^\alpha\right]}{\tau\left(\ln \frac{T}{\tau}\right)^{1-\alpha}} \right|^{p'} d\tau \right)^{1/p'}$;

(iii) OCP B solution at $p \to \infty$ is given by the following expression:

$$u(t) = l \operatorname{sign} \left(\frac{b}{c} \right), \tag{29}$$

where $\alpha > 0$, $t \in (t_0, T^*]$ and T^* can be found from equation $\left|\frac{c}{b}\right| \frac{a}{E_\alpha\left[a\left(\ln \frac{T}{t_0}\right)^\alpha\right] - 1} = l$;

(iv) OCP B solution at $1 < p < \infty$ is given by the following expression:

$$u(t) = l \left| \frac{b}{ct} \left(\ln \frac{T}{t} \right)^{\alpha-1} E_{\alpha,\alpha} \left[a \left(\ln \frac{T}{t} \right)^\alpha \right] \right|^{p'-1} \operatorname{sign} \left(\frac{b}{c} \right), \tag{30}$$

where $\alpha > 1/p$, $t \in (t_0, T^*]$ and T^* can be found from equation $\left|\frac{c}{b}\right| \frac{1}{K_1(t_0,T)} = l$.

Proof. Let us consider the case $p \to \infty$ and solve the minimization problem (7)-(8) subject to (13)-(14). As we have shown above the problem (7)-(8) reduces to unconditional minimization problem (18). Taking into account (13)-(14) and $p' = 1$ we will obtain:

$$\frac{1}{\lambda} = \int\limits_{t_0}^{T} \left| \frac{bE_{\alpha,\alpha}\left[a\left(\ln \frac{T}{\tau}\right)^\alpha\right]}{c\tau\left(\ln \frac{T}{\tau}\right)^{1-\alpha}} \right| d\tau. \tag{31}$$

In order to calculate integral in (31) we will use Mittag-Leffler function representation by power series [9, p. 42]:

$$E_{\alpha,\beta}(t) = \sum_{k=0}^{\infty} \frac{t^k}{\Gamma(\alpha k + \beta)}. \tag{32}$$

This series converge absolutely on number axis at $\alpha > 0$ and $\beta > 0$ [9, p. 42]. Then we can substitute representation (32) into (31), perform the termwise integration and summarize resulting series:

$$\frac{1}{\lambda} = \int\limits_{t_0}^{T} \left| \frac{bE_{\alpha,\alpha}\left[a\left(\ln \frac{T}{\tau}\right)^\alpha\right]}{c\tau\left(\ln \frac{T}{\tau}\right)^{1-\alpha}} \right| d\tau = \left|\frac{b}{c}\right| \sum_{k=0}^{\infty} \int\limits_{t_0}^{T} \frac{a^k\left(\ln \frac{T}{\tau}\right)^{\alpha k + \alpha - 1}}{\Gamma(\alpha k + \alpha)} \frac{d\tau}{\tau} =$$

$$= \frac{1}{a} \left|\frac{b}{c}\right| \sum_{k=0}^{\infty} \frac{\left[a\left(\ln \frac{T}{t_0}\right)^\alpha\right]^{k+1}}{\Gamma[\alpha(k+1)+1]} = \frac{1}{a}\left|\frac{b}{c}\right| \left(E_\alpha \left[a \left(\ln \frac{T}{t_0} \right)^\alpha \right] - 1 \right).$$

Expressing λ from the last formula we will obtain:

$$\lambda = \left|\frac{c}{b}\right| \frac{a}{E_\alpha\left[a\left(\ln\frac{T}{t_0}\right)^\alpha\right]-1}. \tag{33}$$

Substituting expression (33) to (9) we will obtain formula (27). OCP B solution (29) in this case result from (10) and (11) subject to (33).

Let us consider now the case $1 < p < \infty$. Obviously, the expression (18) subject to (13)-(14) could be rewritten as the following formula:

$$\lambda = \left|\frac{c}{b}\right| \frac{1}{K_1(t_0,T)}. \tag{34}$$

Substituting formula (34) into expression (9) we will obtain OCP A solution (28). From (10) and (11) subject to (34) one can obtain OCP B solution (30). Thus, the theorem is proven.

Remark 3. Optimal controls (27) and (29) have no switching points on considered time half-interval, which corresponds to corollary from Feldbaum's interval theorem [16] for integer-order system of order $[\alpha]+1$.

For single integrator analogous solutions for OCP A and OCP B can be obtained by direct solution of problem (7)-(8) subject to (20), (21) or by passage of expressions (27), (28), (29) and (30) to the limit at $a \to 0$. So, in case of $p \to \infty$ we can obtain the following expression for λ:

$$\lambda = \frac{|c|\Gamma(\alpha+1)}{\left(\ln\frac{T}{t_0}\right)^\alpha}. \tag{35}$$

OCP A solution for studied system will be written by formula:

$$u(t) = \frac{c\Gamma(\alpha+1)}{\left(\ln\frac{T}{t_0}\right)^\alpha}. \tag{36}$$

Optimal control in case of OCP B will be described by formula (10), where minimal time T^* can be found (in general case numerically) from (11) subject to (35). In case of $q^T = 0$ this equation can be solved explicitly and the following formula will take place:

$$T^* = t_0 \exp\left(\frac{|s^0|\alpha}{l}\right). \tag{37}$$

Analogously, we can find for single integrator at $1 < p < \infty$:

$$\lambda = \frac{|c|\Gamma(\alpha)}{K_2(t_0,T)}, \tag{38}$$

where $K_2(t_0,T) = \left(\int\limits_{t_0}^{T}\left|\frac{1}{\tau\left(\ln\frac{T}{\tau}\right)^{1-\alpha}}\right|^{p'} d\tau\right)^{1/p'}$. Solutions of OCP A and OCP B can be found from (9), (10) and (11) subject to (38).

Now we will illustrate the results obtained above. Calculations and visualization of the results carried out in MatLab 7.9. A special module was used for Mittag-Leffler function calculation [17]. Numerical integration was performed by Gauss-Kronrod method [18].

Firstly we will illustrate the case $p \to \infty$ ($p' = 1$ correspondingly). At fig. 1 and 2 the dependencies represented of control norm from differentiation index α, which calculated using

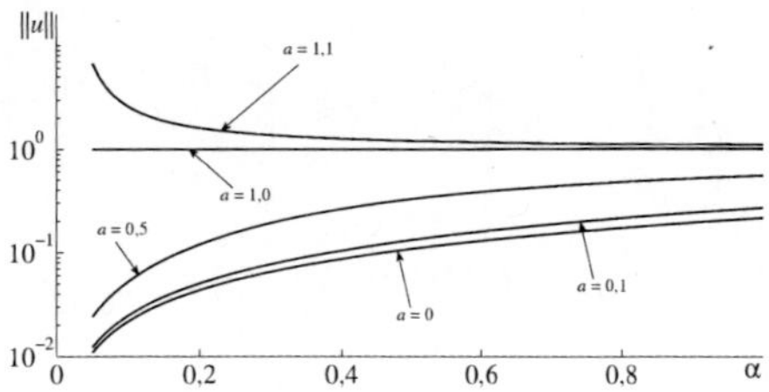

Figure 1. The dependency of control norm from differentiation index α at $q^T = 0$. Logarithmic scale used for ordinates.

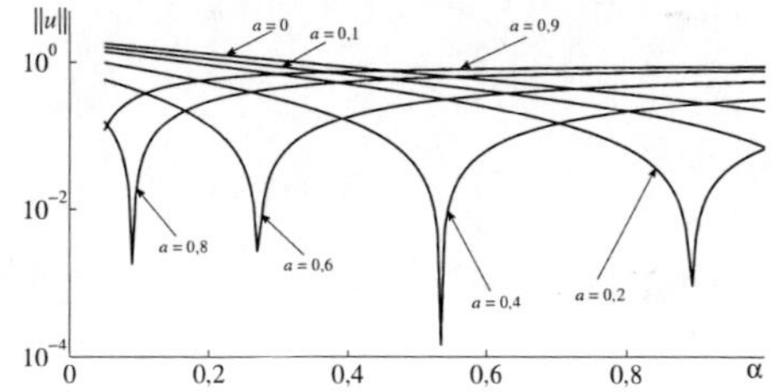

Figure 2. The dependency of control norm from differentiation index α at $q^T = 2$. Logarithmic scale used for ordinates.

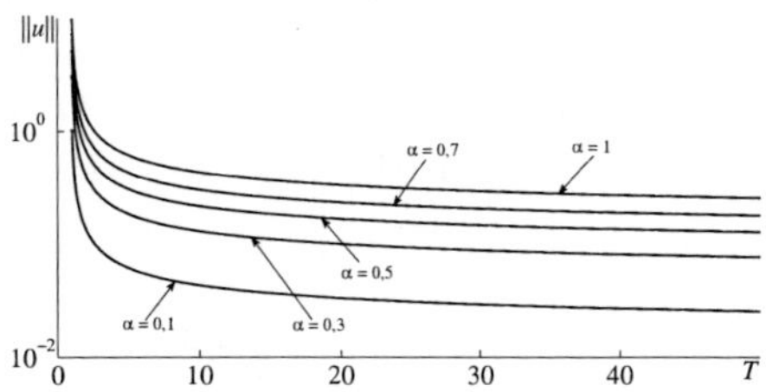

Figure 3. The dependency of control norm from control time T at $p' = 1$, $q^T = 0$ and several values of index α. Logarithmic scale used for ordinates.

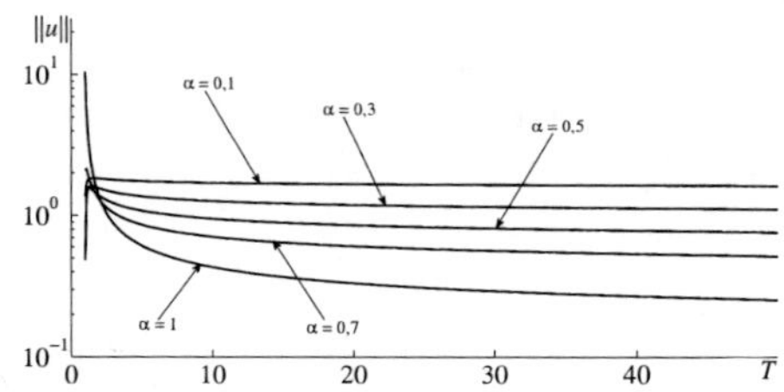

Figure 4. The dependency of control norm from control time T at $p' = 1$, $q^T = 2$ and several values of index α. Logarithmic scale used for ordinates.

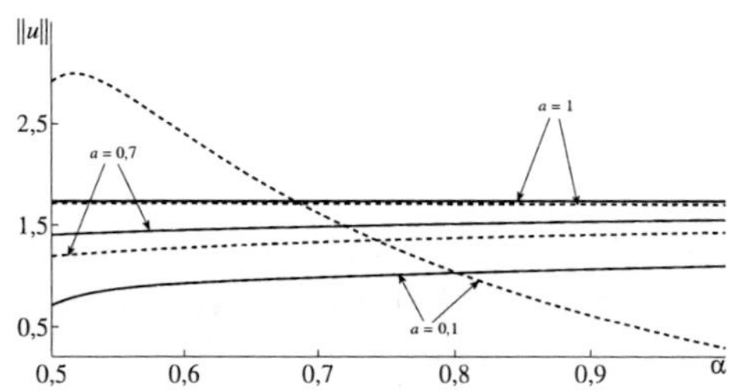

Figure 5. The dependency of control norm from differentiation index α at $p = p' = 2$ and different q^T: $q^T = 0$ (solid lines), $q^T = 2$ (dashed lines).

formulas (33) and (35) for one-dimensional system (1) at different a and for single integrator ($a = 0$ correspondingly). The dependencies obtained for different final coordinates q^T: $q^T = 0$ (fig. 1) and $q^T = 2$ (fig. 2). Other parameters were fixed as follows: $q^0 = 1$, $t_0 = 1$, $T = 100$, $b = 1$. As it seen on fig. 1 the curves differs qualitatively at $a \leq 1$ and at $a > 1$: in first case the monotonic growth obtained (which expressed more slightly in case of $a = 1$, than in case of $a < 1$), and in second case the curve monotonically decreasing. This tendency take place at $a \geq 1$ for $q^T = 2$ (fig. 2), but in this case behaviour of curves changes dramatically at $a < 1$: its demonstrate extremal character at $0.1 < a < 0.9$ and decrease monotonically at $a \leq 0.1$, and monotonic growth present only at $a \geq 0.9$. Figs. 3 and 4 represent the dependencies of control norm from control time T for single integrator at severla fixed values of α. It's seen that,

IOP Conf. Series: Journal of Physics: Conf. Series **918** (2017) 012026 doi:10.1088/1742-6596/918/1/012026

in general, the curves decreasing but in case of $q^T = 2$ at small T the maximum appear and norm behaviour changes in dependence from α: at $q^T = 0$ the norm increases and at $q^T = 2$ it decreases.

At fig. 5 analogical results for control norm dependency from index α represented for $1 < p < \infty$. It's seen that the majority of curves monotonically increasing although at $q^T = 2$ for $a = 0.1$ slight extremum appears at small α.

4.2. Double integrator

Double integrator is represented by system (22) at $N = 2$. Let us consider the case $p \to \infty$ and suppose that at least one of the moments c_1, c_2 (defined by (25)) is non-zero. Then the moment problem is correct and solvable for $\alpha_1 > 0$, $\alpha_2 > 0$ in accordance with Theorem 2. Using expressions (24) at $k = 1, 2$ one can study the problem (7)-(8) and reduce it to the problem of unconditional minimization of the following one-variable function:

$$F(\xi_2) = \frac{1 - \xi_2 c_2}{c_1 \Gamma(\alpha_1 + \alpha_2 + 1)} \left[\left(\ln \frac{T}{t_0} \right)^{\alpha_1 + \alpha_2} - 2 \left(\frac{\Gamma(\alpha_1 + \alpha_2)}{\Gamma(\alpha_2)} \frac{\xi_2 c_1}{\xi_2 c_2 - 1} \right)^{\frac{\alpha_1 + \alpha_2}{\alpha_1}} \right] +$$
$$+ \frac{\xi_2}{\Gamma(\alpha_2 + 1)} \left[\left(\ln \frac{T}{t_0} \right)^{\alpha_2} - 2 \left(\frac{\Gamma(\alpha_1 + \alpha_2)}{\Gamma(\alpha_2)} \frac{\xi_2 c_1}{\xi_2 c_2 - 1} \right)^{\frac{\alpha_2}{\alpha_1}} \right].$$

We can calculate the first derivative of this function and demonstrate that its extrema are roots of the following equation:

$$\left(\ln \frac{T}{t_0} \right)^{\alpha_2} \left[\frac{1}{\Gamma(\alpha_2 + 1)} - \frac{c_2}{c_1} \frac{\left(\ln \frac{T}{t_0} \right)^{\alpha_1}}{\Gamma(\alpha_1 + \alpha_2 + 1)} \right] -$$
$$- \frac{2}{\Gamma(\alpha_2)} \left(\frac{\Gamma(\alpha_1 + \alpha_2)}{\Gamma(\alpha_2)} \frac{\xi_2 c_1}{\xi_2 c_2 - 1} \right)^{\frac{\alpha_2}{\alpha_1}} \left[\frac{1}{\alpha_2} - \frac{1}{\alpha_1 + \alpha_2} \frac{\xi_2 c_2}{\xi_2 c_2 - 1} \right] = 0. \quad (39)$$

Equation (39) at arbitrary values of indices α_1 and α_2 have no explicit solution analogously to the same cases for Caputo [5, 7] and Riemann-Liouville [8] double integrators. But explicit solution of this equation can be obtained in some particular cases, for example, in case of $c_2 = 0$. Then we can solve equation (39) analytically and to find ξ_2^*:

$$\xi_2^* = - \frac{2^{-\frac{\alpha_1}{\alpha_2}}}{c_1} \frac{\Gamma(\alpha_2)}{\Gamma(\alpha_1 + \alpha_2)} \left(\ln \frac{T}{t_0} \right)_1^{\alpha}. \quad (40)$$

From (8) at $c_2 = 0$ we can find:

$$\xi_1 = \frac{1}{c_1}. \quad (41)$$

Using (41), (40) and (24) at $k = 1, 2$ one can obtain:

$$\xi_1^* g_1(t) + \xi_2^* g_2(t) = \frac{1}{c_1 t} \frac{\left(\ln \frac{T}{t} \right)^{\alpha_2 - 1}}{\Gamma(\alpha_1 + \alpha_2)} \left[\left(\ln \frac{T}{t} \right)_1^{\alpha} - 2^{-\frac{\alpha_1}{\alpha_2}} \left(\ln \frac{T}{t_0} \right)_1^{\alpha} \right]. \quad (42)$$

Function (42) reverse sign at the following point:

$$t' = T \left(\frac{T}{t_0} \right)^{-2^{-\frac{1}{\alpha_2}}}. \quad (43)$$

Substituting (42) into (7) and using (43) one can calculate:

$$\lambda_2 = \frac{\Gamma(\alpha_1+\alpha_2+1)}{1-2^{-\frac{\alpha_1}{\alpha_2}}}\frac{|c_1|}{\left(\ln\frac{T}{t_0}\right)^{\alpha_1+\alpha_2}}. \tag{44}$$

OCP A solution result from (9) subject to (42) and (44):

$$u(t) = \frac{\Gamma(\alpha_1+\alpha_2+1)}{1-2^{-\frac{\alpha_1}{\alpha_2}}}\frac{c_1}{\left(\ln\frac{T}{t_0}\right)^{\alpha_1+\alpha_2}}(\text{sign})\left[\left(\ln\frac{\mathrm{T}}{\mathrm{t}}\right)_1^{\alpha} - 2^{-\frac{\alpha_1}{\alpha_2}}\left(\ln\frac{\mathrm{T}}{\mathrm{t}_0}\right)_1^{\alpha}\right]. \tag{45}$$

OCP B solution result from (10) subject to (42):

$$u(t) = l\text{sign}\left[\frac{\left[\left(\ln\frac{T^*}{t}\right)_1^{\alpha} - 2^{-\frac{\alpha_1}{\alpha_2}}\left(\ln\frac{T^*}{t_0}\right)_1^{\alpha}\right]}{c_1}\right], \tag{46}$$

where T^* can be found numerically from (11) subject to (44). In some special cases one can obtain an explicit solution, for example, in case of $s_2^0 = q_1^T = 0$ we will have:

$$T^* = t_0 \exp\left(\frac{|s_1^0|\Gamma(\alpha_1+\alpha_2+1)}{l\Gamma(\alpha_1)\left(1-2^{-\frac{\alpha_1}{\alpha_2}}\right)}\right)^{\frac{1}{\alpha_2+1}}. \tag{47}$$

Remark 4. Optimal controls (45) and (46) have one switching point, which corresponds to the corollary from Feldbaum's interval theorem [16] for integer-order system of order $[\alpha_1]+1+[\alpha_2]+1$.

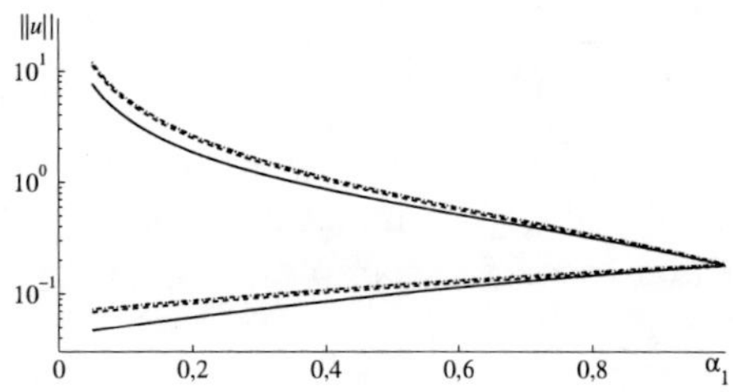

Figure 6. The dependency of control norm from index α_1 at α_2 fixed for double integrator at $q_1^T = 0$ (three upper curves) and at $q_1^T = 2$ (three lower curves). Dependencies calculated for several values of index α_2: $\alpha_2 = 0.2$ (solid lines), $\alpha_2 = 0.5$ (dashed lines) and $\alpha_2 = 0.9$ (dash-dotted lines). Logarithmic scale used for ordinates.

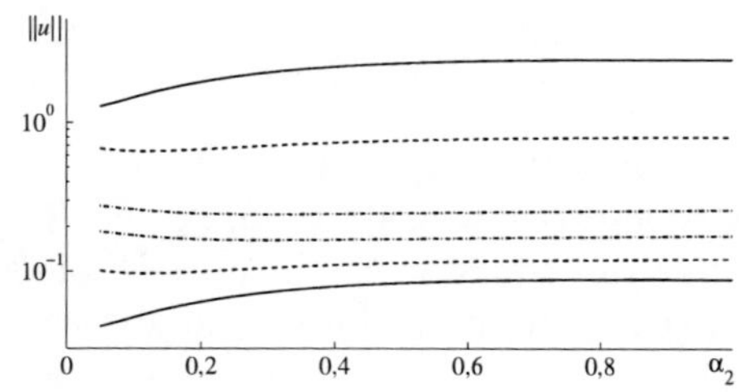

Figure 7. The dependency of control norm from index α_2 at α_1 fixed for double integrator at $q_1^T = 0$ (three upper curves) and at $q_1^T = 2$ (three lower curves). Dependencies calculated for several values of index α_1: $\alpha_1 = 0.2$ (solid lines), $\alpha_1 = 0.5$ (dashed lines) and $\alpha_1 = 0.9$ (dash-dotted lines). Logarithmic scale used for ordinates.

Now we will illustrate results obtained in this subsection. At fig. (6) and (7) represented dependencies of control norm (44) from each of indices α_1 and α_2 at other index fixed. It's seen that dependency from α_1 is more expressive. It's also demonstrated that different values q^T lead to qualitatively different dependencies. At fig. 8 shown dependencies of minimal control time T^* from index α_1 at α_2 fixed, which demonstrate increasing character. Dependencies of T^* from index α_2 at α_1 fixed are analogous to the curves at fig. 8.

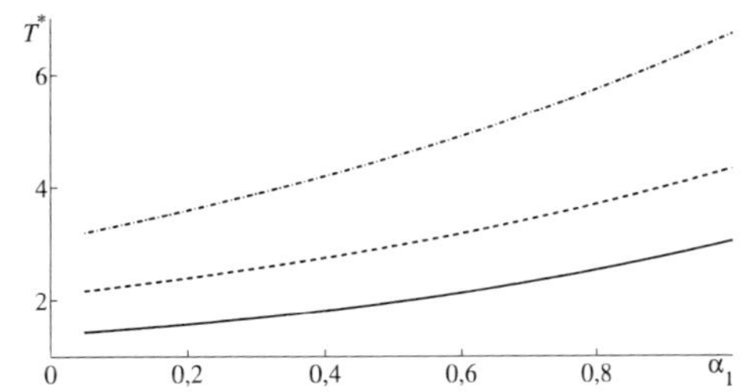

Figure 8. The dependency of minimal control time from index α_1 at α_2 fixed for double integrator at $q_1^T = 0$. Dependencies calculated for several values of index α_2: $\alpha_2 = 0.2$ (solid lines), $\alpha_2 = 0.5$ (dashed lines) and $\alpha_2 = 0.9$ (dash-dotted lines).

5. Some features of phase trajectories of Hadamard double integrator

In this section we continue to study Hadamard double integrator with control $u(t) \in L_\infty(t_0, T]$ and focus on two basic aspects: boundary trajectories calculation and calculation of system phase trajectories in optimal control mode.

Definition 2. We will calling boundary trajectories the system phase trajectories, which correspond to boundary values of control $u(t) = \pm l$.

Remark 5. In case of $\alpha_1 = \alpha_2 = 1$ the boundary trajectories are identical to the boundaries of integral vortex of differential inclusion [19], which correspond to considered system. Boundary trajectories bound the phase plane region containing the all admissible trajectories of considered system.

To find boundary trajectories we suppose $u(t) = \pm l$ in the system (23) at $N = 2$. After calculations of corresponding integrals we obtain:

$$\begin{cases} q_1^{\pm l}(t) = \frac{s_1^0}{\Gamma(\alpha_1)}\left(\ln\frac{t}{t_0}\right)^{\alpha_1-1} + \frac{s_2^0}{\Gamma(\alpha_1+\alpha_2)}\left(\ln\frac{t}{t_0}\right)^{\alpha_1+\alpha_2-1} \pm \\ \qquad \pm\frac{l}{\Gamma(\alpha_1+\alpha_2+1)}\left(\ln\frac{t}{t_0}\right)^{\alpha_1+\alpha_2}, \\ q_2^{\pm l}(t) = \frac{s_2^0}{\Gamma(\alpha_2)}\left(\ln\frac{t}{t_0}\right)^{\alpha_2-1} \pm \frac{l}{\Gamma(\alpha_2+1)}\left(\ln\frac{t}{t_0}\right)^{\alpha_2}. \end{cases} \tag{48}$$

In general case we can't eliminate time from equations (48) and obtain obvious dependence $q_2^{\pm l}(q_1^{\pm l})$, although it's possible to make in some special cases. For example, at $s_2^0 = 0$ we can express time from the second equation in (48) and , substituting it to the first equation, obtain finally:

$$q_1^{\pm l} = \frac{s_1^0}{\Gamma(\alpha_1)}\left(\frac{\Gamma(\alpha_2+1)}{\pm l}q_2^{\pm l}\right)^{\frac{\alpha_1-1}{\alpha_2}} \pm \frac{l}{\Gamma(\alpha_1+\alpha_2+1)}\left(\frac{\Gamma(\alpha_2+1)}{\pm l}q_2^{\pm l}\right)^{\frac{\alpha_1}{\alpha_2}+1}.$$

In case of $\alpha_2 = 1$ one can analogously obtain the following expression:

$$q_1^{\pm l} = \frac{s_1^0}{\Gamma(\alpha_1)}\left(\frac{q_2^{\pm l} - s_2^0}{\pm l}\right)^{\alpha_1-1} + \frac{s_2^0}{\Gamma(\alpha_1+1)}\left(\frac{q_2^{\pm l} - s_2^0}{\pm l}\right)^{\alpha_1} \pm \\ \pm\frac{l}{\Gamma(\alpha_1+2)}\left(\frac{q_2^{\pm l} - s_2^0}{\pm l}\right)^{\alpha_1+1}.$$

At fig. 9 and 10 some examples of boundary trajectories shown, which correspond to the following initial states: $s_1^0 = 1$, $s_2^0 = 0$. Other parameters were fixed as follows: $t_0 = 1$, $T = 100$, $l = 1$.

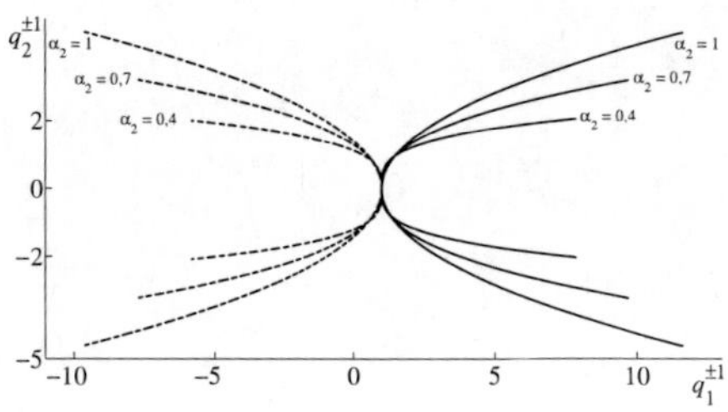

Figure 9. Boundary trajectories of Hadamard double integrator at $\alpha_1 = 1$ and several fixed values of α_2. Solid lines correspond to $l = 1$, dashed lines — $l = -1$.

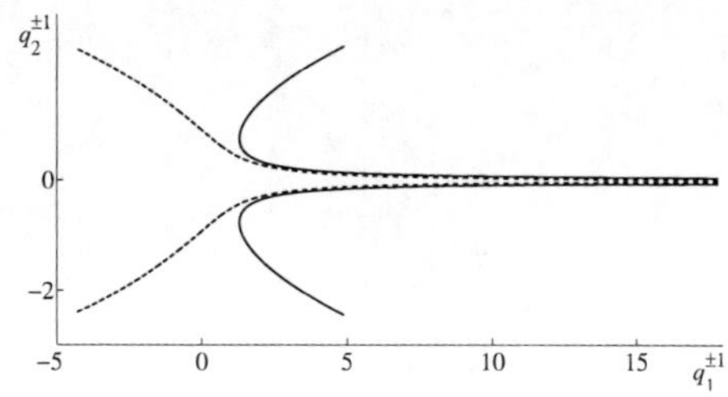

Figure 10. Boundary trajectories of Hadamard double integrator at $\alpha_1 = \alpha_2 = 0.5$. Solid lines correspond to $l = 1$, dashed lines — $l = -1$.

It's seen from fig. 9 that in case of $\alpha_1 = 1$ boundary trajectories are similar to the same for double integrator of integer order [19], Caputo double integrator [5, 7] and Riemann-Liouville double integrator [8]. At $\alpha_2 = 1$ these trajectories bound maximal area on phase plane which decrease with $alpha_2$ decreasing. Boundary trajectories at $\alpha_1 < 1$ (fig. 10) differs qualitatively from the same for integer-order double integrator, Riemann-Liouville and Caputo double integrators. The regions between solid and dashed curves in upper and lower half-planes at fig. 9 and 10 are inaccesible for the system. And decreasing of α_2 corresponds to decreasing of area, bounded by boundary trajectories. Note also that at fig. 10 in the region bounded by dashed lines phase trajectories localized either in upper half-plane, or in lower half-plane.

Let us calculate now phase trajectories of Hadamard double integrator in optimal control mode. Substituting control (45) in system (23) at $N = 2$ we can derive:

$$\begin{cases} q_1(t) = \frac{s_1^0}{\Gamma(\alpha_1)}\left(\ln\frac{t}{t_0}\right)^{\alpha_1-1} + \frac{s_2^0}{\Gamma(\alpha_1+\alpha_2)}\left(\ln\frac{t}{t_0}\right)^{\alpha_1+\alpha_2-1} + \\ \quad + \frac{c_1}{1-2^{-\frac{\alpha_1}{\alpha_2}}}\frac{1}{\left(\ln\frac{T}{t_0}\right)^{\alpha_1+\alpha_2}}\times \\ \quad \times\left[\left(\ln\frac{t}{t_0}\right)^{\alpha_1+\alpha_2} - 2\left(\ln\frac{t}{t'}\right)^{\alpha_1+\alpha_2}\Theta(t-t')\right], \\ q_2(t) = \frac{s_2^0}{\Gamma(\alpha_2)}\left(\ln\frac{t}{t_0}\right)^{\alpha_2-1} + \\ \quad + \frac{c_1}{1-2^{-\frac{\alpha_1}{\alpha_2}}}\frac{1}{\left(\ln\frac{T}{t_0}\right)^{\alpha_1+\alpha_2}}\frac{\Gamma(\alpha_1+\alpha_2+1)}{\Gamma(\alpha_2+1)}\times \\ \quad \times\left[\left(\ln\frac{t}{t_0}\right)^{\alpha_2} - 2\left(\ln\frac{t}{t'}\right)^{\alpha_2}\Theta(t-t')\right], \end{cases} \tag{49}$$

where $\Theta(t - t')$ is the Heaviside function, t' — control switching point (43). It can be verified directly that formulas (49) satisfy final conditions (3) (taking into account $c_2 = 0$).

At fig. 11 and 12 represented some examples of phase trajectories, calculated using formulas (49). Initial state values and other parameters fixed identically to trajectories calculation for fig. 9 and 10. Final states fixed as follows: $q_1^T = q_2^T = 0$. It's seen that at $\alpha_1 = 1$ (fig. 11) decreasing of index α_2 lead to switching point shift to final state. In case of $\alpha_1 < 1$ (fig. 12) the overcontrol effect take place: phase trajectory go beyond final point, switches and return to final point. This behaviour is similar to the same for Caputo [5, 7] and Riemann-Liouville [8] double integrators.

6. Discussion

In this section we will compare obtained results with the same for integer-order systems and Caputo and Riemann-Liouville systems.

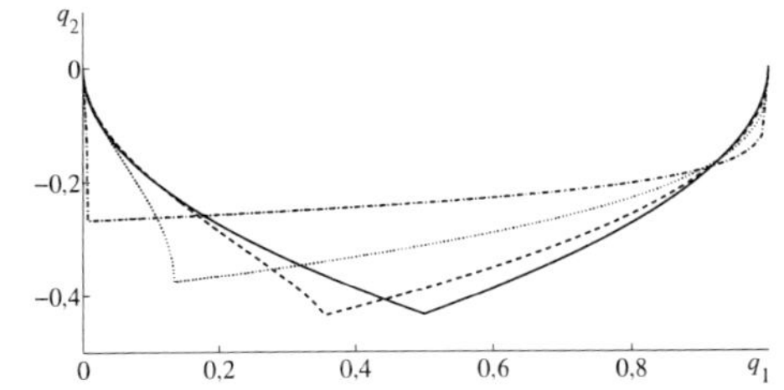

Figure 11. Phase trajectories of Hadamard double integrator in optimal control mode at $\alpha_1 = 1$ and several fixed values of α_2: $\alpha_2 = 0.2$ (dot-dashed line), $\alpha_2 = 0.5$ (dotted line), $\alpha_2 = 0.8$ (dashed line) and $\alpha_2 = 1$ (solid line).

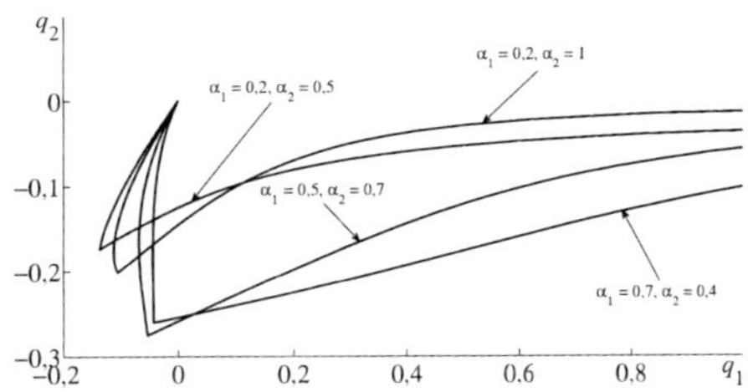

Figure 12. Phase trajectories of Hadamard double integrator in optimal control mode at some fixed values of indices α_1 and α_2.

6.1. Optimal control behaviour at integer values of differentiation indices

Note that for Hadamard systems the functions $g_i(t)$, obtained in this paper and defined by (13), (20) and (24), at $\alpha_1 = 1$, $\alpha_2 = 1$ doesn't reduce to analogical functions for integer-order systems. Analogous situation take place for moments, defined by (14), (21) and (25) (saving moments c_N in (25)). This fact is qualitative difference between Hadamard systems and Caputo and Riemann-Liouville systems, illustrated below by example.

Let us consider single integrator. If in this case the left side of equation (1) contains only classic differentiation operator of first order, then its solution subject to initial condition (2) can be written as follows:

$$q(t) = s^0 + \int_{t_0}^{t} u(\tau) d\tau. \tag{50}$$

Solution (19) for Hadamard single integrator reduces at $\alpha = 1$ to the following expression:

$$q(t) = s^0 + \int_{t_0}^{t} \frac{u(\tau)}{\tau} d\tau. \tag{51}$$

At $t = T$ both of solutions (50) and (51) can be rewritten as a moment problem (4) with the same moments but different functions $g(\tau)$. So, motion laws for ordinary single integrator and Hadamard single integrator of first order will differ from each other. It's shown at fig. 13 for boundary values of control $u(t) = \pm l$.

It's known [4, Ch. 2] that in case of $u(t) \in L_\infty(t_0, T)$ OCP A solution for ordinary single integrator can be expressed by the following formula (here and below it's taking into account that for considered in this paper systems initial conditions established not at $t = 0$ but at $t = t_0 > 0$):

$$u^1(t) = \frac{q^T - s^0}{T - t_0}, t \in (t_0, T]. \tag{52}$$

OCP B solution for this case represented by expressions [4, Ch. 2]:

$$u^1(t) = l \text{sign}(q^T - s^0), t \in (t_0, T^*], \tag{53}$$

$$T_1^* = t_0 + \frac{|q^T - q^0|}{l}. \tag{54}$$

IOP Conf. Series: Journal of Physics: Conf. Series **918** (2017) 012026 doi:10.1088/1742-6596/918/1/012026

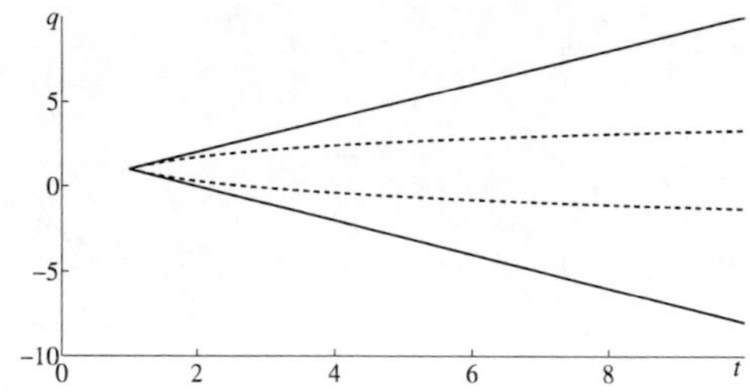

Figure 13. Motion laws for ordinary single integrator (solid lines) and for Hadamard single integrator of first order (dashed lines) at $u(t) = \pm l$. Other parameters were fixed as follows: $t_0 = 1$, $T = 10$, $l = 1$, $s^0 = 1$.

If we fix $\alpha = 1$ at expression (36) subject (21) then we can obtain the following formula, which obviously differs from (52):

$$u(t) = \frac{q^T - s^0}{\ln \frac{T}{t_0}}. \tag{55}$$

Expression (10) subject to (21) at $\alpha = 1$ is identical to expression (53) but expression (37) reduces to the following formula, which didn't match with (54):

$$T^* = t_0 \exp\left(\frac{|s^0|}{l}\right). \tag{56}$$

Let us consider the classical double integrator of first order. At arbitrary initial and final conditions solution of conditional minimization problem (7)-(8) lead to the following quadratic equation for ξ_2 [4, Ch. 2]:

$$(c_1 + Ac_2)c_2\xi_2^2 - 2(c_1 + Ac_2)\xi_2 + A = 0, \tag{57}$$

where $A = A^{cl} = \left(\frac{c_2}{c_1}\frac{T-t_0}{2} - 1\right)(T - t_0)$. Equation (39) at $\alpha_1 = \alpha_2 = 1$ reduces to equation (57) with another A: $A = A^H = \left(\frac{c_2}{2c_1}\ln\frac{T}{t_0} - 1\right)\ln\frac{T}{t_0}$.

OCP A solution for double integrator of first order at $s_2^0 = q_2^T = 0$ is given by the following formula [4, Ch. 2]:

$$u^2(t) = \frac{4(q_1^T - s_1^0)}{(T - t_0)^2}\text{sign}\left(\frac{T - t_0}{2} - \frac{t - t_0}{2}\right). \tag{58}$$

OCP A solution (45) for Hadamard double integrator subject to $s_2^0 = q_2^T = 0$, $\alpha_1 = \alpha_2 = 1$ and (25) at $k = 1, 2$ reduces to the following expression:

$$u^2(t) = \frac{4(q_1^T - s_1^0)}{\left(\ln\frac{T}{t_0}\right)^2}\text{sign}\left(\ln\frac{T}{t} - \frac{1}{2}\ln\frac{T}{t_0}\right). \tag{59}$$

OCP B solution for classical double integrator of first order at $q_1^T = 0$ and $s_1^0 > 0$ given by the following formulas [4, Ch. 2]:

$$u^2(t) = l\text{sign}\left(\frac{t - t_0}{s_1^0} - \frac{1}{\sqrt{ls_1^0}}\right), t \in (t_0, T_2^*], \tag{60}$$

$$T_2^* = t_0 + 2\sqrt{\frac{s_1^0}{l}}. \tag{61}$$

At the same assumptions formulas (46) and (47) for Hadamard double integrator at $\alpha_1 = \alpha_2 = 1$ reduces to the following expressions:

$$u(t) = l\text{sign}\left(\frac{1}{s_1^0}\ln\frac{t}{t_0} - \frac{1}{\sqrt{ls_1^0}}\right), \tag{62}$$

$$T^* = t_0 \exp\left(\frac{4|s_1^0|}{l}\right)^{\frac{1}{2}}. \tag{63}$$

Thus, we have demonstrated in detail that for Hadamard systems solutions of optimal control problem and motion laws doesn't reduce to the same for integer-order systems when differentiation indices fixed to be equal to 1. It differs qualitatively Hadamard systems from analogous Caputo and Riemann-Liouville systems. And there is one interesting feature take place: all of the formulas obtained for Hadamard systems reduce to the same for corresponding integer-order systems when we change the value $T - t_0$ or function $t - \tau$ by the value $\ln\frac{T}{t_0}$ or function $\ln\frac{t}{\tau}$.

On the other hand, expressions (48) for boundary trajectories of Hadamard double intergrator at $\alpha_1 = \alpha_2 = 1$ reduced to the same equation, which can be obtained for classic double integrator of first order [19, §35]:

$$q_1^{\pm l} = s_1^0 \pm s_2^0\frac{q_2^{\pm l} - s_2^0}{l} \pm \frac{1}{2l}\left(q_2^{\pm l} - s_2^0\right)^2.$$

To the same equation reduced expresses for boundary trajectories of Caputo [5, 7] and Riemann-Liouville [8] double integrator at $\alpha_1 = \alpha_2 = 1$.

6.2. *Comparison of optimal control features for Hadamard, Caputo and Riemann-Liouville systems*

As noted above, today optimal control problems are studied most carefully for Caputo and Riemann-Liouville systems. Hadamard systems, investigated in this paper, are more similar to the systems of second type: both of Hadamard and Riemann-Liouville differentiation operators represented by first derivative from the convolution of considering function with some kernel; the statement of Cauchy problem need the nonlocal initial conditions for both types of systems. These facts allow to make not only qualitative but also quantitative comparison for results concerning Hadamard and Riemann-Liouville systems. Comparison of results obtained for Hadamard systems with the same for Caputo systems will be more qualitative than quantitative and will be make more briefly.

As for integer-order systems we can derive all the expressions for Riemann-Liouville systems from the same expressions for Hadamard systems by performing the change of value $\ln\frac{T}{t_0}$ by value $T - t_0$ and of function $\ln\frac{t}{\tau}$ by function $t - \tau$ correspondingly.

Let us consider firstly the single integrator in case of $u(t) \in L_\infty(t_0, T]$. For the Hadamard system at $q^T = 0$ expression (35) subject to (21) reduces to the following formula:

$$\lambda^H = \frac{|s^0|\alpha}{\ln\frac{T}{t_0}}. \tag{64}$$

The same formula for Riemann-Liouville single integrator obtained in [8] (formula (32) subject to (31)) and can be written here as follows:

$$\lambda^{RL} = \frac{|s^0|\alpha}{T - t_0}. \tag{65}$$

Thus, the dependency of control norm from differentiation index is linear for both types of single integrator and differ from each other only by scale multiplier. The value of control norm can vary significantly for different types of integrator, for example, as follows from (64) and (65), at $T >> t_0 \geq 1$ and identical initial conditions the control norm for Hadamard integrator is distinctly more than for Riemann-Liouville integrator. Additionally, it results from (64)-(65) that ratio of control norms corresponding to Hadamard and Riemann-Liouville integrators doesn't depend on differentiation index and defines only by time interval value:

$$\frac{\lambda^H}{\lambda^{RL}} = \frac{T - t_0}{\ln \frac{T}{t_0}}.$$

It can be obtained analogously that dependencies for minimal control time from differentiation index also differ qualitatively for Hadamard and Riemann-Liouville integrators: in the first case exponential dependency take place and linear — in the second case.

In more general case of linear one-dimrnsional system at arbitrary final condition control norm for Hadamard system defines by formula (33), which can be written subject to (14) as follows:

$$\lambda^H = \frac{a}{|b|} \frac{\left| q^T - s^0 \left(\ln \frac{T}{t_0}\right)^{\alpha-1} E_{\alpha,\alpha}\left[a\left(\ln \frac{T}{t_0}\right)^{\alpha}\right]\right|}{E_\alpha\left[a\left(\ln \frac{T}{t_0}\right)^{\alpha}\right] - 1}. \tag{66}$$

For the same Riemann-Liouville system the following expression take place (it result from formula (24) subject to expression (23) in [8]):

$$\lambda^{RL} = \frac{a}{|b|} \frac{\left| q^T - s^0 (T - t_0)^{\alpha-1} E_{\alpha,\alpha}\left[a(T - t_0)^{\alpha}\right]\right|}{E_\alpha\left[a(T - t_0)^{\alpha}\right] - 1}. \tag{67}$$

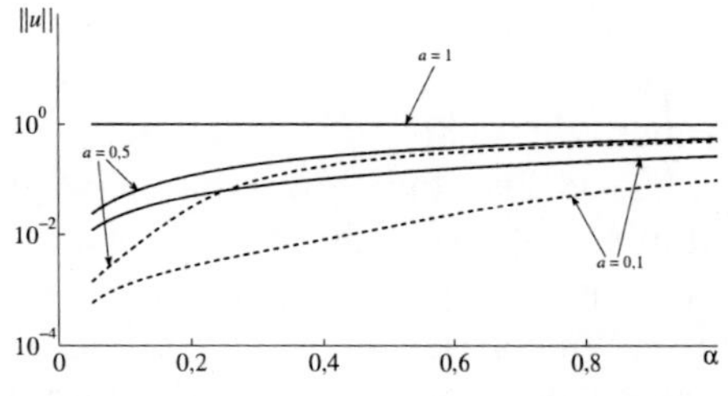

Figure 14. The dependency of control norm from index α for one-dimensional Hadamard (solid lines) and Riemann-Liouville (dashed lines) systems at $q^T = 0$ and different values of parameter a. Logarithmic scale used for ordinates.

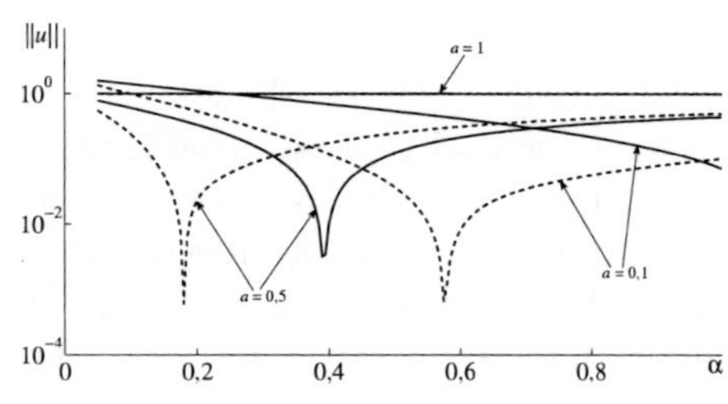

Figure 15. The dependency of control norm from index α for one-dimensional Hadamard (solid lines) and Riemann-Liouville (dashed lines) systems at $q^T = 2$ and different values of parameter a. Logarithmic scale used for ordinates.

At the same initial conditions and other parameters and at $q^T = 0$, $T >> t_0 \geq 1$ it result from (66)-(67) that control norm for Hadamard system is more than for Riemann-Liouville one (see fig. 14). And difference between control norms decreasing with parameter a growth. In case of $q^T \neq 0$ this difference reveal non-monotonic behaviour for $a < 1$ and eliminates for $a = 1$ (see fig. 15).

In case of double integrator expressions for control norm for Hadamard and Riemann-Liouville systems also differ qualitatively from each other. In first case control norm can be calculated using (44) subject to (25) at $k = 1$:

$$\lambda_2^H = \frac{\Gamma(\alpha_1+\alpha_2+1)}{1-2^{-\frac{\alpha_1}{\alpha_2}}} \frac{\left|q_1^T - \frac{s_1^0}{\Gamma(\alpha_1)}\left(\ln\frac{T}{t_0}\right)^{\alpha_1-1}\right|}{\left(\ln\frac{T}{t_0}\right)^{\alpha_1+\alpha_2}}. \tag{68}$$

For Riemann-Liouville double integrator control norm expressed by formula (which result from formula (40) subject to (37), obtained in [8]):

$$\lambda_2^{RL} = \frac{\Gamma(\alpha_1+\alpha_2+1)}{1-2^{-\frac{\alpha_1}{\alpha_2}}} \frac{\left|q_1^T - \frac{s_1^0}{\Gamma(\alpha_1)}(T-t_0)^{\alpha_1-1}\right|}{(T-t_0)^{\alpha_1+\alpha_2}}. \tag{69}$$

It's interesting that ratio of control norms (68) and (69) at $q_1^T = 0$ and the same initial conditions and other parameters depend on α_2 and time interval value only:

$$\frac{\lambda_2^H}{\lambda_2^{RL}} = \left(\frac{T-t_0}{\ln\frac{T}{t_0}}\right)^{\alpha_2+1}.$$

Let us compare now the motion laws and phase trajectories for Hadamard and Riemann-Liouville double integrators. Boundary trajectories of the Hadamard system in general case are expressed by formulas (48). Analogous formulas for the Riemann-Liouville system obtained in [8] (formulas (52)):

$$\begin{cases} q_1^{\pm l}(t) = \frac{s_1^0}{\Gamma(\alpha_1)}(t-t_0)^{\alpha_1-1} + \frac{s_2^0}{\Gamma(\alpha_1+\alpha_2)}(t-t_0)^{\alpha_1+\alpha_2-1} \pm \\ \qquad \pm \frac{l}{\Gamma(\alpha_1+\alpha_2+1)}(t-t_0)^{\alpha_1+\alpha_2}, \\ q_2^{\pm l}(t) = \frac{s_2^0}{\Gamma(\alpha_2)}(t-t_0)^{\alpha_2-1} \pm \frac{l}{\Gamma(\alpha_2+1)}(t-t_0)^{\alpha_2}. \end{cases} \tag{70}$$

Expressions (70) differ from (48) but in special cases, when we can eliminate time and obtain an explicit expression for $q_2^{\pm l}(q_1^{\pm l})$, these expression is the same for both of Hadamard and Riemann-Liouville systems. Particularly, formulas obtained above from (48) at $s_2^0 = 0$ and $\alpha_2 = 1$ are identical to formulas obtained from (70) at the same conditions (see formulas (53) and (54) in [8]). Comparison of expressions (48) and (70) shown that boundary trajectories for Riemann-Liouville system envelop more area at phase plane that boundary trajectories for Hadamard system at the same time interval and other parameters (fig. 16).

In optimal control mode phase trajectories of the Hadamard double integrator at $c_2 = 0$ described by formulas (49) and analogous expressions for the Riemann-Liouville system were obtained in [8] (formulas (55)) and can be written as follows:

$$\begin{cases} q_1(t) = \frac{s_1^0}{\Gamma(\alpha_1)}(t-t_0)^{\alpha_1-1} + \frac{s_2^0}{\Gamma(\alpha_1+\alpha_2)}(t-t_0)^{\alpha_1+\alpha_2-1} + \\ \quad + \frac{c_1^{RL}}{1-2^{-\frac{\alpha_1}{\alpha_2}}} \frac{1}{(T-t_0)^{\alpha_1+\alpha_2}} \times \\ \quad \times \left[(t-t_0)^{\alpha_1+\alpha_2} - 2(t-t'_{RL})^{\alpha_1+\alpha_2}\Theta(t-t'_{RL})\right], \\ q_2(t) = \frac{s_2^0}{\Gamma(\alpha_2)}(t-t_0)^{\alpha_2-1} + \\ \quad + \frac{c_1^{RL}}{1-2^{-\frac{\alpha_1}{\alpha_2}}} \frac{1}{(T-t_0)^{\alpha_1+\alpha_2}} \frac{\Gamma(\alpha_1+\alpha_2+1)}{\Gamma(\alpha_2+1)} \times \\ \quad \times \left[(t-t_0)^{\alpha_2} - 2(t-t'_{RL})^{\alpha_2}\Theta(t-t'_{RL})\right], \end{cases} \tag{71}$$

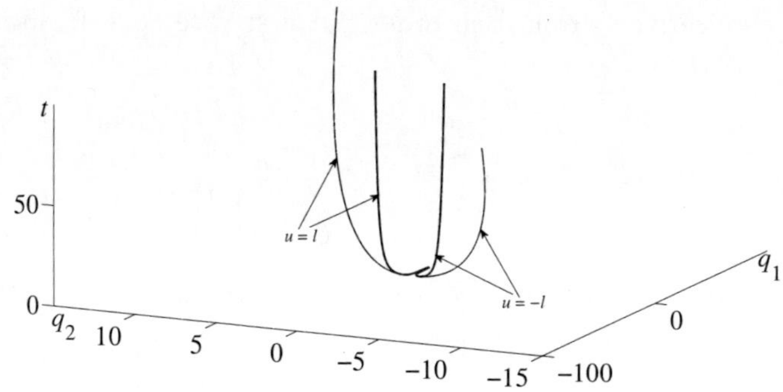

Figure 16. Three-dimensional representation of boundary trajectories for Hadamard (heavy lines) and Riemann-Liouville (thin lines) double integrators at $\alpha_1 = \alpha_2 = 0.5$, $s_1^0 = 1$, $s_2^0 = 0$, $t_0 = 1$, $T = 100$, $l = 1$.

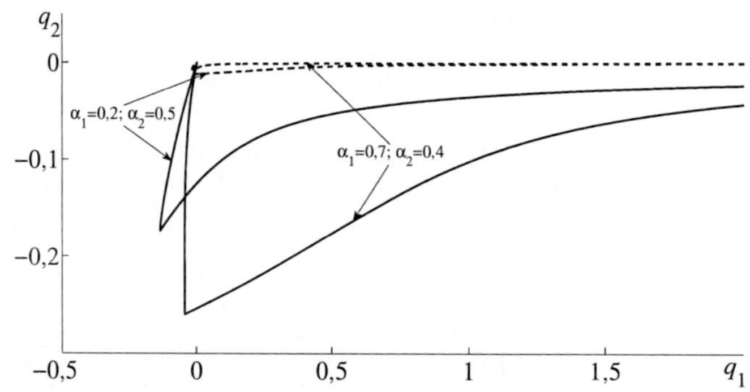

Figure 17. Phase trajectories for Hadamard (solid lines) and Riemann-Liouville (dashed lines) double integrators at some values of indices α_1 and α_2 and $s_1^0 = 1$, $s_2^0 = 0$, $t_0 = 1$, $T = 100$.

where $c_1^{RL} = q_1^T - \frac{s_2^0(T-t_0)^{\alpha_1+\alpha_2-1}}{\Gamma(\alpha_1+\alpha_2)} - \frac{s_1^0(T-t_0)^{\alpha_1-1}}{\Gamma(\alpha_1)}$. For switching point of optimal control for Riemann-Liouville double integrator in this case the following formula valid:

$$t'_{RL} = T\left(1 - 2^{-\frac{1}{\alpha_2}}\right) + 2^{-\frac{1}{\alpha_2}} t_0. \tag{72}$$

Comparison of calculations by formulas (49) and (71) demonstrates that in case of Hadamard double integrator overcontrolling effect expressed more significantly (fig. 17).

At fig. 18 the dependencies of switching point of optimal control from index α_2 represented for both types of double integrators studied here. It's seen that for the Hadamard system switching of control happens earlier than for the Riemann-Liouville system. And for both of systems dependencies decreasing with α_2 growth which correspond to switching point shifting to the left boundary of time interval (it also take place for Caputo systems [5, 7]).

7. Conclusion

In this paper investigation of the optimal control problem carried out for the linear fractional-order systems, which described by equations with Hadamard derivative. One-dimensional system

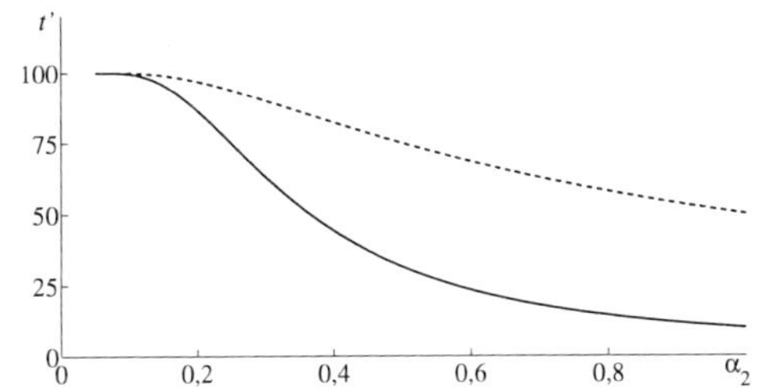

Figure 18. The dependence of switching point of optimal control from index α_2 for Hadamard (solid lines) and Riemann-Liouville (dashed lines) double integrators at $t_0 = 1$, $T = 100$.

of general view and N-fold integrator studied. The optimal control problem reduced to the l-problem of moments. For the last problem conditions of correctness and solvability obtained. Analytical solutions of the studied problem obtained and its features analyzed. Comparison of results performed with the same results for integer-order systems and fractional-order systems, described by equations with Riemann-Liouville and Caputo derivative.

Obtained results can be used for optimal control investigation concerning fractional-order systems and for model and parameter choosing of particular control systems.

Acknowledgements.
This paper dedicated to the memory of my teacher and co-author V.A. Kubyshkin.

References

[1] Butkovskii A G, Postnov S S and Postnova E A 2013 *Automn. Remote Contr.* **74** 725.
[2] Kamocki R 2014 *Math. Meth. Appl. Sci.* **37** 1668.
[3] Kamocki R and Majewski M 2014 *Optimal Control Appl. Meth.* **36** 10.1002/oca.2150.
[4] Butkovskii A G 1969 *Distributed Control Systems* New York: American Elsevier.
[5] Kubyshkin V A and Postnov S S 2014 *Automn. Remote Contr.* **75** 805.
[6] Kubyshkin V A and Postnov S S 2015 *Automn. Remote Contr.* **76** 1302.
[7] Kubyshkin V A and Postnov S S 2015 *Discontinuity, Nonlinearity and Complexity* **4** 429.
[8] Kubyshkin V A and Postnov S S 2016 *J. Control Science and Engineering* **2016** 4873083.
[9] Kilbas A A, Srivastava H M and Trujillo J J 2006 *Theory and applications of fractional differential equations* Amsterdam: Elsevier.
[10] Almeida R and Torres D F M 2009 *Appl. Math. Lett.* **22** 1816.
[11] Heymans N and Podlubny I 2006 *Rheol. Acta* **45** 765.
[12] Tarasov V E 2010 *Fractional Dynamics* Berlin: Springer.
[13] Kolmogorov A N and Fomin S V 1999 *Elements of the Theory of Functions and Functional Analysis* Mineola, N.Y.: Dover Publications Inc..
[14] Akhiezer N I 1965 *The classical moment problem and some related questions in analysis* New York: Hafner Publishing Co..
[15] Krein M G and Nudelman A A 1977 *The Markov moment problem and extremal problems. Ideas and problems of P. L. Chebyshev and A. A. Markov and their further development* Transl. Math. Monograph. **50** Rhode Island: Am. Math. Soc. Providence.
[16] Feldbaum A A 1965 *Optimal control systems* New York: Academic Press.
[17] `http://www.mathworks.com/matlabcentral/fileexchange/8738-mittag-leffler-function`.
[18] Shampine L F 2008 *J. Comp. Appl. Math.* **211** 131.
[19] Butkovskiy A G 1991 *Phase portraits of control dynamical systems* Dordrecht, Boston: Kluwer Academic Publishers.
[20] Budak B M and Fomin S V 1973 *Multiple Integrals, Field Theory and Series* Moscow: Mir Publishers.

Optimization of the matrices structural parameters for the synthesis of restricted access media (RAM) adsorbents for chromatography

S Yu Bogoslovskii, V V Echeistov and V S Boldyrev
Department of Chemistry, Bauman Moscow State Technical University, Moscow, Russia

E-mail: b.su@bmstu.ru

Abstract. The geometric parameters of a matrix for the synthesis of restricted access media adsorbents are optimized, using mathematical modeling based on the basis of a globular model with tetrahedral and octahedral pores. For calculations were used the distribution curves of pore sizes obtained on the basis of the desorption branches of isotherms obtained via low temperature nitrogen adsorption. In contrast to the model of cylindrical pores, the calculation model of globular pores gives realistic values of the surface area of the silica with grafted hexadecylsilyl groups, due to the change of ratio of the amounts of tetrahedral and octahedral pores in the model. The obtained value of the surface area of the silica with grafted hexadecylsilyl groups is in good agreement with elemental analysis data. The obtained refined size of the mouth of the pores makes it possible to choose correctly the particle diameter of the external protective coating.

Most restricted access media (RAM) adsorbents are chemically improved modified silica obtained through the covalent binding of molecules on hydroxylated silica surfaces and specially selected to be effective for the required application tasks [1]. The most significant chromatographic feature of RAM is their ability to differentially interact with components of a chromatographic sample depending on the size of its molecules.

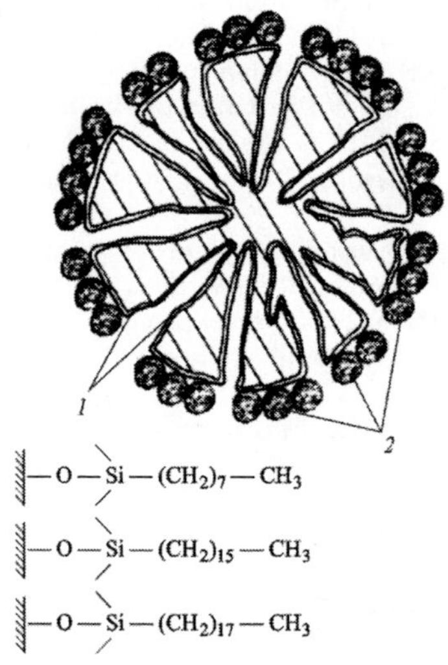

Fig. 1. Schematic structure of RAM adsorbent: (*1*) internal surface (alkyl groups); (*2*) external surface with a protective coating

The surface accessible to the macromolecules of the sample is inert to ensure the exclusion of macromolecules without separation and at a speed close to the speed of movement of the mobile phase. They quickly leave the chromatographic column, since they do not practically interact with the groups grafted on their external surface, while other components of the sample separation are implemented by one or more mechanisms through various interactions with the groups grafted on the inner surface, inaccessible to macromolecules. This allows us to inject complex samples, whose components differ substantially in molecular masses, into a chromatographic column [2,3].

About 70-80% of routine methods of analysis in high performance liquid chromatography (HPLC) are based on dominant method reversed-phase chromatography [1]. To use the existing methods it is necessary that the inner surface of RAM is allowed to work in the mode of reversed-phase chromatography, and the influence of external surface on the retention of low molecular weight components is minimal. It can be achieved through optimization of the structural parameters of the original silica matrix, which must be done as accurately as possible taking into account the chain length of grafted alkyl groups [2-4].

The structural and geometric characteristics of RAM adsorbents were controlled using a method of low-temperature nitrogen adsorption. Despite high informativeness this method, the data obtained should be used with care, since measurement conditions (gas phase; tempreture of 77 K) are far from conditions of RAM synthesis and use. Fig. 2 shows the adsorption isotherm of nitrogen on the silica with grafted hexadecylsilyl groups (*1*) and final RAM (*2*).

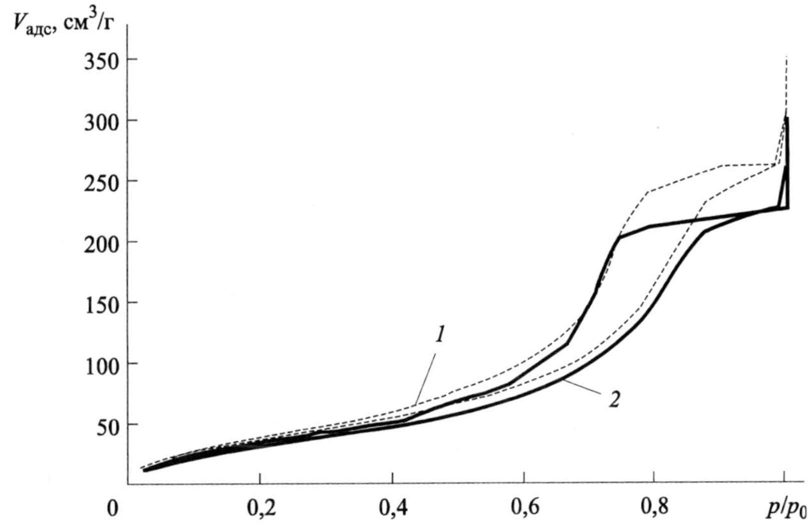

Fig. 2. Adsorption–desorption isotherms were obtained via low temperature nitrogen adsorption: (*1*) silica with grafted hexadecylsilyl groups; (*2*) final RAM adsorbent with human serum albumin on the external surface

There are three areas on the isotherm: a low-pressure area with no hysteresis; a medium-pressure area (0.4–0.9) with a hysteresis and a final area where adsorption increased considerably. The point near the origin of the hysteresis loop corresponds to the onset of capillary condensation in the narrowest pores. The lower closure point of hysteresis for nitrogen, corresponding to $p/p_0 = 0.4$, is responsible for cylindrical pore diameter of about 3.6

nm. We can see that both isotherms (Fig. 2) are almost identical up to values of $p/p_0 = 0.75$. Isotherms therefore differ less if better silica is used. The changes in high-pressure area are associated with the influence of the wide pores geometry coating on the RAMs external surface.

Commonly used to estimate the structural parameters of the silica is the model of cylindrical pores. Fig. 3 shows three curves of pore size distribution, obtained from the desorption branches of the isotherms; pore size distributions were found with the capillary condensation equation for cylindrical pores.

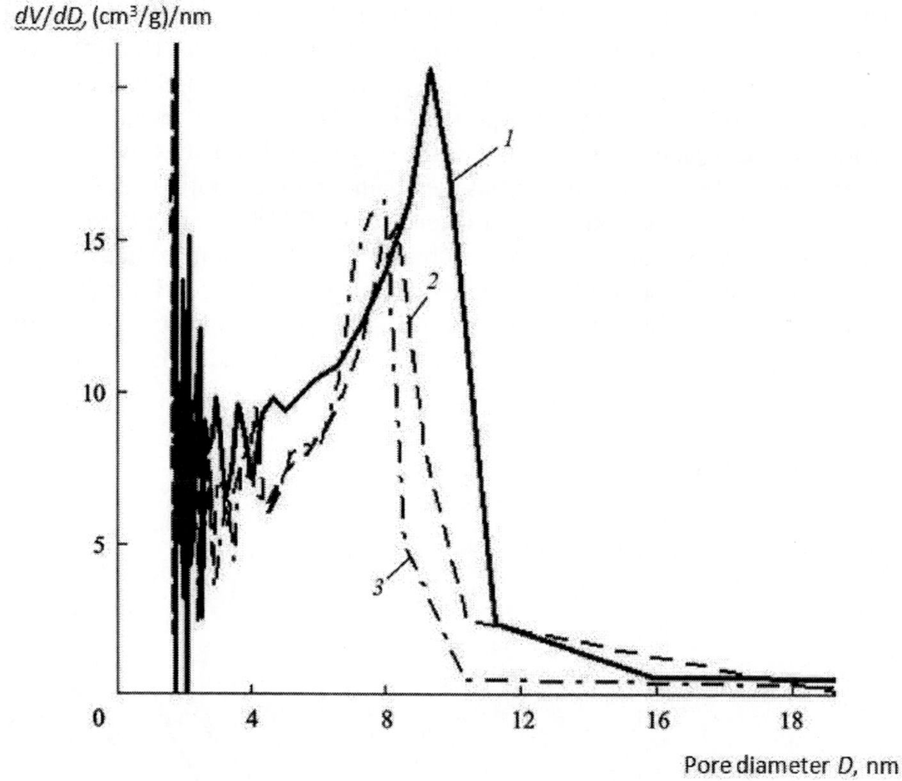

Fig. 3. Differential curve of pore diameter distribution: (*1*) original silica; (*2*) silica with grafted hexadesylsilyl groups; (*3*) RAM

However, it does not convey the true geometry of the silica matrix. Silica is known to have globular structure and space of the pores formed by the voids between the globules of fused silica [1]. Therefore, the use of the model of cylindrical pores could hinder the correct selection of a source of silica and lead to a significant deterioration of the chromatographic properties of the finished RAM.

Application of a globular model, in contrast to the model of cylindrical pores, gives the real size of the mouths of the pores, effectively allowing you to compare the silica matrix parameters with the size of inert macromolecules to sample particles. Fig. 4 shows a representation of the octahedral (*1*, *2*) and tetrahedral (*3*, *4*) pores. The surface (*A*), marked in blue, is not modified silica, available for the nitrogen molecules. Between the blue and green areas there is a modified surface (*B*), partially accessible for the nitrogen molecules. In this region the layers of the modifier are closed, impeding the diffusion of adsorbate. In red (*D*) it is shown the mouth of the pores at the contact of the globules of silica; green (*C*) represents modified surface of silica globules fully accessible to the adsorbate.

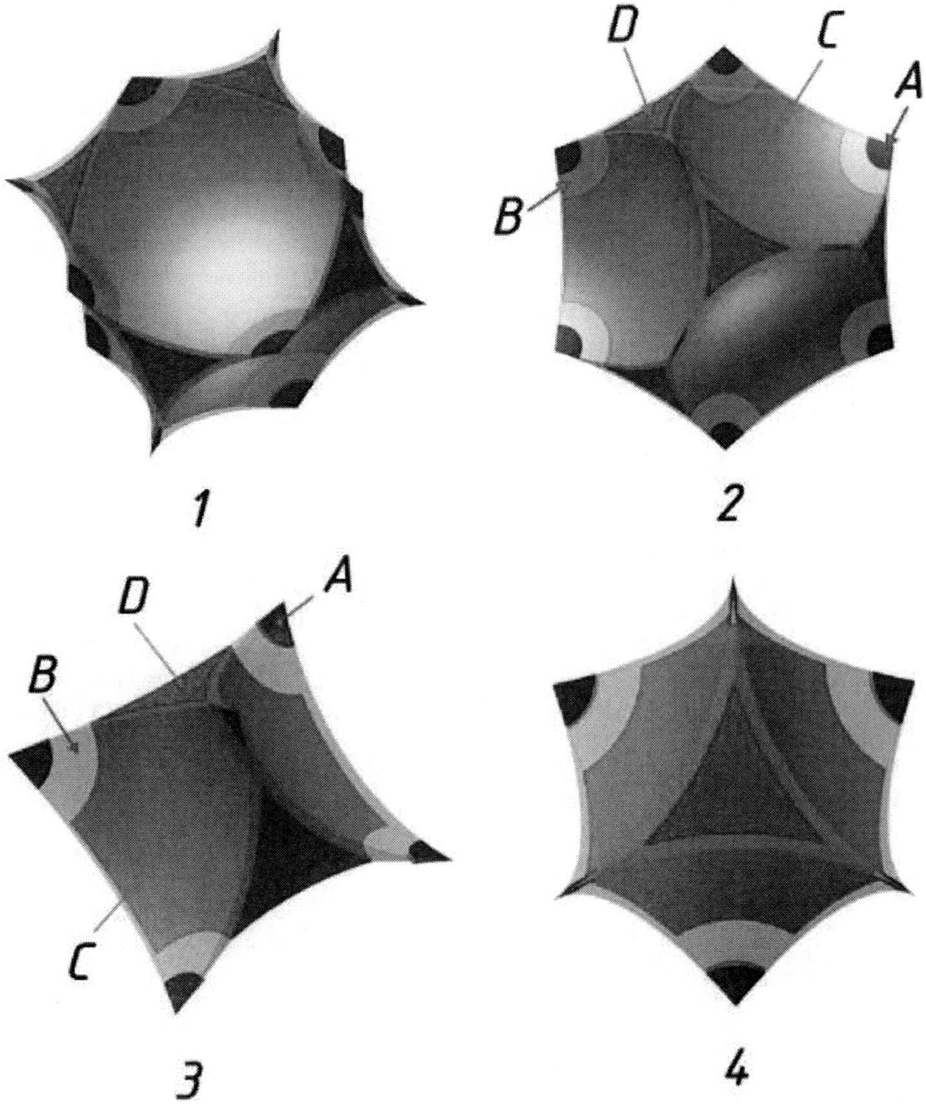

Fig. 4. Representation of the octahedral (*1*, *2*) and tetrahedral (*3*, *4*) pores. Blue color (*A*) shown is not modified silica, (*B*) shows modified surface, partially accessible for the nitrogen molecules; in red (*D*) is shown the mouth of the pores; green (*C*) represents modified surface of silica globules

Globular proteins (e.g., albumin [2-4], which is modeled by ellipsoid is about 8.5 nm long [5]) or particles of inorganic nature with a modified surface modeled by a sphere of known diameter can be used as shielding particles. The shielding coating must not penetrate into the mouth of tetrahedral and octahedral pores of silica matrix formed by the contact of the globules of the same diameter. It allows us to calculate the size of the shielding particles on the basis of the diameter of the globules of silica. For example, for shielding the external surface with human or bovine serum albumin the desired maximum distribution of the pore sizes for the matrix obtained is close to 5 nm.

It is optimal to use a method of low-temperature nitrogen adsorption for the experimental evaluation of the structural parameters of unmodified and modified matrix. The best match between estimated and actual values of the diameters of the orifices of the pores is obtained by using the distribution curves of pore sizes obtained on the basis of the desorption branches of isotherms. The presence of one maximum on the curve makes it easy to interpret it on the basis of a globular model with tetrahedral and octahedral pores. By changing the ratio of the amounts of tetrahedral and octahedral pores in the model it is possible to achieve satisfactory agreement with the experimental shape of the distribution curve. In contrast to the model of cylindrical pores, the calculation model of globular pores gives realistic values of the surface area of the silica with grafted hexadecylsilyl groups: 131 m^2/g instead of 183 m^2/g. Film thicknesses were calculated from the obtained surface area for initial and modified silica and determined the number of layers of the modifier. The obtained value 1.41 is in good agreement with elemental analysis data.

References

[1] Lisichkin G V, Fadeev A Y, Cerdan A A 2003 Khimiya privitykh poverkhnostnykh soedineniy *Fizmatlit Moscow* P 592

[2] Serdan A A, Bogoslovskiy S Yu, Nesterenko P N 1991 Retention of certain medicinal preparations on diphilic sorbent as a function of pH and eluent ionic strength *Russ. J. Phys. Chem.* Vol **65** Issue 10 pp 1396-1399

[3] Bogoslovskiy S Y Serdan A A Nesterenko P N 2015 Column packed with heterosurface adsorbent for direct chromatographic analysis of pharmaceutical in biological fluids Fundamental research. No **2-21** pp 4643-4647; URL: https://fundamental-research.ru/ru/article/view?id=38043

[4] Bogoslovskiy S Yu Serdan A A 2015 Optimizing heterosurface adsorbent synthesis for liquid chromatography *Russ. J. Phys. Chem.* Vol **90** Issue 3 pp 661-664 **DOI:** 10.1134/S0036024416030067
URL: https://link.springer.com/article/10.1134/S0036024416030067

[5] Kiselev M.A., et al. 2001 The size of Hyman Serum Albumin Molecules in Solution *Biophysics*, Vol **46**, No 3 pp 402–405

IOP Conf. Series: Journal of Physics: Conf. Series **918** (2017) 012028 doi:10.1088/1742-6596/918/1/012028

Parameter optimization of electrolytic process of obtaining sodium hypochlorite for disinfection of water

S Yu Bogoslovskii, N N Kuznetsov and V S Boldyrev

Department of Chemistry, Bauman Moscow State Technical University, Moscow, Russia

E-mail: b.su@bmstu.ru

Abstract.
Electrochlorination parameters were optimized in flowing and non-flowing modes for a cell with a volume of 1 l. At a current density of 0.1 A/cm^2 in the range of flow rates from 0.8 to 6.0 l/h with a temperature of the initial solution below 20°C the outlet temperature is maintained close to the optimal 40°C. The pH of the solution during electrolysis increases to 8.8 ÷ 9.4. There was studied a process in which a solution with a temperature of 7-8 $^{\circ}$C and a concentration of sodium chloride of 25 and 35 g/l in non-flowing cell was used. The dependence of the concentration of active chlorine on the electrolysis time varies with the concentration of the initial solution of sodium chloride. In case of chloride concentration of 25 g/l virtually linear relationship makes it easy to choose the time of electrolysis with the aim of obtaining the needed concentration of the product.

Sodium hypochlorite belongs to a new generation of reagents used for water disinfection as alternative to gaseous chlorine, and can be applied either in the form of commodity (technical) hypochlorite (is a highly concentrated strong alkaline solution produced in chemical plants with a content of active chlorine 14%) or produced at the place of use in the required quantity by the electrolysis of sodium chloride. This method of obtaining hypochlorite directly into the sodium chloride solution with the help of electric current is called electrochlorination. Due to high concentration commodity hypochlorite (CH) decomposes faster than the one obtained directly on the spot in the cell (EH).
This decomposition of hypochlorite into chlorate leads both to loss of product and the appearance of undesirable by-product in the water. The EH is made at the place of consumption as needed, thus reducing the storage period before use. CG has a high pH value (pH=12) to ensure stability of the product. Therefore, when introducing it in water with a high content of cations of calcium and magnesium there may form insoluble carbonates and hydroxides. It leads to the need for additional filtering of water prior to use. Finally, CH contains a potentially carcinogenic bromate, the concentration of which must not exceed 10 μg/1 in the EU. EH with a low content of bromate can be obtained through the use of sodiumchloride with a low content of bromide. Therefore, the electrochlorination systems are increasingly used instead of technical hypochlorite batching operations.

Sea water can be used for process water instead of sodium chloride solution [1]. Chlorine is produced on the anode in the cell of the electrolyzer; water decomposes on the cathode with formation of hydrogen and hydroxyl ions. In the open reactor reagents freely mix and react with the formation of the target product. At concentrations less than 1% sodium hypochlorite is classified as non-hazardous reagent, being at the same time effective disinfection, so it is advisable to choose electrochlorination parameters to generate a concentration from 0,2% to 0.8%. Besides, there present practically no chlorate at these concentrations of hypochlorite. In case of 14% TG 0.6% of active chlorine is converted to chlorate within the first days of storage at 22 ^{o}C [2], so during transportation and storage its ratio can be significantly increased.

Optimization of parameters of electrochlorination was carried out with the help of flow installation with a reactor volume equal to 1 liter, containing four titanium electrodes with oxide-ruthenium anodes, each with an area of 60 centimeters. The concentration of active chlorine was determined by iodometric method [3]. Oxide-ruthenium anode is known to work best of all at temperatures not below 18 ^{o}C. On the other hand, excessive heating of the electrolyte adversely affects the yield of sodium hypochlorite, therefore, the optimal temperature is 40-45 ^{o}C [4]. In case of flow through electrolytic cell the lower the flow rate the higher electrolyte temperature and its residence time in the reactor. Other conditions being equal, heating of the electrolyte also increases with decrease of concentration of sodium chloride. According to the data obtained in a wide range of flow rates from 0.8 to 6.0 1/h there occurs no excess of temperature if the initial solution does not exceed 20 ^{o}C.

The pH of the solutions during electrolysis increases to a value $pH = 8,8 \div 9,4$ with the increase of time of electrolysis and the concentration of salt due to the reduction of water molecules. Conductivity of the solutions changes slightly in the range of $3 \div 5\%$ of the original. Consumption ensuring the achievement of active chlorine concentration of 0.2% at a current density of $I = 9.1$ A/cm^2 is from 6 to 7.5 1/h depending on the concentration of chloride. The concentration of active chlorine of 0.4% is achieved at the flow rate $2.5 \div 2.7$ 1/h.

We also studied the dependence of the temperature and concentration of sodium hypochlorite on the time of electrolysis in a non-flowing cell. In this experiment we simulated the use of sea water in winter as raw material. Initial temperature of sodium chloride solution was 7° to 8°C (Fig. 1,2). The concentration of sodium chloride was 35 g/l and 25 g/l. It is seen that the temperature of the stagnant electrolyte in a cell increases almost linearly. The dependence of the concentration of hypochlorite on time in the case of more concentrated solution has a more pronounced deviation (Fig. 2). Each experiment was conducted three times. The obtained results show that by choosing the time of electrolysis, it is possible to obtain a product of needed concentration if we use sea water as raw material in winter time.

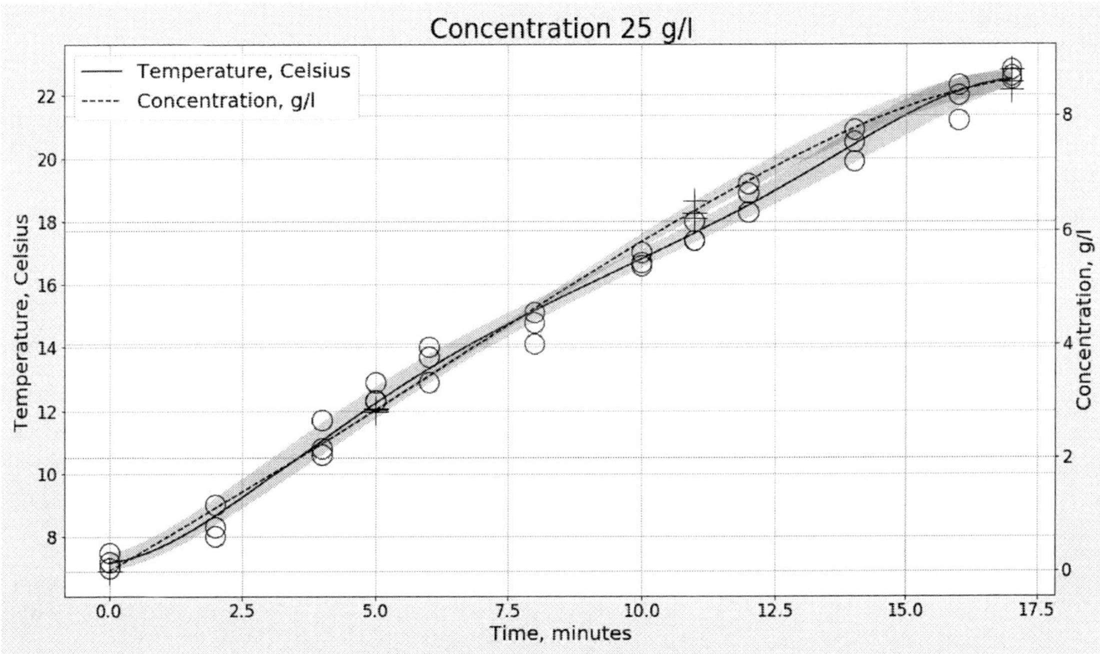

Fig. 1. The dependence of the temperature (left scale, circles) and concentration of active chlorine (right scale, cross) on time in a non-flowing cell for concentration of sodium chloride 25 g/l.

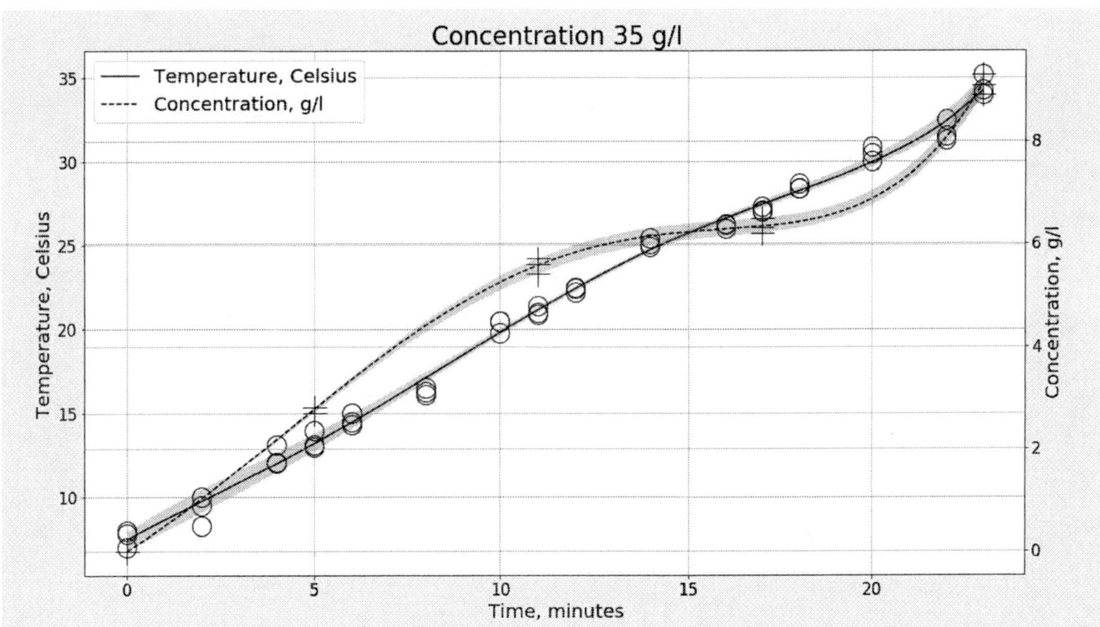

Fig. 2. The dependence of the temperature (left scale, circles) and concentration of active chlorine (right scale, cross) on time in a non-flowing cell for concentration of sodium chloride 35 g/l.

References

[1] Muhammad Saleem *et al* 2012 Int. J. Electrochem. Sci. **7** pp 3929 – 3938
[2] White's Handbook of Chlorination and Alternative Disinfectants 2010 *Hoboken: John Wiley & Sons* P 452—571
[3] Standard Methods for the Examination of Water & Waste Water, 2005
[4] DOI: 10.1007/s12613-012-0596-0

IOP Conf. Series: Journal of Physics: Conf. Series **918** (2017) 012029 doi:10.1088/1742-6596/918/1/012029

Photoluminescence and polariton dispersion law in terbium nitrate hydrate crystals

V S Gorelik[1,2] and M G Burdanova[2]

[1]Lebedev Physical Institute of the Russian Academy of Sciences, Moscow, Russia
[2]Bauman Moscow State Technical University, Moscow, Russia

E-mail: gorelik@sci.lebedev.ru

Abstract. The article examines the dispersion law of polariton waves in terbium nitrate hydrate crystals. We have calculated the values of the unitary polaritons frequencies corresponding to the refractive index equal to unit as well as the velocity and the effective mass of the unitary polaritons. The diagram of terms splitting in the low symmetric crystalline field is constructed. It is shown that the abnormal decrease of the polariton waves' group velocity occurs due to the splitting of the terbium ions spectral terms in the crystalline field. We consider the conditions for the essential increase of the photoluminescence intensity and decrease of the lasing threshold in terbium nitrate hydrate crystals.

1. Introduction

Electromagnetic radiation in the dielectric medium interacts with the crystal lattice polar vibrations as well as with the exciton states of a semiconductor or a dielectric. These interactions are the most efficient when the frequencies and the wave vectors of the electromagnetic and exciton waves are close, that is when the resonance takes place. Therefore, in such conditions the hybrid (excitonic polaritons) waves with the typical dispersion law $\omega(k)$ are formed [1, 2]. The dispersion of polaritons has found numerous applications in many fields of science and engineering. By changing the polaritons velocities it is possible to design new devices such as filters [3], delay lines [4-7], scintillators [8] etc. varying the crystal absorption by means of injecting the active ions. The recent researches show the most perspective fields to use the polaritons effects for designing optical sources based on the dielectric crystals [9-12]. The polaritons theory allows calculating the refractive index and some characteristics of the electromagnetic waves such as the group velocity, the effective mass of polaritons, the spectral thermal energy density and others. The rare earth is interesting for investigating because of its properties. The photoluminescence spectra of the rare earth ions are based on the electron transitions between the levels of the inner 4f-shell, which is screened from the external influences of outer electrons (usually p- and s-shells). It allows the appearance of the exciton terms and excitonic polaritons in bulky materials even at room temperature. The inner-shell 4f-4f transitions comprise both visible and near-infrared ranges [13]. They are sharp and easily recognizable. Since these transitions are forbidden by the selection rules, the lifetimes of the excited states are long [14]. In addition, they are almost independent of the chemical environment. They have a high photoluminescence quantum yield [15]. Only a limited number of researches include the information about the polaritons dispersion characteristics of rare earth ions in the dielectric medium. In our earlier research we have already investigated one group of rare earth ions: erbium, ytterbium and thulium [16, 17]. In this step of our research we have obtained information about terbium ions, which are prospective for the laser generating. The examination of paramagnetic terbium ions Tb^{3+} is particularly important because it is necessary to develop the models of solid laser in the visible (green) spectral range with high efficiency and power of laser emission together with its high spatial characteristics. Terbium ions are mostly used as the impurity centers in dielectric mediums. This allows creating population inversion of levels, which leads to laser transitions. Thereto, terbium ions have a set of terms which will be used to realize a three-level scheme of laser generating in the visible (green) range that has not yet been realized. So far there was a tendency in the solid laser technology to focus so much on the research near the infrared range. It is partially related to the fact that for a long time powerful radiation sources capable to pump the active medium in ultraviolet and visible ranges were not developed. Only for the last few decades the new types of diodes have been created in line with the mentioned goals. Moreover, the investigation of terbium optical properties is significantly important in the case of creating the scintillator. This paper presents a method for assessing the optical properties of terbium nitrate hydrate. The excitonic polaritons spectrum of terbium ions was investigated. As a result, the refractive index and the dispersion characteristics of terbium nitrate hydrate were obtained in the wide frequency range. We provide theoretical and experimental analysis by using Maxwell's equations with regard to the external field and oscillations. In particular, we have calculated the velocity of the unitary polaritons, for which the refractive index is unity. It allowed us to present a principal scheme of the terbium laser. Our main goal was to provide the possibility of lowering the lasing threshold by reducing the group velocity when splitting in the crystal field.

2. Experimental research

The research involved using the polycrystalline samples of terbium nitrate hydrate ($Tb(NO_3)_3 \cdot 6H_2O$). The analysis of the samples optical characteristics was described by recording the spectra of transmission and reflection. The scheme of the experimental setup for observing the photoluminescence spectra is illustrated in Figure 1. As the sources of excitation radiation we used the laser Nd^{3+}: YAG, the nitrogen laser and the ultraviolet (UV) light-emitting diodes (1) with the wavelengths of λ= 266, 337, 369 and 385 nm correspondingly. This radiation was directed with the help of the fiber (3) into the cuvette (6) with the samples placed on the table (4) by means of fastening (5). The absorption and photoluminescence spectra were registered on the spectrometer (2). The focusing lens system (7) was used to prevent the divergence of the beam. The experimental data was input to the analog-to-digital converter of the computer for the final processing. Figure 2 illustrates the photoluminescence spectra excited by different light sources. In Figure 2a the photoluminescence spectrum of terbium nitrate hydrate for the diodes pump radiation is given. It should be noted that these relations are normalized according to the maximum experimentally obtained data (curve 1 – pump radiation, curve 2 – photoluminescence spectrum). In Figure 2b the photoluminescence spectrum of terbium nitrate hydrate for the nitrogen laser pump radiation (λ=266 nm) is given. These relations are normalized according to the maximum experimentally obtained data as well. The low-symmetry crystal field resulted in the appearance of splitting and, consequently, the sub-levels $^7F_6$ and $^5D_4$ of the paramagnetic ion Tb^{3+} multiplets (see Figure 3). In this case the nitrogen laser (λ=337 nm) is used as the source of light.

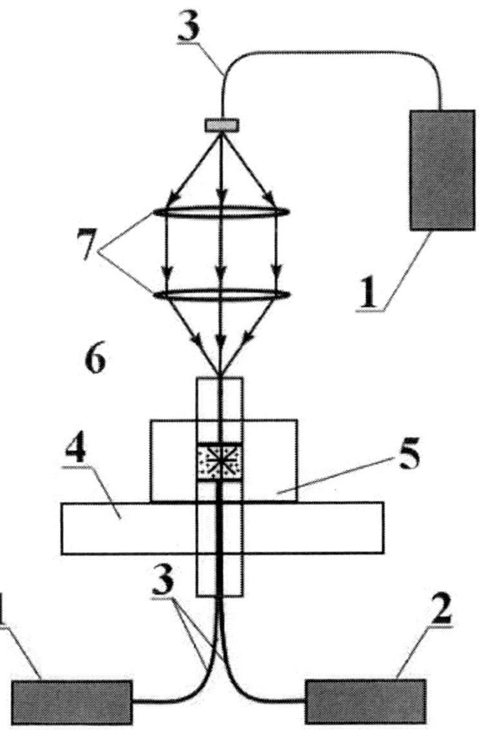

Figure 1. The schematic of the experimental setup for observing the photoluminescence spectrum of erbium nitrate hydrate: 1 – light-emitting diodes or laser; 2 – spectrometer; 3 – fiber; 4 – table, 5 – adapter; 6 – cuvette, 7 – lens system.

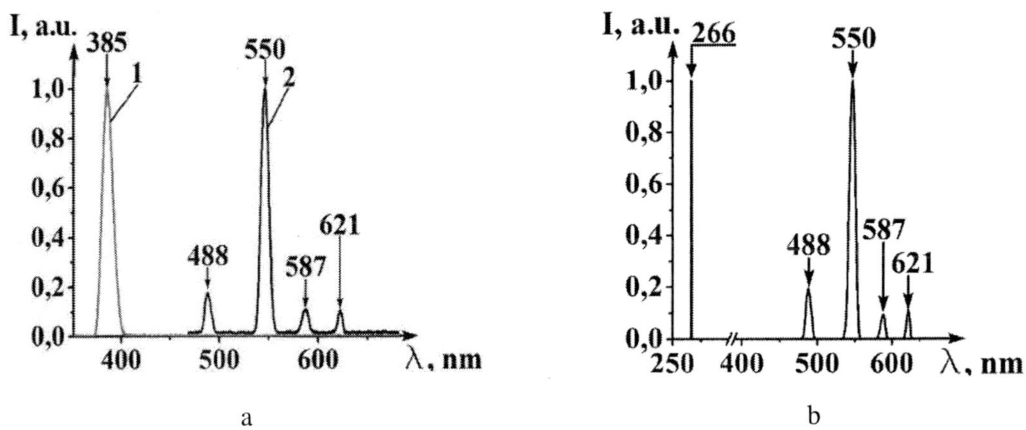

Figure 2. The spectra of the experimental setup for observing the photoluminescence spectrum of terbium nitrate hydrate with the wavelengths of pump radiation a) 385 nm (diode) b) 266 nm (laser).

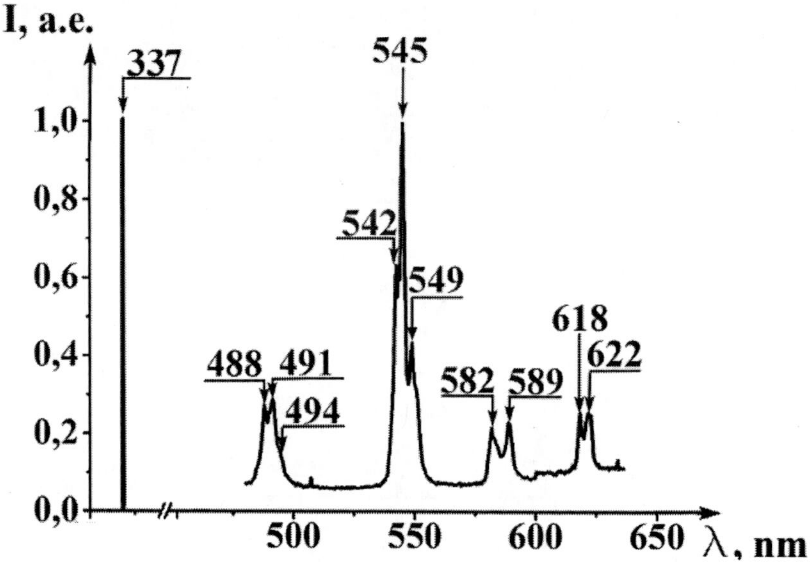

Figure 3. The photoluminescence spectrum of terbium nitrate hydrate at room temperature under the nitrogen laser excitation (λ = 337 nm).

When excited by different sources a number of lines corresponding to the transitions between different energy levels in terbium ions were revealed. Pumping by different sources, however, has shown the same results: the most intensive peak was at the wavelength of λ=550 nm; the second intensive peak corresponded to the wavelength of 488 nm. In our experimental setup we used different diodes. All of these diodes had the same capacities. However, when we used the λ=385 nm diode the intensity of the photoluminescence spectrum was the highest (see Figure 2a). The photoluminescence spectra of terbium ions in the optical range are based on the electron transitions between the levels of the inner 4f-shell, which is screened from the external influences of outer electrons (usually p- and s-shell). This screening leads to the appearance of the fluorescence narrow spectral lines.

3. Discussion

The free ion of erbium Tb^{3+} has 11 electrons in its 4f shell. The ground-state manifold for terbium ions is $^7F_6$. After the excitation a transition between the ground-state manifold and another manifold is realized, which can be nonradiative or fluorescent. The most complete description of the free ions spectrum was obtained in [18-20]. These researches provide the information about a wide spectral range – from the ultraviolet wavelength region to the infrared one, with the additional calculation of splitting the levels in the magnetic field. All energy levels present in the photoluminescence spectrum are accounted in Table 1. By comparing the data from [18-20] with the data obtained from our experiment, we defined the wavelengths corresponding to the quantum transitions in the 4f-shell free terbium ions in terbium nitrate hydrate. The wavelength λ = 488 nm corresponds to the transition from $^5D_4$ to $^7F_6$ ground state level. The lines at wavelengths λ = 550 nm, 587 nm, 621 nm and 655 nm correspond to the transitions from $^5D_4$ level to the closely placed levels of $Tb^{3+.}$ The corresponding transitions between the energy levels are shown in Figure 4. Thus, by using the diode and laser we realized the pumping scheme for the inverse population between $^5D_4$ and $^7F_6$ levels. This diode is prospective for realizing laser generation with the wavelength of 550 nm corresponding to the green spectral range. When we use the diode (λ = 385 nm) the resonant pumping of $^5D_3$ term takes place with relaxation to $^5D_4$. As a result the superliminescent emission from $^5D_4$ level to the first excited state $^7F_5$ may be observed (see Figure 2a). The same thing happens if a pulsed UV laser (266 nm) is used (see Figure 2b). Using the pulsed nitrogen laser (λ = 337nm) as the exciting source of light results in more complicated photoluminescence spectrum (see Figure 3). Thus, in all these cases the superluminescence and lasing at fairly strong pumping is observed.

Table 1. Calculated and experimentally observed energy levels for terbium nitrate hydrate.

Energy level	Free ions [18]		Terbium nitrate hydrate	
	Tb^{3+}		$Tb(NO_3)_3 \cdot 6H_2O$	
	E, cm^{-1}	λ, nm	E, cm^{-1}	λ, nm
$^7F_6$	0	∞	0	∞
$^7F_5$	2137	4679	2037	4909
$^7F_4$	3457	2893	3256	3071
$^7F_3$	4525	2210	4362	2293
$^7F_2$	5120	1953	5044	1983
$^7F_1$	5672	1763	5478	1825
$^7F_0$	5244	1907	-	-
$^5D_4$	20634	485	20488	488
$^5D_3$	26383	379	-	-

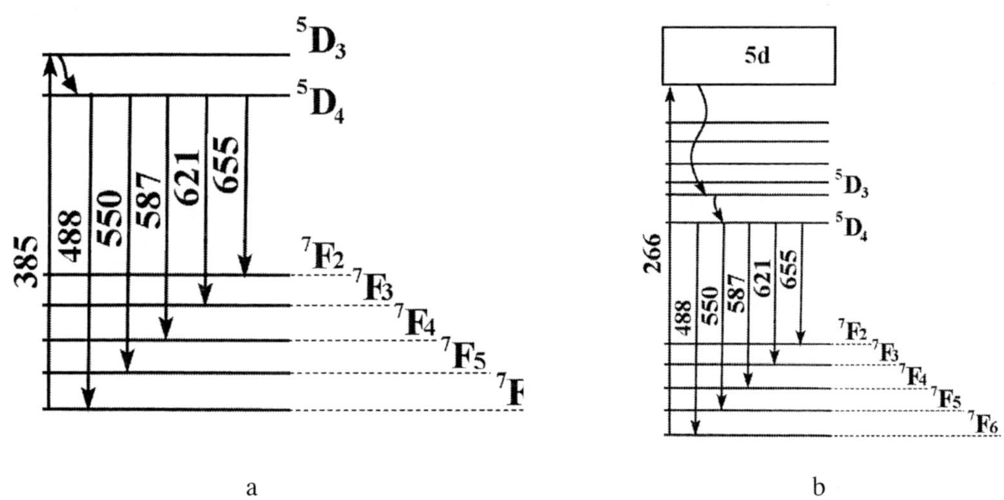

Figure 4. The energy diagram of terbium nitrate hydrate pumping at the wavelengths of: a) λ = 385 nm, b) 266 nm.

4. Theoretical considerations

The theoretical modeling of the electromagnetic wave propagation in erbium nitrate hydrate relies on the Maxwell's equation [6, 7]:

$$\begin{cases} rot\vec{E} = -\dfrac{\partial \vec{B}}{\partial t}; \\ div\vec{B} = 0; \\ rot\vec{H} = \dfrac{\partial \vec{B}}{\partial t}; \\ div\vec{D} = 0. \end{cases} \tag{1}$$

The material equations are needed to close the Maxwell's equations. The material equations link the electromagnetic quantities with each other:

$$\begin{cases} \vec{D} = \varepsilon_0 \varepsilon \vec{E} = \varepsilon_0 \vec{E} + \vec{P}; \\ \vec{B} = \mu_0 \mu \vec{H} = \mu_0 \vec{H} + \mu_0 \vec{M}. \end{cases} \tag{2}$$

If we want to solve (1), we have the following conditions and materials:

$$\rho(r) = 0,\ j(r) = 0$$

Here, the electric charge density and electric current density are assumed to be zero.

After applying a few vector operations we get the wave equation for the electromagnetic field E:

$$rot\, rot\vec{E} - grad\, div\vec{E} = -\mu_0\, \mu \frac{\partial}{\partial t} rot\vec{H} = -\mu_0\, \mu\, \varepsilon_0 \varepsilon \frac{\partial^2 \vec{E}}{\partial t^2}. \tag{3}$$

The simplest possible solution to the Maxwell's equations is the plane wave. For monochromatic plane waves in the dielectric medium:

$$(\nabla^2 - \frac{\varepsilon\mu}{c_0^2})\vec{E} = 0. \tag{4}$$

Finally, the dispersion law for polaritons waves has the following form:

$$\omega^2 = \frac{c_0^2 k^2}{\varepsilon(\omega)\mu(\omega)};\ \mu(\omega) = 1;\ \omega^2 = \frac{c_0^2 k^2}{\varepsilon(\omega)}; \tag{5}$$

$$i\varepsilon\varepsilon_0 k\vec{E}_0 \exp\left[ik\vec{r} - \omega t\right] = 0. \tag{6}$$

For the dipole moment and the polarization vector we obtain:

$$\vec{p}_0 = \frac{e^2 F}{m(\omega_0^2 - \omega^2)} \vec{E}_0; \tag{7}$$

$$\vec{P}_0 = \frac{e^2 F}{mV_0(\omega_0^2 - \omega^2)} \vec{E}_0. \tag{8}$$

The equation of motion for the polarization vector is represented as:

$$\ddot{\vec{P}} = -\omega_0^2 \vec{P} + \frac{e^2 F}{mV_0}\vec{E}; \quad \vec{P} = \frac{e\sqrt{F}\vec{u}}{V_0}\vec{P}_0 \exp i(\vec{k}\vec{r} - \omega t). \tag{9}$$

Introducing the plasma frequency ω_p, from the equation (9) we obtain the relation:

$$\ddot{\vec{P}} = -\omega_0^2 \vec{P} + \omega_p^2 \vec{E}; \ \omega_p^2 = \frac{e^2 F}{mV_0}. \tag{10}$$

For the electric displacement we have:

$$\vec{D}_0 = \varepsilon_0 \vec{E}_0 + \vec{P}_0 = \varepsilon_0 [1 + \frac{e^2 F}{mV_0(\omega_0^2 - \omega^2)}]\vec{E}_0 = \varepsilon_0 \varepsilon(\omega)\vec{E}_0. \tag{11}$$

Therefore, the dispersion of the electromagnetic waves can be described as:

$$\begin{cases} \varepsilon(\omega) = 1 + \dfrac{e^2 F}{mV_0(\omega_0^2 - \omega^2)} = 1 + \dfrac{\omega_p^2}{\omega_0^2 - \omega^2} = \dfrac{\omega_l^2 - \omega^2}{\omega_0^2 - \omega^2}; \\ \omega_l^2 = \omega_0^2 + \omega_p^2; \\ \omega_p^2 = \dfrac{e^2 F}{mV_0}. \end{cases} \tag{12}$$

With regard for the electronic polarizability the dielectric permittivity equation becomes:

$$\varepsilon(\omega) = \varepsilon_\infty \frac{\omega_l^2 - \omega^2}{\omega_0^2 - \omega^2}; \ \varepsilon_\infty = n_\infty^2. \tag{13}$$

Here $\varepsilon_\infty = n_\infty^2$ is the high-frequency permittivity, ω_l and ω_0 are the frequencies of the corresponding longitudinal and transverse waves. Accordingly, the dispersion law of polaritons in the dielectric media is written as:

$$\omega^2 = \frac{c_0^2 k^2}{\varepsilon(\omega)\mu(\omega)} = \frac{c_0^2 k^2(\omega_0^2 - \omega^2)}{\varepsilon_\infty(\omega_l^2 - \omega^2)} = \frac{c^2 k^2(\omega_0^2 - \omega^2)}{(\omega_l^2 - \omega^2)}; \ c^2 = \frac{c_0^2}{\varepsilon_\infty}. \tag{14}$$

Solving (14) gives two frequency-dependent solutions for the wave vector k as a function of ω since the left-hand side is quadratic in k. Close to the absorption two solutions exist. This means that near the resonance the dispersion of the light cannot be considered independently from the dispersion of the excitations. They both form a composite new entity propagating in the medium. The corresponding quantum excitations may be classified as polaritons, resemble to lattice polaritons found in such crystals as NaCl, $LiNbO_3$ and others.

In this case the dielectric function may be presented as:

$$\varepsilon(\omega) = \varepsilon_\infty + \sum_{j=1}^{j=n} \frac{q_j N}{m_j \varepsilon_0(\omega_{0j}^2 - \omega^2)}. \tag{15}$$

Finally, we obtain the permittivity description known as the Kurosawa relation:

$$\varepsilon(\omega) = \varepsilon_\infty \prod_{j=1}^{j=n} \frac{\omega_{lj}^2 - \omega^2}{\omega_{0j}^2 - \omega^2}. \tag{16}$$

With regard to the solution of the wave equation we come to the polariton dispersion law written as:

$$\omega^2 = \frac{c_0^2 k^2}{\varepsilon_\infty \prod_{j=1}^{j=n} \frac{\omega_{lj}^2 - \omega^2}{\omega_{0j}^2 - \omega^2}}. \tag{17}$$

Here, for simplicity we consider the magnetic permeability to be equal to unity. For the unitary polaritons, for which the refraction index n is unity, we obtain the following expression:

$$n^2 = \varepsilon\mu = \frac{c_0^2 k^2}{\omega^2} = 1 \tag{18}$$

$$\frac{c_0^2 k^2}{\omega^2} = \frac{\varepsilon_\infty c_0^2 k^2 \prod_{j=1}^{j=n} \frac{\omega_{lj}^2 - \omega^2}{\omega_{0j}^2 - \omega^2}}{c_0^2 k^2} = \varepsilon_\infty \prod_{j=1}^{j=n} \frac{\omega_{lj}^2 - \omega^2}{\omega_{0j}^2 - \omega^2} = 1 \tag{19}$$

The group velocity of polaritons waves in the material can be found from the known relation:

$$V(\omega) = \frac{d\omega}{dk} = \left(\frac{dk}{d\omega}\right)^{-1} \tag{20}$$

For the effective mass of quasi-particles in the isotropic media we use the formula:

$$m(\omega) = \left(\frac{d^2E}{dp^2}\right)^{-1} = \hbar^2 \left(\frac{d^2\varepsilon}{dk^2}\right)^{-1} = \hbar\left(V(\omega)\frac{dV(\omega)}{d\omega}\right)^{-1} \tag{21}$$

The work [11] shows that the longitudinal frequencies are of the same order as the transverse frequencies. To make a more exact calculation that will agree more closely with the results of the experiment we must consider $\omega_{LO} = (1,1 \div 1,2)\omega_{TO}$ for the high frequency and $\omega_{LO} = (1,01 \div 1,02)\omega_{TO}$ for the low frequency. Figures 5 and 6 illustrate the calculation dispersion law $\omega(k)$ for the electromagnetic waves in the visible range with and without splitting.

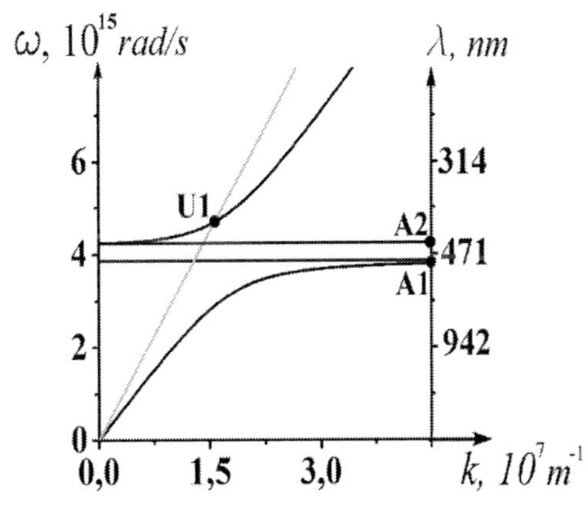

a

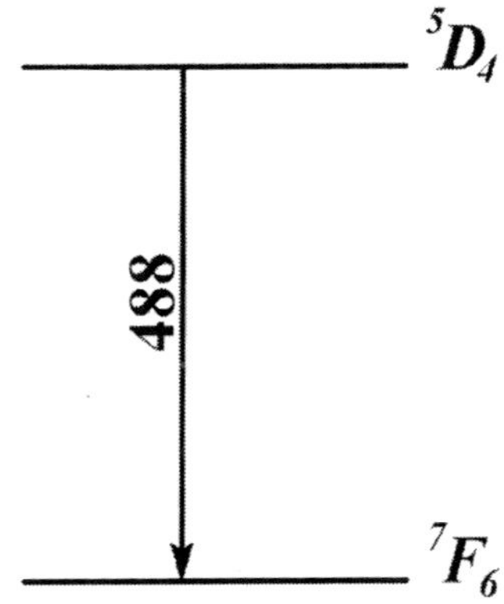

b

Figure 5. a) The dispersion curves for the terbium nitrate hydrate in the visible range. The corresponding wavelengths: A1 – 488 nm, A2 – 444 nm, U1 – nm, b) The corresponding transition between the energy levels.

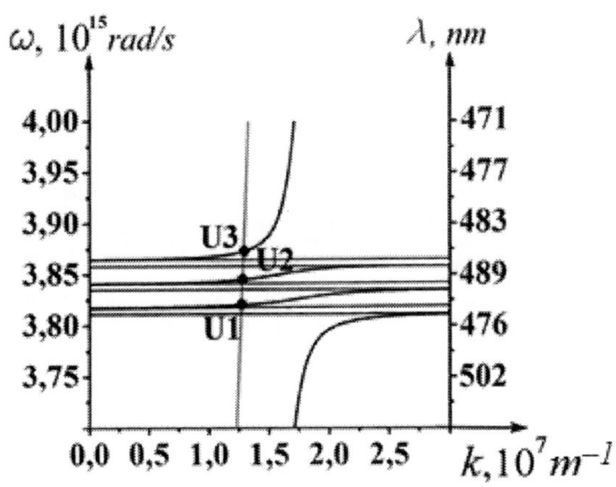

Figure 6. The dispersion curves for the terbium nitrate hydrate in the visible range with splitting.

The parameters of the electromagnetic waves at the points corresponding to the unitary polaritons, for which the refractive index is unity, are shown in Tables 2 and 3. In those points the reflection coefficient R at normal incidence is zero, i.e. the material is characterized by high transparency for the incident radiation.

Table 2. The parameters of the electromagnetic waves at the singular points in the visible range without splitting.

Singular points	$\omega, 10^{15}$rad/s	λ, nm	$k, 10^7 m^{-1}$	n	R	v, 10^8m/s
U1	4.69	401	1.57	1.00	0.00	1.75
A1	3.86	488	-	$-\infty$	1.00	0.00

A2	4.24	444	-	∞	1.00	0.00

Table 3. The parameters of the electromagnetic waves at the singular points in the visible range with splitting.

Singular points	ω,10^{15}rad/s	λ,nm	k,10^7m^{-1}	n	R	v, 10^8m/s
U1	3.81	494	1.27	1.00	0.00	0.09
U2	3.84	491	1.28	1.00	0.00	0.17
U3	3.87	487	1.29	1.00	0.00	0.77

5. Applications

Nowadays one of the main applications of the rare earth elements is to use them as the active medium in fiber optics and lasers. Nevertheless, industrial lasers based on terbium do not currently exist. As it has already been mentioned this is because the first high-power UV LED, allowing for pumping of the active medium in the visible range, appeared only at the end of the last century. Due to its high sensitivity and intensity the phosphors based on the rare earth elements have a high quantum yield. It is possible to create a more efficient high-energy radiation detector even at low intensity. The experimental and theoretical works presented in this paper allow using terbium as the phosphors in such detectors. The experimental setup presented in Figure 1 may be used for the superluminescence and lasing with λ = 550 observation [16, 17] in the dielectrics doped with the terbium ions. The high efficiency inverse population pumping is reached due to the "stopping of light" when the frequency of the exciting emission coincides with the spectral positions of the unitary polariton excitation (points U1, U2, and U3 in Figure 6). The group velocities of the corresponding polariton waves are very small and the probability of the exciting emission absorbance by the dielectric crystals becomes very high. For the unitary polaritons (points U1,U2 ,U3 in figure 6) the energy density of the laser electric field inside the discussed dielectrics essentially increases (on several orders) and lasing may be reached at fairly small exciting light intensity.

6. Conclusions

Summarizing, in this paper we have presented and analyzed the photoluminescence spectrum of terbium nitrate hydrate. The theoretical and experimental results for the energy level of free ions of Tb^{3+} and terbium nitrate hydrate levels have been obtained. The dispersion dependencies of the polaritons waves in terbium nitrate hydrate have been established. The positions of the unitary polaritons, for which the refraction index is unity, have been determined. It was shown that the group velocities of the unitary polaritons are thirty times less than the speed of light in the vacuum. The increase of the efficiency of the laser generating is justified. The experimental schemes for the terbium laser have been proposed. The possibility of laser generating at the wavelength of 550 nm has been shown.

Acknowledgments

This work was partially supported by Russian Foundation for Basic Research grants 15-02-02882, 16-52-00026.

References

[1] Hopfield J J 1958Theory of the Contribution of Excitons to the Complex Dielectric Constant of

IOP Conf. Series: Journal of Physics: Conf. Series **918** (2017) 012029 doi:10.1088/1742-6596/918/1/012029

Crystals *Phys. Rev.* **112** 1555–67
[2] Henry C H and Hopfield JJ1965 Raman Scattering by Polaritons *Phys. Rev. Lett.* **15** 964–6
[3] SukhoivanovIA and GuryevIV 2009*Photonic Crystals. Physics and Practical Modeling*(Heidelberg: Springer) p 242
[4] Vizoso B, Vazquez C, Civera R, Lopez-Amo M and Muriel M A 1994 Amplified fiber-optic recirculating delay lines *Journal of Lightwave Technology***12** 294–305
[5] Gorelik V S, Ivicheva S N, Lepnev L S, LitvinovaA O and Moiseenko V N 2015 Emission of opal photonic crystals filled with europium and terbium *Inorg Mater***51** 525–531
[6] Li M, Zhang P, Li J, Zhou J, Alexander S, Abramova V V, Klimonsky S O and Tretyakov Y D 2007 Directional emission from rare earth ions in inverse photonic crystals *Appl. Phys.*B **89** 251–262
[7] Pollard M E, Knight K J, Parker G J, Hewak D W and Charltonet M D B 2012 Fabrication of photonic crystals in rare-earth doped chalcogenide glass films for enhanced upconversion *SPIE OPTO. International Society for Optics and Photonics* 223–239
[8] Lee Y-S 2009 Principles of Terahertz Science and Technology (New York: Springer US)p 340
[9] Dzedolik I V and Pereskokov V 2014 Variation of spectral properties of dielectric ionic crystal in the terahertz range due to the polariton absorption *Applied Optics***53** 3318–27
[10] Tasolamproul A C, Zografopoulos D C and Kriezis E E 2011Liquid crystal-based dielectric loaded surface plasmon polariton optical switches *J. Appl. Phys.***110** 093102–093102-9
[11] Wu L, Jiang L, Ding C, Sheng Q, Ding X and Yao J 2015 Effect of Optical Pump on the Dielectric Properties of $LiTaO_3$ in Terahertz Range J *Infrared Milli Terahz Waves***36** 1–6
[12] Mendonça J T 2006 Photon acceleration and polariton wakefields in dielectric crystals *New Journal of Physics***8** 35–41
[13] Bünzli J C 2010 Lanthanide Luminescence for Biomedical Analyses and Imaging *Chemical Reviews***110** 2729–55
[14] Selvin P R 2002 Principles and Biophysical Applications of Lanthanide-Based Probes *Annual Review of Biophysics and Biomolecular Structure***31** 275–302
[15] Andradea A A , Pillaa V, Lourençob S A, Silvac A C A and Dantasc N O 2012 Fluorescence quantum efficiency dependent on the concentration of Nd^{3+} doped phosphate glass *Chemical Physics Letters***547** 38–41
[16] Gorelik V S and Burdanova M G 2016 Electromagnetic waves in optical fibres in a magnetic field *Laser Phys.* **26** 035001
[17] Gorelik V S and Burdanova M G 2014 Photoluminescence and polariton dispersion in erbium nitrate hydrateю*Laser Phys.***24** 125001
[18] Meinders E and Klinkenberg P F A Structure of the spectrum of neutral terbium, Tb I: Part IV*Physica* **32** 1617–32
[19] Klinkenberg P F A Structure of the spectrum of neutral terbium, Tb I: Part III *Physica***37** 197–214
[20] Dieke G H 1968 Spectra and Energy levels of Rare Earth Ions in Crystal (New York: Int. Pub.: A division of John Wiley & Sons) p 401

IOP Conf. Series: Journal of Physics: Conf. Series **918** (2017) 012030 doi:10.1088/1742-6596/918/1/012030

On the nature of crystal plasticity under compression

M N Magomedov
Chief Researcher, Institute for Geothermal Problems of Dagestan Scientific Center of Russian Academy of Sciences, Makhachkala, Russia
E-mail: mahmag4@mail.ru

Abstract. The nanostructuring of crystals into domains under uniform compression, the plasticity of a solid nanostructure under pressure, and the bimodal distribution of domains size are explained based on the dependence of the surface energy and surface pressure from the shape, size, and density of nanocrystals.

It is known that as the uniform (hydrostatic) compression grows, fluctuations in a thermodynamically equilibrium single crystal of an elementary (singlecomponent) substance are decreased. Based on these considerations, it seems that upon compression of the crystal its structure must become more homogeneous. However, experiments show that this is true only if the uniform compression is low ($v/v_0 > 0.85$), whereas at a high level of compression ($v/v_0 < 0.6$), a single crystal of an elementary substance acquires a nanocrystalline domain structure. A domain is a region of a crystal that differs from neighboring ones by the direction of the displacement vector of crystal lattice and has an intercrystalline surface. Here, v/v_0 is the ratio of the specific volume (per atom: $v = V/N$) of the crystal at pressure P and temperature T and volume when $P = 0$ and $T = 0$ K.

It was noted [1] that the distribution histograms for the number of megaplastic-deformation-induced domains plotted against their characteristic linear size have two maxima, d_{s1} and d_{s2} with $d_{s1} < d_{s2}$. Here, $d_s = \Sigma_s^{1/2}$, where Σ_s is the area of the electron-microscopic image of the domain. That is, domains resulting from nanostructuring have a bimodal size distribution.

Theoretical methods that studying the influence of nanostructuring on the strength and plasticity of materials assume that the intercrystalline surface in a crystal exists a priori. But the question why the crystal, starting from a certain pressure, experiences fragmentation (nanostructuring) and acquires a domain structure with a clear-cut interdomain surface remains unclear. When a crystal becomes nanostructured, the appearance of an interdomain surface (Σ) seems to make the system less stable because of the contribution of surface energy: $\Sigma\cdot\sigma$, where σ is the specific surface energy (energy per unit area). Then, why does the baric nanostructuring of the crystal take place? Why does a nanostructured solid substance under compression become plastic? Why do the distribution histograms of the number of domains resulted from nanofragmentation have a bimodal form? To answer these questions, one should know how the surface energy and surface pressure depend on the shape and size of the nanocrystal at different densities and temperatures.

It is shown (Fig. 1 top graph) that the nature of the baric structuring of the crystal is that the surface energy of the crystal becomes negative ($\sigma < 0$) at $P > P_{fr}$ [2, 3]. At this, as was shown in [4], the intercrystalline surface energy of domain (σ_d) in structured solid is related to the surface energy of freestanding nanocrystal by formula: $\sigma_d = \chi\cdot\sigma$, where the coefficient χ depends on the crystallographic indices of the contacting faces of the domains: $1 > \chi > 0$.

Therefore, if performed σ < 0, then we have σ_d < 0 too.

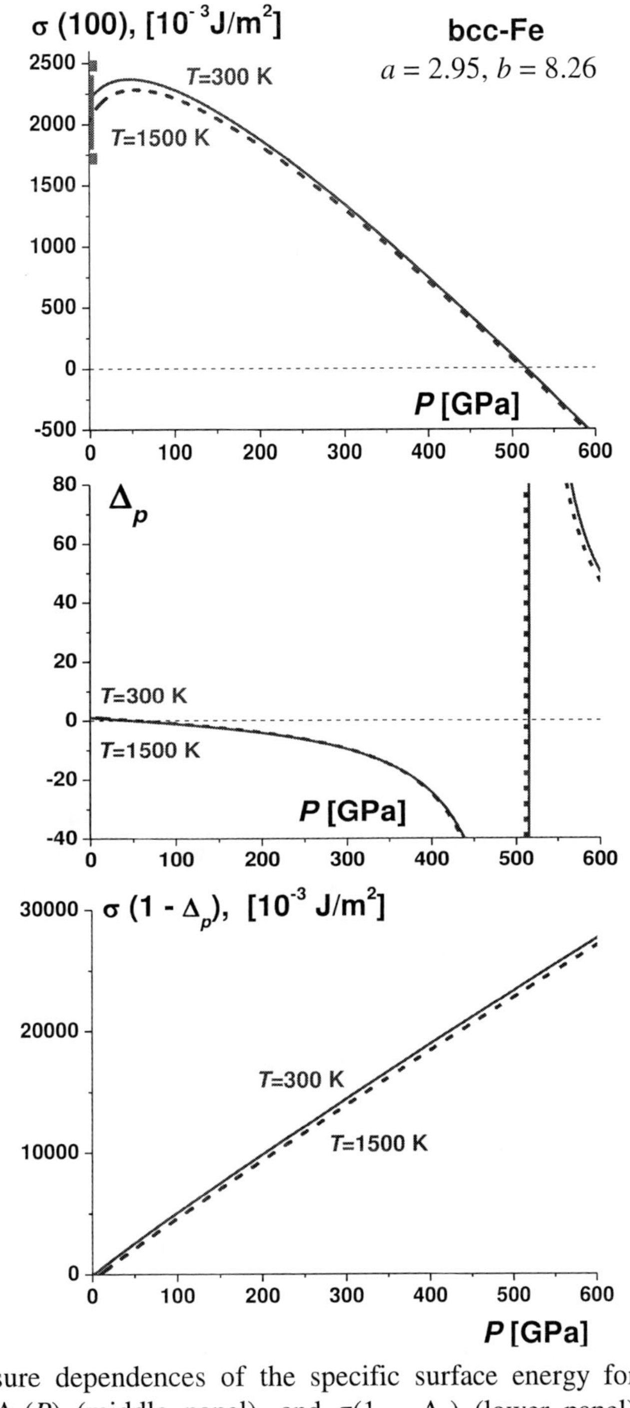

Figure 1. Pressure dependences of the specific surface energy for the (100) face (upper panel), $\Delta_p(P)$ (middle panel), and $\sigma(1 - \Delta_p)$ (lower panel) for BCC iron. Continuous lines, $T = 300$ K; dashed line, $T = 1500$ K.

In addition, it is shown (Fig. 1 bottom graph), that the structured at compression solid substance becomes plastical due to the growth of surface pressure P_{sf} in the nanocrystal. For the surface pressure was obtained the following expression [5]:

$$P_{sf} = \left[\frac{\partial(\sigma\Sigma / N)}{\partial v}\right]_{T,N} = P_{ls}\cdot(1-\Delta_p)\,, \tag{1}$$

where P_{ls} is the Laplace pressure, which is determined by the change of surface area at the change of volume of nanosystem:

$$P_{ls} = \sigma\cdot\left[\frac{\partial(\Sigma / N)}{\partial v}\right]_{T,N}\,, \tag{2}$$

the expression for the function Δ_p is given by: $\Delta_p = -\,[\partial\ln(\sigma)/\partial\ln(\Sigma/N)]_{T,N}$.

In Fig. 1 are shown the calculated by method from [5] for the macrocrystalline body-centered cubic (BCC) iron the isothermal dependencies from pressure (in GPa) of the next functions: σ(100) is the specific surface energy (in 10^{-3} J/m^2) for (100) face – top graph, $\Delta_p(P)$ – the middle graph, $\sigma\cdot(1-\Delta_p)$ – the lower graph.

It was assumed that the atoms interact via the pairwise potential of the Mie-Lennard-Jones:

$$\varphi(r) = \frac{D}{(b-a)}\left[a\left(\frac{r_o}{r}\right)^b - b\left(\frac{r_o}{r}\right)^a\right]. \tag{3}$$

Here, D and r_o are the depth and the coordinate of the potential minimum, $b > a > 1$ are the numerical parameters. Parameters of the Mie-Lennard-Jones potential (3) for BCC-Fe defined in [3] are: $r_o = 2.4775\cdot10^{-10}$ m, $D/k_B = 12576.7$ K, $a = 2.95$, $b = 8.26$.

Based on these parameters, the normal specific volume, the Debye temperature and the first Gruneisen parameter at $v/v_o = 1$ are [3]: $v_o = [\pi/(6k_p)]r_o^3 = 7.0494$ cm^3/mole, $\Theta(1) = 408.073$ K, $\gamma(1) = 1.702$. Here, k_B is the Boltzmann constant, k_p is the structure packing coefficient (for BCC structure $k_p = 0.6802$). The state equation for macrocrystalline BCC iron was calculated in [3], where the good agreement with experimental data was obtained.

In Fig. 1 the solid line is the isotherm of 300 K, the dashed line represents the isotherm of 1500 K. The vertical line in the top graph is the area of the scatter of the experimental data for σ(100) at $P = 0$ [2, p. 416]: σ(100)/[10^{-3} J/m^2] = 1720 – 2480.

Since the inequality $\sigma \le 0$ is fulfilled for $P \ge P_{fr}$, the Laplace pressure will be negative in accordance with (2) ($P_{ls} \sim \sigma < 0$), i.e. P_{ls} will be tensile. However, when nanocrystals or macrocrystals are compressed, function $\sigma(1-\Delta_p)$ varies at the fragmentation point continuously, as it shown in Fig. 1 for BCC-Fe. Therefore, surface pressure according to (1) $P_{sf} = P_{ls}\,(1-\Delta_p) \sim \sigma(1-\Delta_p) > 0$ continuously grows with the density of the nanocrystal under compression. A domain resulted from compression will experience elevated surface pressure P_{sf}. The higher the outer pressure, the higher pressure P_{sf} (Fig. 1). The fragmentation-induced pressure P_{sf} squeezes the surface bounded domain still greater. As a result, fragmentation-induced domains shrink, leaving a space in between. Therefore, the material nanostructured by baric fragmentation will resemble a fluid material: like any liquid phase, it will take the shape of a vessel containing it.

The nature of the bimodal distribution of domain over sizes, it was felt that the maximum size of the d_{s1} corresponds to the domains of deformation origin, and the second maximum size $d_{s2} > d_{s1}$ corresponds to recrystallized domains [1]. However, this bimodality can be explained on the basis of the dependence of energy parameters of a nanocrystal from the shape of its surface. As it shown in [2, 6], using the RP-model of the nanocrystal at high pressures (at $P \ge P_{fr}$) formed under of the plastic fragmentation the domain is energetically advantageous to have a noncubic (i.e. platelike or rodlike) forms. This leads to the bimorphism: within the framework of RP model at constant values of k_p, T, v and N the noncubic nanocrystal with equal probability can have either platelike or rodlike form. Both these shapes of nanocrystal have equal values Σ and σ, but different values of the diameter (the distance between the most distant atoms): the diameter of platelike shape is less than rodlike. A bimodal distribution of the formed under the "megaplastic" deformation of domains on size, discovered by the authors of [1] can be explained by means of the bimorphism.

At $P > P_{fr}$ the noncubic shapes of the nanocrystal are energetically more favorable than the cubic shape of the same isomers. Therefore, if the pressure is reduced drastically (so that the system would have no time to relax into a single-crystalline state) at $P > P_{fr}$,the compressed nanostructured substance is expected to disintegrate into fragments with a noncubic surface. The dispersion of solids under sudden reset of hydrostatic pressure was discovered in [7]. In that work was shown that, when a compression pressure exceeds some material-specific critical value and at the fast pressure drop (about 2 GPa/ms) not only inorganic materials but also teflone, rubber, paraffin, etc., were dispersed. The substances fragments, which were observed in [7], possess a typical splintery (oblate or prolate) habitus.

Conclusions

Thus, it was found that (i) the baric nanostructuring of a crystal takes place when function σ becomes negative, (ii) a nanostructured solid becomes the ductile under compression because of the growth of the surface pressure, which compresses the domains, and (iii) the bimodal distribution for domains which produced by baric nanofragmentation is explained by the fact that under such conditions it is energetically favorable for the domains to have a maximal specific (per atom) surface area.

Therefore, the compression-induced exothermic transition of the crystal to a nanostructural state leads to a sharp rise in the domain density, on the one hand, and to a sharp decrease in the viscosity of the resulting nanostructured medium on the other. Domains produced by such exothermic baric fragmentation exhibit a splintery (platelike or rodlike) habitus. A crystal nanostructured by exothermic baric fragmentation becomes ductile (passes to a fluid state) and takes the shape of a vessel. If the pressure is sharply drop, the nanostructured medium shatters like sprayed liquid.

This work was funded by the Russian Foundation for Basic Research (grant no. 16-03-00041_a) and by a program of the Russian Academy of Sciences Presidium (project no. 1. 11P(1)).

References

[1] Glezer A M *et al* 2014 *Doklady Physics: Journal of the Russian Academy of Sciences* **59** 360
[2] Magomedov M N 2010 *Study of Interatomic Interaction, Vacancy Formation and Self-Diffusion in Crystals* (Moscow: Fizmatlit) [in Russian]
[3] Magomedov M N 2015 *Technical Physics* **60** 1619
[4] Zadumkin S N and Karashaev A A 1965 in *Surface Phenomena in Melts and Solid Phases Developing from Them* (Nal'chik: Kabardino-Balkarskoe Knizhnoe Izd.) pp. 85–88 [in Russian]
[5] Magomedov M N 2015 *Nanotechnologies in Russia* **10** 89
[6] Magomedov M N 2016 *Technical Physics* **61** 722
[7] Fedorov V T and Khokonov Kh B 1988 *Soviet Physics Doklady* **33** 463

IOP Conf. Series: Journal of Physics: Conf. Series **918** (2017) 012031 doi:10.1088/1742-6596/918/1/012031

On the polymorphism of silicon and germanium

M N Magomedov
Chief Researcher, Institute for Geothermal Problems of Dagestan Scientific Center of Russian Academy of Sciences, Makhachkala, Russia
E-mail: mahmag4@mail.ru

Abstract. Based on the pairwise interatomic Mi-Lennard-Jones potential the state equations and the baric dependencies of the properties for various polymorphic modifications of silicon and germanium were obtained. It is shown that for the semiconductor phase the best results are obtained using the interatomic potential of the elastic type, and for metallic phases the best results are obtained using the potential of the plastic type that has smaller depth of the potential well.

It is known [1-5] that under compression of the silicon (Si) or germanium (Ge) a phase transition (PT) of the first kind is observed, in which a diamond structure (*cd*-phase) converted in a body-centered tetragonal structure type β-Sn (*tin*-phase). At this the inter-atomic bonds are converted from the covalent type to the metal ones. At further compression, the structure of β-Sn is converted in other more dense polymorphs. Although there are many theoretical methods for calculation of equations of state of various crystalline modifications of Si and Ge were proposed [3-5], but the nature of the baric metallization is not entirely clear. Especially unclear the nature of irreversibility of PT from the diamond structure (*cd*-phase) in the structure β-Sn (*tin*-phase).

In this paper, the method for calculation of the dependence of pressure (P) versus of volume (V) and temperature (T) of the crystal is proposed, and on the basis of this method the irreversibility of polymorphic PT of the first kind that occur in Si and Ge under compression is studied.

It was assumed that the atoms interact via the pairwise potential of the Mie-Lennard-Jones:

$$\varphi(r) = \frac{D}{(b-a)}\left[a\left(\frac{r_o}{r}\right)^b - b\left(\frac{r_o}{r}\right)^a\right]. \qquad (1)$$

Here, D and r_o are the depth and the coordinate of the potential minimum, $b > a > 1$ are the numerical parameters.

At normal conditions (T = 300 K and P = 0) the crystals of Si and Ge have the diamond structure (*cd*-phase) with $k_n = 4$ and $k_p = 0.3401$, where k_n is the first coordination number, k_p is the structure packing coefficient. The values of r_o, b and a for *cd*-phase of Si and Ge were determined in [6], and presented in the table. With regard to D in [6] it was shown that for covalent crystals the value D can be defined both from the elastic modulus (B_0) and from the specific energy of atomization (L_0) at $T = 0$ K and $P = 0$ by the equations:

$$D_b = \frac{18 B_0 V_0}{k_n ab N} = D_s + \Delta D, \qquad D_s = \frac{L_0}{k_n/2}. \qquad (2)$$

For metals these two formulas give the same result, but for the covalent crystals the depth of interatomic potential that is restored from the elastic modulus (D_b) is approximately twice bigger than the magnitude, which follows from the energy of atomization of the crystal (D_s).

Moreover, the difference: $\Delta D = D_b - D_s$, for crystals of elements of the carbon subgroup (diamond, Si, Ge, α-Sn, Pb) decreases at increasing of the atomic mass m, and disappears for lead: ΔD(Pb) $\approx$ 0 [6].

Table. The parameters of interatomic potential (1) for the diamond structure of Si and Ge.

Crystal	r_o [6] nm	b	a	B_0 [6] GPa	D_b eV	L_0 [6] eV	D_s eV	ΔD eV
Si	**0.2351**	4.0	2.48	**97.7**	5.54	**4.64**	2.32	3.22
Ge	**0.2450**	4.3	2.75	**74.9**	4.03	**3.88**	1.94	2.09

The feature of this interatomic bond is that for its destruction is enough to break the weakest of the two links. It was shown [6, 7] that the energy of whole interatomic bond (D_b) is working under the elastic (reversible) deformation, and the energy of weak bond (D_s) is working at the plastic (irreversible) deformation of the covalent crystal. From D_b it is necessary to calculate parameters that not lead to the interatomic bonds rupture while measurement: the sound velocity, the Debye temperature, the thermal expansion coefficient. From D_s are calculated parameters, associated with the bonds ruptures, such as energies of sublimation, vacancy formation, self-diffusion and the surface energy. Under the plastic (irreversible) deformation of the covalent crystal only weak links are breaking, and the depth of the potential (1) is determined by the value D_s.

When the parameters of potential (1) were defined, we used the proposed in [8] method for the calculating of the pressure P versus the volume V dependence for Si and Ge by means of elastic (i.e. with D_b) and plastic (i.e. with D_s) types of potential (1).

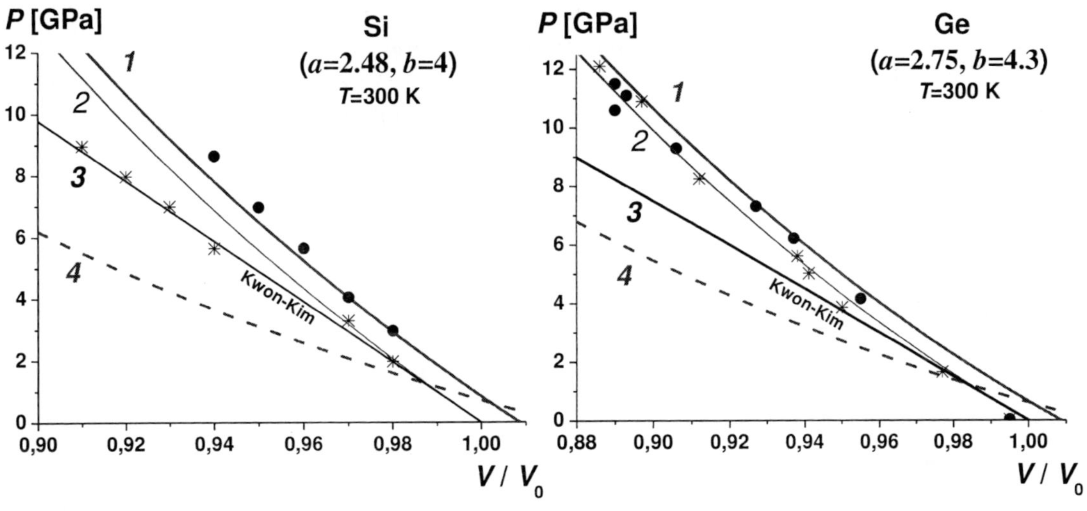

Figure 1. State equation $P(V/V_0)$ for *cd*-phase of silicon (left) and germanium (right).

In Fig. 1 left shows the dependence of $P(V/V_0)$ for *cd*-phase of Si, where $V_0 = N\pi r_o^3/(6\ k_p)$. Symbols are the experimental data from [5]: points are Bridgman data, stars are X-ray data. Curve *2* was calculated by means of the Murnaghan–Birch equation in the following form:

IOP Conf. Series: Journal of Physics: Conf. Series **918** (2017) 012031 doi:10.1088/1742-6596/918/1/012031

$$P(x) = 1.5B_0 \frac{(1-x^2)}{x^7}\left[1+\frac{3}{4}(B_0'-4)\frac{(1-x^2)}{x^2}\right], \qquad x = \left(\frac{V}{V_0}\right)^{1/3}. \tag{3}$$

Curve *3* is the result of calculations by means of the Kwon–Kim equation, which is given by [5]:

$$P(V/V_0) = B_0[1-(V/V_0)]. \tag{4}$$

In Eqs. (3) and (4) for *cd*-phase of Si there were taken $B_0 = 97.8$ GPa and $B_0' = 4.1$ [5].

In Fig. 1 right shows the dependencies of $P(V/V_0)$ for the *cd*-phase of Ge. Dots and stars are the results of measurements [1] by the methods of ungasketted cell and gasketted cell, respectively. Curve *2* is obtained by the Eq. (3), curve *3* was calculated by the Kwon–Kim equation (4). In Eqs. (3) and (4) for *cd*-phase of Ge were taken B_0= 74.9 GPa, B'_0= 4.35 [1].

Our calculations at 300 K are shown by curves *1* and *4* for elastic and plastic types of the potential (1), respectively. As it is seen from Fig. 1, our dependencies $P(V/V_0)$ with the elastic type of the potential (i.e. with D_b – curve *1*) describe well the experimental data for the *cd*-phase of Si and Ge. Along with this, the calculations with the plastic type of the potential (i.e. with D_s – curve *4*) are yield the values of pressure much less than the experimental data.

It was found that the dependence of $P(V)$ for the metallized phases of Si and Ge can also be calculated by the method [8], but with the parameters of interatomic potential of the plastic type (i.e. with D_s). Herewith the values r_o i.e. the coordinates of the minimum of potential (1) for β-Sn and HCP structures were determined by means of the known dependence of interatomic distance from the crystal structure [9, p. 141]: if at the polymorphic transformation the value of first coordination number k_n increases in the sequence:

$$3\ (\text{structure A9}) \rightarrow 4\ (\text{A4}) \rightarrow 6\ (\text{A5}) \rightarrow 8\ (\text{A2}) \rightarrow 12\ (\text{A1}),$$

then the length of interatomic bond increases in the sequence:

$$1.00 \rightarrow 1.02 \rightarrow 1.09 \rightarrow 1.11 \rightarrow 1.14.$$

For structures β-Sn and HCP the first coordination number and the packing coefficient are equal [9, p. 167]: $k_n = 6$, $k_p = 0.524$ and $k_n = 12$, $k_p = 0.7405$, respectively. Then for the coordinates of the interatomic potential minimum we can obtained foloving values:

$$r_o(k_n{=}6) = 0.25123 \text{ nm и } r_o(k_n{=}12) = 0.26276 \text{ nm – for Si}, \tag{5}$$

$$r_o(k_n{=}6) = 0.26181 \text{ nm и } r_o(k_n{=}12) = 0.27382 \text{ nm – for Ge}. \tag{6}$$

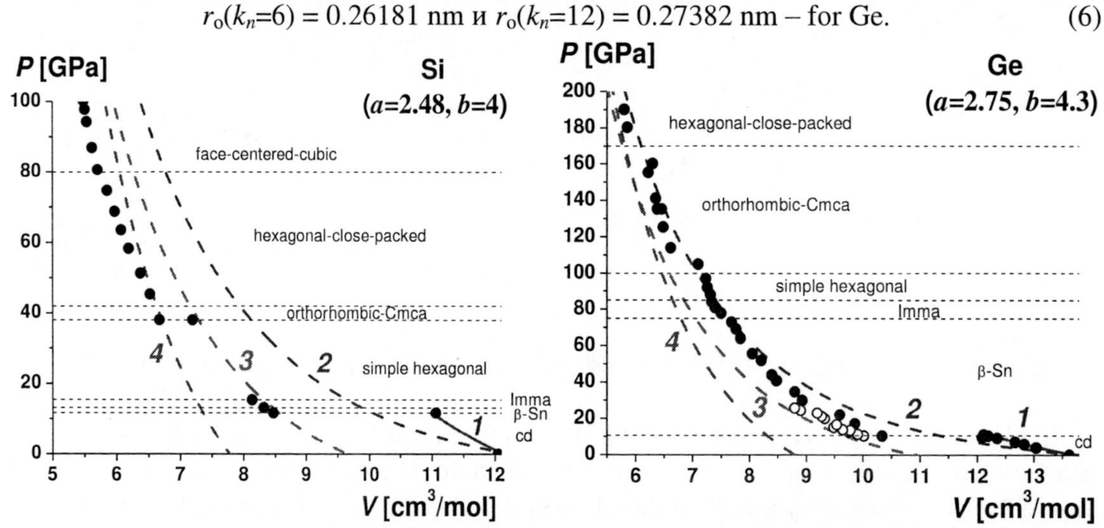

Figure 2. State equation $P(V)$ for silicon (left) and germanium (right).

In Fig. 2 shows the dependence $P(V)$ for different structures of Si and Ge. Horizontal lines show the pressure boundary for the polymorphic phases [3, Tabl. 1].

For silicon are obtained (the pressure of the phase transition P_t is given in [GPa]):

cd→(**11.7**)→β-Sn→(**13.2**)→Imma→(**15.4**)→sh→(**~38**)→Cmca→(**~42**)→hcp→(**79-80**)→fcc

For germanium are obtained the following estimates for P_t:

cd → (**10.6**) → β-Sn → (**~75**) → Imma → (**~85**) → sh → (**~100**) → Cmca → (**~170**) → hcp

Curves show our calculations of $P(V)$ along the isotherm 300 K: *1* and *2* are the state equations for *cd*-phase, which are obtained by means of the elastic and the plastic types of potential (see Fig. 1); *3* and *4* are the state equations for the structure of β-Sn (*tin*-phase) and for the hexagonal closely packed (HCP) structure, which were calculated with the plastic type of potential (i.e. with D_s). The points are the experimental dates from works [1-5].

As can be seen from Fig. 2, the obtained dependencies for Si are in good agreement with experimental data from [2], and for Ge from [1, 4]. The 300 K isotherms calculations for structures β-Sn and the HCP with the elastic type of potential show that calculated dependence of $P(V)$ lies much higher than the experimental data.

The transition from D_b to D_s indicates that baric metallization of Si and Ge is the same irreversible process, as the plastic deformation of the crystal. For this reason, after removing of the pressure the metallizing state is not returned to the original covalent *cd*-phase, and for the obtaining of the covalent diamond structure from the baric metallized state it is necessary to anneal it under $P = 0$ and at high (500 – 900 K) temperatures [1-3].

By the method presented in [8], the baric dependencies of the lattice properties for the *cd*- and *tin*-phases of silicon and germanium were calculated. The following properties were calculated: Θ is the Debye temperature and $\Theta'(P) = (\partial\Theta/\partial P)_T$ is the derivative of Θ with respect to pressure along an isotherm, the first ($\gamma = -[\partial\ln(\Theta)/\partial\ln(V)]_T$), the second ($q = [\partial\ln(\gamma)/\partial\ln(V)]_T$) and the third ($z = -[\partial\ln(q)/\partial\ln(V)]_T$) Gruneisen parameters, $B_T = -[\partial P/\partial\ln(V)]_T$ is the isothermal bulk modulus and it's pressure derivative $B'(P) = (\partial B_T/\partial P)_T$, $\alpha_p = [\partial\ln(V)/\partial T]_P$ is the thermal expansion coefficient and $\alpha_p'(P) = (\partial\alpha_p/\partial P)_T$, $C_v = \alpha_p\cdot B_T\cdot V/\gamma$ and $C_p = C_v(1 + \gamma\cdot\alpha_p\cdot T)$ are the isochoric and isobaric heat capacity, $C_v'(P) = (\partial C_v/\partial P)_T$ and $C_p'(P) = (\partial C_p/\partial P)_T$, the product: $\alpha_p\cdot B_T = (\partial P/\partial T)_V = (\partial S/\partial V)_T$, σ is the specific (per unit area) the surface energy for the face (100), and also $\sigma'(T) = (\partial\sigma/\partial T)_V$ and $\sigma'(P) = (\partial\sigma/\partial P)_T$.

From the calculation of the isotherm 300 K it follows that at *cd-tin* PT the next properties are abruptly increased: q, z, α_p, C_v, C_p, $|\alpha_p'(P)|$, $\alpha_p\cdot B_T$, and the next ones abruptly reduced: Θ, γ, B_T, $B'(P)$, $|C_v'(P)|$. At *cd-tin* PT the thermal expansion coefficient undergoes the greatest change: 80.9 % for Si, and 42.7 % for Ge.

This work was funded by the Russian Foundation for Basic Research (grant no. 16-03-00041_a) and by a program of the Russian Academy of Sciences Presidium (project no. 1. 11P(1)).

References

[1] Menoni C S et al 1986 *Phys. Rev.* B **34** 362
[2] Duclos S J et al 1990 *Phys. Rev.* B **41** 12021
[3] Mujica A et al 2003 *Rev. Modern Phys.* **75** 863
[4] Chen X et al 2011 *Phys. Rev. Lett* **106** 135502
[5] Pandya T C et al 2012 *Journal of Physics: Conference Series* **377** 012097

[6] Magomedov M N 2004 *Russian Journal of Inorganic Chemistry* **49** 1906
[7] Magomedov M N 2013 *Technical Physics* **58** 1789
[8] Magomedov M N 2015 *Technical Physics* **60** 1619
[9] Batsanov S S 2000 *Structural Chemistry* (Moscow: Dialog-MGU) [in Russian]

IOP Conf. Series: Journal of Physics: Conf. Series **918** (2017) 012032 doi:10.1088/1742-6596/918/1/012032

Projector approach to the general linear equation with variables from the subring of the rational functions and a factorable coefficient

[1]T G Voytik, [2]G S Poletaev, [1]S A Yatsenko
[1] Odessa National Maritime University, Odessa, Ukraine
[2] Odessa State Academy of Buildings and Architecture, Odessa, Ukraine
poletayev_gs@ukr.net

Abstract. The problem of finding two rational, linearly related functions with poles in the half-planes is considered. A new approach to this theory is proposed. The problem with factorable rational coefficients is formulated and solved. The method is based on the theory that is derived from previous research by the second author. The result is used for the corresponding abstract equations in the ring, with a special factorization pair of subrings. The formulas of solutions and illustrative examples are given. The procedure does not involve Fourier integral, Cauchy integral theory, or the Hölder function.

Introduction. The paper follows on research in [1]. The approaches to equations theory that are based on the application of functions of a complex variable and the Cauchy-type integrals may lead to significant analytic difficulties, which cannot be always justified. In particular, it concerns equations and boundary conditions simulated in Riemann-Hilbert problem and related problems [2-7]. New ideas and results of new approaches to this kind of problem were first proposed in [2]. That research involved alternative general assumptions and did not involve Hölder functions. The aim of the present work is to develop a new projector approach, simplifying a similar type theory of Riemann-Hilbert-Privalov problem. This problem is formulated further in the text. Namely, it is the problem of finding two rational functions with poles belonging to different semi-planes using a related general linear equation with a factorable coefficient on the closed real axis. The goal is achieved through using the general results of [8-14] and the theorem of existence with formulas of solutions from [15]. The basic provisions, notation and definitions are taken from [1].

1. Problem defenition and the formula of solution

Problem: "For the given rational functions $A(x), B(x)$, $-\infty < x < \infty$, find a pair of rational functions $X^+(z) \in \Re_r^+, Y_-(z) \in \Re_{r-}$ belonging to $\Re_r$, such that all the poles (if existing) of the first and the second functions are located in the lower and the upper semiplane respectively. The poles on the closed real axis conform to the following general linear equation:

$$\alpha A(x)X^+(x) + \beta Y_-(x) = \gamma B(x); x \in \{-\infty;\infty\}; \alpha\beta \neq 0; \alpha,\beta,\gamma \in C". \qquad (1)$$

where all known functions are defined on the closed real axis, and it is assumed that for $x = +\infty$ and $x = -\infty$ each has matching end values, equal to the corresponding limits.

When solving the problem in $\Re_r$, we start from the obvious assumption that each function, and therefore the entire ratio (1) can be projected onto the entire complex plane [1]. This can be done by replacing the real variable x with a complex variable z, while remaining in the appropriate subclass of rational functions. Thus, instead of (1), we get the equation:

$$\alpha A(z)X^{+}(z)+\beta Y_{-}(z)=\gamma B(z); z\in(C\cup\infty);\alpha\beta\neq 0;\alpha,\beta,\gamma\in C. \tag{2}$$

where $A(z), B(z)\in\Re_{r}$, $z\in C$ are assumed to be known functions; while $X^{+}(z)\in\Re_{r}^{+}, Y_{-}(z)\in\Re_{r-}$ are unknown functions. Any pair of rational functions $X^{+}(z)\in\Re_{r}^{+}, Y_{-}(z)\in\Re_{r-}$ being the solution to (2) generates the desired solution to (1) and to the problem under consideration through the restriction onto the closed real axis. Since the results from [8, 9] (see also [15]) can be implemented in the ring R = $\Re_{r}$ with the factorization pair ($\Re_{r}^{+},\Re_{r}^{-}$) and owing to the obvious transformations under appropriate assumptions, we obtained the following formulas to solve the equations (2), (1) and the problem under consideration [1, 8-15]:

$$X^{+}(z)=\frac{\gamma}{\alpha}\cdot\Gamma^{+}(z)S^{0}P^{+}[T^{-}(z)B^{+}(z)], Y_{-}(z)=\frac{\gamma}{\beta}\{B_{-}(z)+[T^{-}(z)]^{-1}\cdot P_{-}[T^{-}(z)B^{+}(z)]\}; \tag{3}$$

$$X^{+}(x)=X^{+}(z)\downarrow_{z=x},\ Y_{-}(x)=Y_{-}(z)\downarrow_{z=x},\ -\infty<x<\infty. \tag{4}$$

where

$$A^{-1}(z)=\Gamma^{+}(z)S^{0}T^{-}(z); z\in C.$$

2. Examples

Let us solve the problem set by the boundary condition on the closed real axis. The boundary condition is given by equation (1):

1) when

$$A(x)=\frac{x^{2}+9}{x^{2}+1};\quad B(x)=\frac{x^{2}+1}{(x^{2}+4)(x+5i)};\alpha=500,\beta=250,\gamma=10000.$$

We find

$$A^{-1}(z)=\frac{z+i}{z+3i}\cdot 1\cdot\frac{z-i}{z-3i}.\qquad \text{Hence, } S^{0}=1;$$

$$\Gamma^{+}(z)=\frac{z+i}{z+3i};T^{-}(z)=\frac{z-i}{z-3i};(T^{-}(z))^{-1}=\frac{z-3i}{z-i}.$$

B(z) is expanded into the form:

$$B(z)=-\frac{1}{4(z+2i)}+\frac{3}{28(z-2i)}+\frac{8}{7(z+5i)}.$$

Therefore:

$$B^{+}(z)=\frac{1}{28}\cdot\frac{25z+29i}{(z+2i)(z+5i)};\qquad B_{-}(z)=\frac{3}{28(z-2i)}.$$

Implementing the formulas (3), we find the solution representation in the form:

$$X^{+}(z)=\frac{1}{7}\cdot\frac{99z^{2}+234zi-135}{z^{3}+10z^{2}i-31z-30i}, Y_{-}(z)=\frac{2}{7}\cdot\frac{41z-67i}{z^{2}-3zi-2}. \tag{5}$$

2) When,

$$A(x)=\frac{x^{2}+1}{x^{2}+9};\ B(x)=10;\alpha=5;\beta=2;\gamma=1,$$

It should be:

$$5\cdot\frac{z^2+1}{z^2+9}X^+(z)+2\cdot Y_-(z)=10\,.$$

Similarly to section 1), we consistently find:

$$[A^{-1}(z)]^{-1}=\frac{z+3i}{z+i}\cdot 1\cdot\frac{z-3i}{z-i};\; X^+(z)=\frac{1}{5}\cdot\frac{z+3i}{z+i}\cdot 1\cdot P^+\left[\frac{z-3i}{z-i}\cdot 10\right];$$

$$X^+(z)=\frac{10}{5}\cdot\frac{z+3i}{z+i}\cdot P^+\left[1-\frac{2i}{z-i}\right]=2\cdot\frac{z+3i}{z+i}.\; Y_-(z)=\frac{1}{2}\cdot\{0+\frac{z-i}{z-3i}P_-\left[\frac{z-3i}{z-i}\cdot 10\right]\}=$$

$$=5\cdot\{\frac{z-i}{z-3i}P_-\left[1-\frac{2i}{z-i}\right]\}=5\cdot\{\frac{z-i}{z-3i}\left[-\frac{2i}{z-i}\right]\}=-\frac{10i}{z-3i}.$$

Thus,

$$X^+(z)=2\cdot\frac{z+3i}{z+i};Y_-(z)=-10\cdot\frac{i}{z-3i}. \qquad (6)$$

The solutions corresponding to the constant in the right-hand side play an important role in the theory of such type equations [8].

Conclusions and prospects.

Using the assumptions made, a transition can be made from Riemann-Hilbert-Privalov problem to a related problem. The sought-for functions are assumed to belong to the respective subsets of the rational functions. The related problem with the factorable rational coefficient has been formulated and solved. The solution is constructed explicitly using the assumptions. The constructions are simpler and exempt from the Cauchy-type integral, the notion of index and the Hölder condition dating back to F.D. Gahov and others ([3-7], compare to [2]). They are based on the results obtained by the second author for the corresponding equations in the ring with the factorization pair of the subrings [8-14]. We note finally that a database of specific examples will lay the groundwork for studying the properties of the problem and its solutions further on.

References

[1] Voytik T G, Poletaev G S and Yatsenko S A 2015 *Trudyi vosmoy Vserossiyskoy konferentsii (Moscow)* part II (Izd. MGTU im. N.E. Baumana) 125-9
[2] Krein M G 1958 *Uspehi matematicheskih nauk* **5 (83)** pp 3-120.
[3] Gahov F D and Cherskii Yu I 1978 *Uravnenie tipa svertki* (Moscow: Nauka) p 296
[4] Popov G Ya, Kerekeshcha P V and Kruglov V E 1976 Ed. Popov G Ya *Metod faktorisatsii i ego chislennaia realisatsiia* (Odessa: Odessa State University) p 82
[5] Mhitarian S M 1968 *Isvestiia AN Armianskoi SSR Mehanika,* **XXI** *№ 5-6* pp 3-20.
[6] Akopian V N and Dashchtoian L L 2013 *Sovremennye problemy mehaniki deformiruemogo tverdogo tela, differentsial'nyh i integral'nyh uravnenii.* Odessa p 12
[7] Cherskii Yu I, Kerekeshcha P V and Kerekeshcha D P 2010 *Metod sopryazheniya analiticheskih funktsiy s prilozheniyami* (Odessa: Astroprint) p 552
[8] Poletaev G S 1988 Ob uravneniiah i sistemah odnogo tipa v kol'tsah s faktorisatsionnymi parami. *Preprint /AN USSR. Institut matematiki: 88.31* (Kiev) p 20
[9] Poletaev G S 2000 Ob odnoproektornyih vtorogo poryadka uravneniyah s pravilno faktorizuemyimi koeffitsientami v koltse s faktorizatsionnoy paroy *Vestnik Hersonskogo gos. tehnich. un-ta, № 2 (8)* 191–195.

[10] McNabb A, Schumitzky A 1972, March. Factorization of operators – I: Algebraic theory and examples. *Journal of Functional Analysis, Vol. 9, № 3*, 262–295. *Doi:10.1016/0022-1236(72)90002-x*

[11] Poletaev G S 1991, September. Abstract analogue of a dual equation of convolution type in a ring with a factorization pair. *Ukrainian Mathematical Journal, Vol. 43, № 9*, 1124–1135. *Doi:10.1007/bf01089213*

[12] Poletaev G S, Voytik T G, Yatsenko S A 2013 Nahozhdenie dvuh ratsionalnyih funktsiy s polyusami iz poluploskostey po lineynomu uravneniyu s pravilno faktorizuemyim koeffitsientom // *GlushkovskI chitannya. NTUU «KPI», Kiev,* 74-77.

[13] Poletaev G S 2015 Podvid dvuproektornyih pervogo poryadka uravneniy s pravilno faktorizuemyim koeffitsientom v koltse s faktorizatsionnoy paroy // *XYII International Conference DYNAMICAL SYSTEM MODELLING AND STABILITY INVESTIGATION. MODELLING & STABILITY. ABSTRACTS OF CONFERENCE REPORTS. Kiev, Ukraine. May 27-29, 2015. VISNIK Kiyivskogo natsIonalnogo unIversitetu im. Tarasa Shevchenka,* 46.

[14] Poletaev G S 2016 Metod reshcheniia abstraktnyh uravnenii s dvumia neisvestnymi is podkolets faktorisatsionnoi pary. *Materialy mizhnarodnoi naukovo-praktychnoi konferentsii «Matematyka v suchasnomu tekhnichnomu universyteti», 24-25 hrudnia 2015 r. Kiev*, 85–88.

[15] Voytik T G, Poletaev G S, Yatsenko S A 2016 Method nahozhdeniia ratsional'nyh funktsii s poliusami is rasnyh poluploskostei po uravneniiu s pravil'no faktorisuemym koeffitsientom. *Naukovi*

IOP Conf. Series: Journal of Physics: Conf. Series **918** (2017) 012033 doi:10.1088/1742-6596/918/1/012033

Pseudoscalar lattice modes in the amino acid crystals and DNA

V S Gorelik[1,2], V P Sverbil[1], B P Gorshunov[3,4], A I Seredin[2]

[1]Lebedev Physical Institute of the Russian Academy of Sciences, Moscow, Russia
[2]Bauman Moscow State Technical University, Moscow, Russia
[3]Moscow Institute of Physics and Technology, Dolgoprudny, Russia
[4]Prokhorov General Physics Institute of the Russian Academy of Sciences, Moscow, Russia

gorelik@sci.lebedev.ru

Abstract. Intense sharp lines corresponding to the librational modes were found in the low-frequency Raman scattering spectra of the glycine, lysine, asparagine and tyrosine aminoacids as well as in the dry DNA crystal lattices. According to the group-theoretical analysis such modes were assigned to the pseudoscalar type of symmetry.

1. Introduction

The effective methods for studying the vibrational spectra of molecules and crystals are infrared (IR) spectroscopy and Raman spectroscopy (RS). The vibrational spectra of amino acids were researched in many works using IR spectroscopy [1-2] and Raman spectroscopy [3-4]. In this paper we have carried out the group-theoretic analysis of the normal vibrations of such crystals as glycine, lysine, asparagine and tyrosine and registered the Raman spectra of the corresponding amino acids and dry DNA.

2. Experimental results and discussion

The schematic of the experimental setup for the Raman spectra registration is shown in Figure 1. The argon laser (1) with the wavelength of 514.5 nm and the power of 10 mW was used as the source of pumping radiation. Having passed through the semitransparent mirror (2) the laser radiation was focused into 10 micron spot by the collecting lens (3) on the sample (4) (polycrystalline amino acid or dry DNA of the calf). The scattered radiation was collected on the entrance slit of the triple monochromator Horiba Jobin Yvon T64000 (5). The spectra were detected by the CCD-matrix,

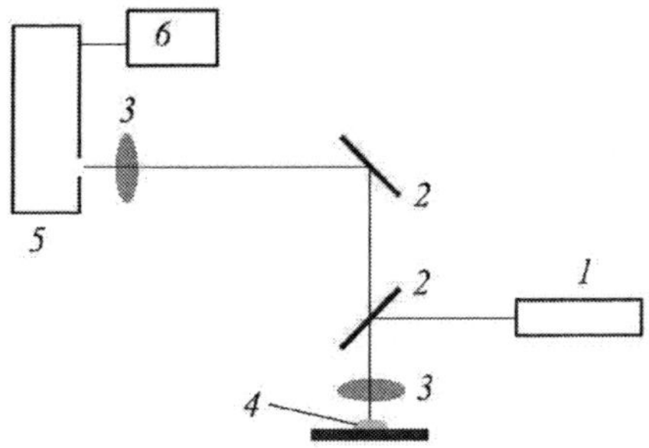

Figure 1. The experimental setup for the Raman spectra registration: 1 - laser; 2 – mirrors; 3 - lenses; 4 - sample; 5 - monochromator; 6 - computer

the data being transmitted and accumulated in the computer (6). The spectral resolution was less than 1 cm^{-1}.

Glycine (H_2NCH_2COOH) is the simplest aliphatic amino acid which has no optical isomers (Figure 2a). L-lysine (figure 2b), D-asparagine (figure 2c) and tyrosine (figure 2d) have more complex structural formulas.

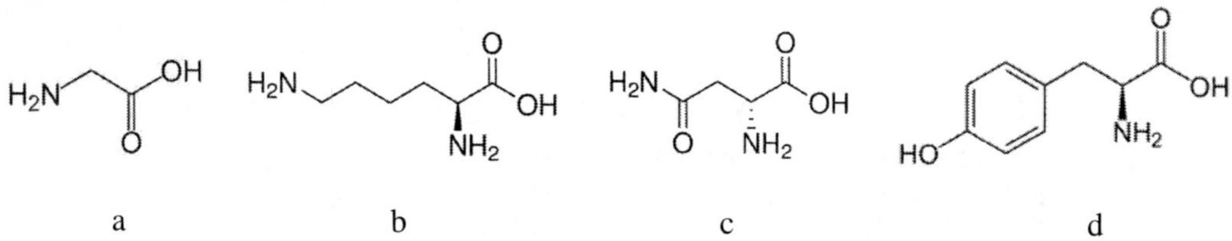

Figure 2. Structural formulas of glycine (a), L-lysine (b), D-asparagine (c) and tyrosine (d)

The α-glycine crystal is monoclinic; the space group of this crystal symmetry is C_{2h}^5 (the structural class is $P2_1/n$) [5,6]; the primitive cell contains 4 molecules (figure 3a). The α-glycine unit cell has the following parameters at room temperature: a = 5.1020 ± 0.0005 Å, b = 11.9709 ± 0.0010 Å and c = 5.4575 ± 0.0009 Å; the volume of the unit cell is equal to 333.3 Å^3 .

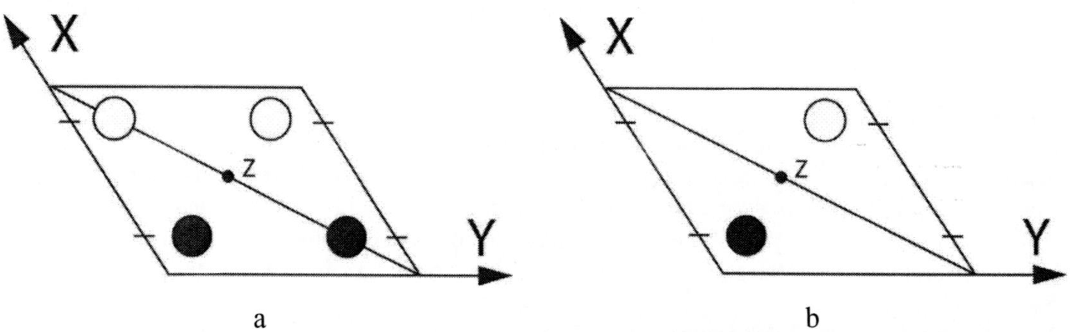

Figure 3. Primitive cells of α-glycine (a), β-glycine, L-lysine and D-asparagine (b). The molecules within the cell are presented by circles; the black circles are in the XY plane, the white circles are shifted along the Z-axis by half-period

In the case of β-glycine the structural class is $P2_1$ (space group C^2_2) [7]; the primitive cell contains two molecules (Figure 3b). L-lysine [8] and D-asparagine [7] crystals belong to the same structural class.

Another example of the amino acid is tyrosine, for which the amino-acid residue is of the form $R=(CH_2)–(C_6H_6)–OH$. The amino acid molecules in the protein are formed as zwitterions $(NH_3)^+–(CH–R)–(COO)^-$. Tyrosine ($C_9H_{11}NO_3$) is aromatic amino acid which exists in two optically isomeric forms - L and D as well as in the form of the racemate DL.

The crystal structure of L-tyrosine is orthorhombic with the space group D_2^4 ($P2_12_12_1$) [9, 10]; the primitive cell contains 4 molecules (Figure 4a). At room temperature the unit cell parameters have the following values: a = 6.913 Å, b = 21.116 Å and c = 5.829 Å; the volume of the unit cell is equal to 850.9 Å^3.

The DL-tyrosine has a space group C_{2v}^{9} (the structural class *Pna2₁*) [7], the primitive cell contains 4 molecules and is presented in Figure 4b.

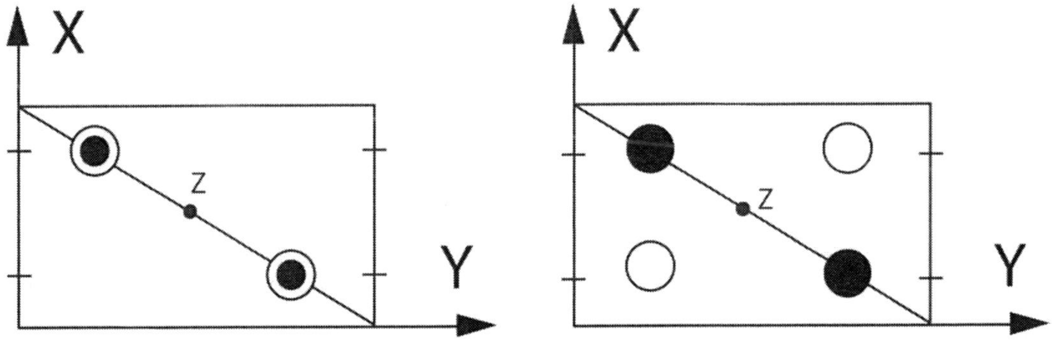

a b

Figure 4. Primitive cells of L-tyrosine (a) and DL-tyrosine (b). The molecules within the cell are presented by circles; the black circles are in the XY plane, the white circles are shifted along the Z-axis by half-period

Based on the carried out group-theoretical analysis [11-13] of the discussed crystals the selection rules were determined in the Raman and IR spectra for the translational (tr) and libration (lib) modes of these crystals (see tables 1-4).

Table 1. The results of the group-theoretic analysis of the spatial symmetry group C_{2h}^{5} $(P2_1/n)$ (α-glycine crystals); a_i, a_{ij} – the components of the vector and symmetric tensor of the second rank respectively; $\chi_V, \chi_{\tilde{V}}, \chi_{tr}, \chi_{lib}$ - the characters of the vector, pseudovector, translational and libration representations, respectively.

	E	$C_2(z)$	σ_h	I	a_i	a_{ij}
$\vec{\alpha}$	0;0;0	$0;0;\frac{1}{2}$	$\frac{1}{2};0;0$	$\frac{1}{2};0;\frac{1}{2}$		
A_g	1	1	1	1		$a_{xx}, a_{yy}, a_{zz}, a_{xy}$
A_u	1	1	-1	-1	a_z	
B_g	1	-1	-1	1		a_{xz}, a_{yz}
B_u	1	-1	1	-1		
χ_V	3	-1	1	-3		$T_{tr} = 2A_u+1B_u$ (IR)
$\chi_{\tilde{V}}$	3	-1	-1	3		$T_{lib} = 3A_u+3B_u$ (IR)
χ_{tr}	9	1	-1	3		$T_{tr} = 3A_g+3B_g$ (RS)
χ_{lib}	12	0	0	0		$T_{lib} = 3A_g+3B_g$(RS)
m_0	4	0	0	0		

IOP Conf. Series: Journal of Physics: Conf. Series **918** (2017) 012033 doi:10.1088/1742-6596/918/1/012033

Table 2. The results of the group-theoretic analysis of the β-glycine, D-asparagine and L-lysine crystals with the space group symmetry C^2_2 $(P2_1)$; a_i, a_{ij} – the components of the vector and symmetric tensor of the second rank respectively; χ_V, $\chi_{\tilde{V}}$, χ_{tr}, χ_{lib} - the characters of the vector, pseudovector, translational and libration representations, respectively.

	E	C_2	a_i	a_{ij}
$\vec{\alpha}$	$0;0;0$	$0;0;\frac{1}{2}$		
A_1	1	1	a_z	a_{xx}, a_{yy}, a_{zz}, a_{xy}
A_2	1	-1	a_x, a_y	a_{xz}, a_{yz}
χ_V	3	-1		$T_{tr}=2A_1+1A_2$ (IR)
$\chi_{\tilde{V}}$	3	-1		$T_{lib}=3A_1+3A_2$ (IR)
χ_{tr}	3	1		$T_{tr}=2A_1+1A_2$ (RS)
χ_{lib}	6	0		$T_{lib}=3A_1+3A_2$ (RS)
m_0	2	0		

Table 3. The results of the group-theoretic analysis of the spatial symmetry group C^9_{2v} $(Pna2_1)$ (DL-tyrosine crystals); a_i, a_{ij} – the components of the vector and symmetric tensor of the second rank respectively; χ_V, $\chi_{\tilde{V}}$, χ_{tr}, χ_{lib} -the characters of the vector, pseudovector, translational and libration representations, respectively.

	E	$C_2(z)$	$\sigma(xz)$	$\sigma(yz)$	a_i	a_{ij}
$\vec{\alpha}$	$0;0;0$	$\frac{1}{2};\frac{1}{2};\frac{1}{2}$	$\frac{1}{2};0;0$	$0;\frac{1}{2};\frac{1}{2}$		
A_1	1	1	1	1	a_z	a_{xx}, a_{yy}, a_{zz}
A_2	1	1	−1	−1		a_{xy}
B_1	1	−1	1	−1	a_x	a_{xz}
B_2	1	−1	−1	1	a_y	a_{yz}
χ_V	3	−1	1	1	$T_{tr}=3A_1+2B_1+2B_2$ (IR)	
$\chi_{\tilde{V}}$	3	−1	−1	−1	$T_{lib}=3A_1+3B_1+3B_2$ (IR)	
χ_{tr}	9	1	1	1	$T_{tr}=3A_1+2A_2+2B_1+2B_2$ (RS)	
χ_{lib}	12	0	0	0	$T_{lib}=3A_1+3A_2+3B_1+3B_2$ (RS)	
m_0	4	0	0	0		

IOP Conf. Series: Journal of Physics: Conf. Series **918** (2017) 012033 doi:10.1088/1742-6596/918/1/012033

Table 4. The results of the group-theoretic analysis of the spatial symmetry group D_2^4 ($P2_12_12_1$) (L-tyrosine crystals); a_i, a_{ij} – the components of the vector and symmetric tensor of the second rank respectively; χ_V, $\chi_{\tilde{V}}$, χ_{tr}, χ_{lib} - the characters of the vector, pseudovector, translational and libration representations, respectively.

	E	$C_2(z)$	$C_2(y)$	$C_2(x)$	a_i	a_{ij}
$\vec{\alpha}$	0;0;0	$\frac{1}{2};\frac{1}{2};0$	$0;\frac{1}{2};\frac{1}{2}$	$\frac{1}{2};0;\frac{1}{2}$		
A	1	1	1	1		a_{xx}, a_{yy}, a_{zz}
B_1	1	1	−1	−1	a_z	a_{xy}
B_2	1	−1	1	−1	a_y	a_{xz}
B_3	1	−1	−1	1	a_x	a_{yz}
χ_V	3	−1	−1	−1	$T_{tr}=1B_1+3B_2+3B_3$	(IR)
$\chi_{\tilde{V}}$	3	−1	−1	−1	$T_{lib}=2B_1+4B_2+4B_3$	(IR)
χ_{tr}	9	−3	1	1	$T_{tr}=2A+1B_1+3B_2+3B_3$	(RS)
χ_{lib}	12	−4	0	0	$T_{lib}=2A+2B_1+4B_2+4B_3$	(RS)
m_0	4	4	0	0		

The Raman spectra of some amino acids were reported in [14-16]. These spectra were registered in the middle (above 400 cm^{-1}) and high wave numbers range [14]. In [15] the Raman spectrum of L-lysine is shown in the 600-1700 cm^{-1} range. In [16] the Raman spectrum of D-asparagine is reported in the 200-3500 cm^{-1} range.

Figure 5 shows the Raman spectrum of the D-asparagine polycrystalline amino acid registered in the wide (10-3700 cm^{-1}) wave numbers range at room temperature.

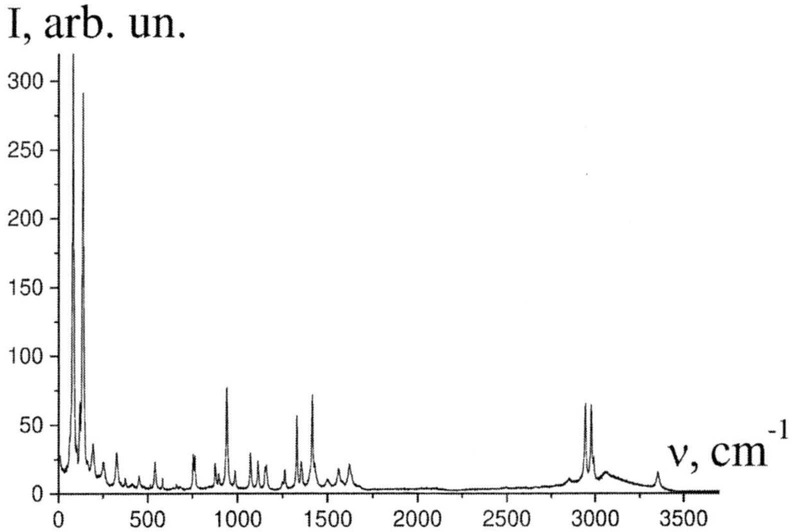

Figure 5. The Raman spectrum of the D-asparagine polycrystalline amino acid in the wide spectral range at room temperature

Figure 6 shows the Raman spectrum of the polycrystalline L-lysine amino acid registered in the wide (10-3700 cm^{-1}) wave numbers range at room temperature.

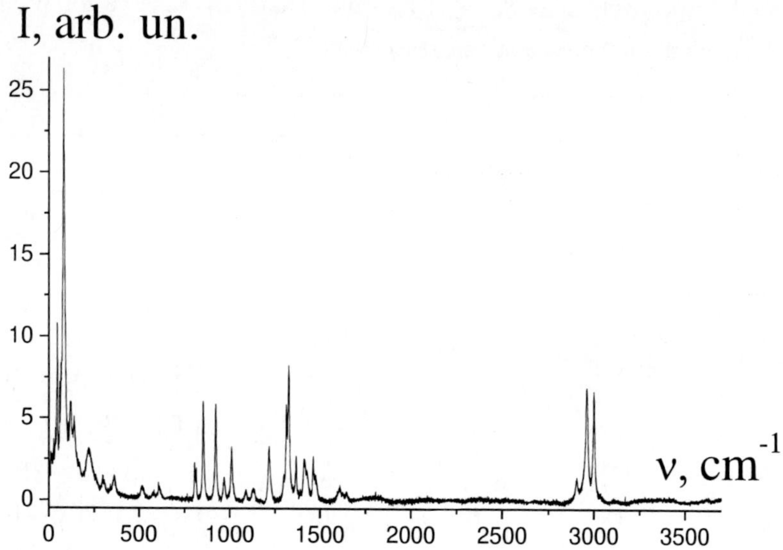

Figure 6. The Raman spectrum of the L-lysine polycrystalline amino acid in the wide spectral range at room temperature

Figure 7 shows the Raman spectrum of the calf dried DNA registered in the wide (10-3700 cm^{-1}) wave numbers range at room temperature.

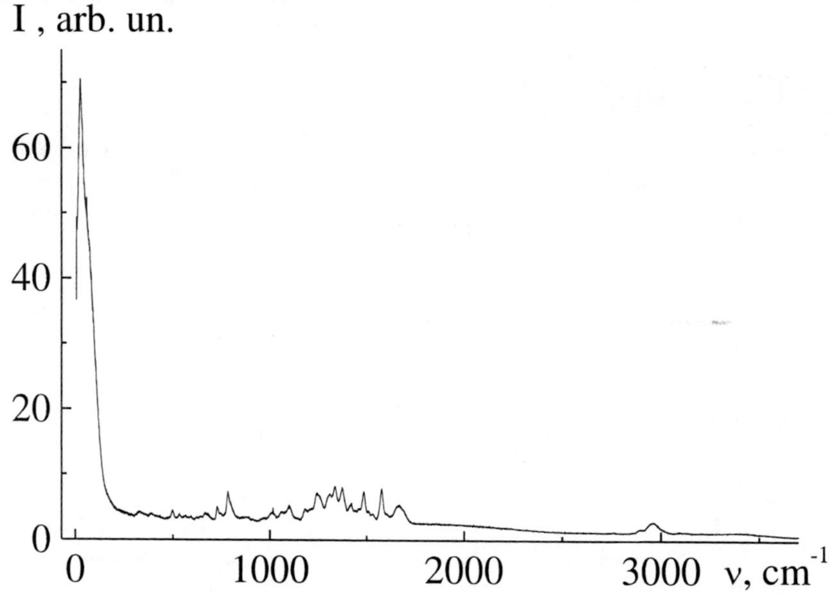

Figure 7. The Raman spectrum of the calf dried DNA in the wide spectral range at room temperature

Figure 8 shows the low-frequency Raman spectra of the polycrystalline amino acids: α-glycine (a), L-tyrosine (b), DL-tyrosine (c), D-asparagine (d) and L-lysine (e) at room temperature. Figure 8(f) shows the comparison between the low-frequency Raman spectra of the polycrystalline L-lysine amino acid (1) and the dried DNA of the calf (2).

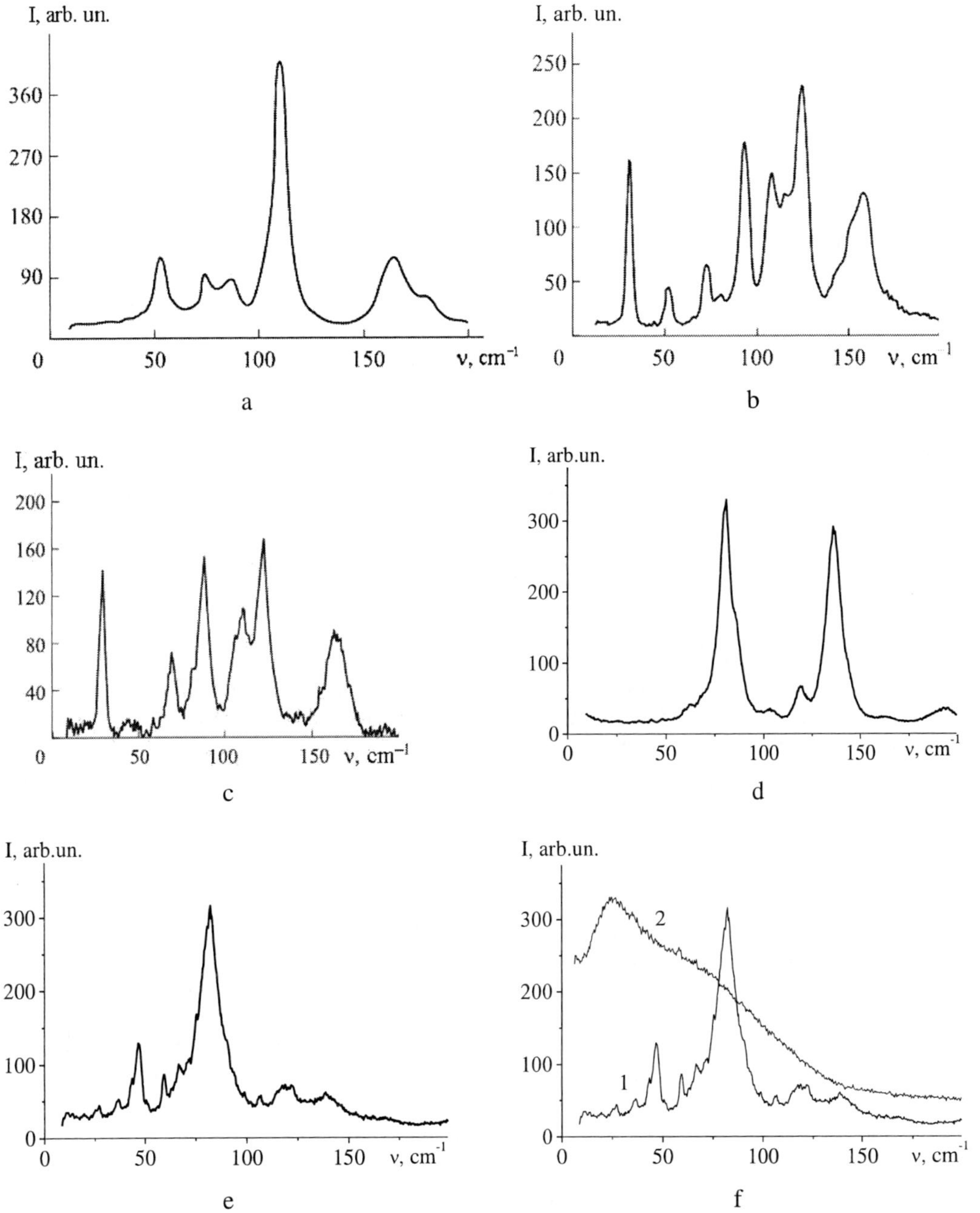

Figure 8. The Raman spectra of the polycrystalline amino acids: α-glycine (a), L-tyrosine (b), DL-tyrosine (c), D-asparagine (d) and L-lysine (e) in the low-frequency region at room temperature, (f) the low-frequency Raman spectra of the polycrystalline L-lysine amino acid (1) and dried DNA of the calf (2)

IOP Conf. Series: Journal of Physics: Conf. Series **918** (2017) 012033 doi:10.1088/1742-6596/918/1/012033

The low-frequency spectra of the polycrystalline amino acids D-asparagine and L-lysine (figures 8d,e) consist of several Raman lines, the most narrow and intense of which falls (figure 8f) into the region of the low-frequency component of the DNA bands.

According to the properties of the lattice modes in the Raman spectra the most intense peaks of the spectra in Figures 5,6,7 should be attributed to the pseudoscalar libration modes corresponding to the turning antiphase motions of two L-lysine or D-asparagine molecules present in the primitive cell of the corresponding crystal and similar to the motions of the nucleobases in DNA. The corresponding peaks in the glycine and tyrosine vibration spectra were also attributed to the pseudoscalar modes [17].

3. Conclusions

Thus, it has been established that at the low-frequency region of the Raman spectra of the α-glycine, L- and DL-tyrosine, D-asparagine and L-lysine amino acids there are very intensive and sharp lines. Such lines were attributed to the pseudoscalar libration modes of the amino acid crystalline lattice. The spectral positions of such libration modes are close to the low-frequency intensive band of the dry DNA molecule that may result in the strong resonant interaction between the amino acids and the DNA molecules in biostructures.

Acknowledgments

This work was supported in part by RFBR grants 15-02-02882, 16-52-00026 and Ministry of Education and Science of the Russian Federation Projects 5top100 and N3.9896.2017/BY.

References

[1] Breen M S, Kemena C, Vlasov P K et al 2012 *Nature* **490** 535
[2] Suzuki S, Ohshima T, Tamiya N et al 1959 *Spectrochim. Acta* **15** 969
[3] Dupuy B, Castinel C, Garrigou-Lagrange C 1969 *Spectrochim. Acta A* **25** 571
[4] Tipping M, Viras K and King T A 1984 *Biopolymers* **23** 2891
[5] Krishnan R S, Sc F A, Balasubramanian K 1958 *Proceedings of the Indian Academy of Sciences- Section A* **48** 55
[6] Stenback H 1976 *Journal of Raman Spectroscopy* 5 49
[7] Boggs R and Donohue J 1971 *Acta Cryst. B* 27 247
[8] Frey M N, Koetzle T F, Lehmann M S and Hamilton W C 1973 *Chem. Phys.* 58 2547
[9] Williams P A, Hughes C E, Harris K D 2015 *Angew Chem Int Ed Engl.* **54** 3973
[10] Görbitz C H 2015 *Crystallography Reviews* **21** 160
[11] Sushchinskii M M 1969 *Raman spectra of molecules and crystals* (Moscow: Nauka)
[12] Landau L D, Lifshitz E M 1977 *Quantum Mechanics (Non-relativistic Theory)* (Amsterdam: Elsevier Butterworth-Heinemann)
[13] Lyubarsky G Y 1958 *Group theory and its applications to physics* (Moscow: Fizmatgiz)
[14] Jenkins A L, Larsen R A and Williams T B 2005 *Spectrochimica Acta Part A* **61** 1585
[15] Zhu G, Zhu X, Fan Q, Wan X 2011 *Spectrochimica Acta Part A* **78** 1187
[16] Casado J, Lopez Navarrete J T and Ramirez F J 1995 *Journal of Raman Spectroscopy* 26 1003
[17] Belyanchikov M A, Gorelik V S, Gorshunov B P and Pyatyshev A Y 2017 *Journal of Experimental and Theoretical Physics* **124** 77

IOP Conf. Series: Journal of Physics: Conf. Series **918** (2017) 012034 doi:10.1088/1742-6596/918/1/012034

QCL spectroscopy combined with the least squares method for substance analysis

D A Samsonov[1], **A S Tabalina**[2], **I L Fufurin**[3]

[1]Department of Physics student, Bauman Moscow State Technical University, Moscow, Russia

[2]Department of Physics student, Bauman Moscow State Technical University, Moscow, Russia

[3]Cand. Sc. (Phys.), Assoc.Professor, Department of Physics, Bauman Moscow State Technical University, Moscow, Russia

Email: blitzar90@gmail.com, ast_295@mail.ru, igfil@mail.ru

Abstract. The article briefly describes distinctive features of quantum cascade lasers (QCL). It also describes an experimental set-up for acquiring mid-infrared absorption spectra using QCL. The paper demonstrates experimental results in the form of normed spectra. We tested the application of the least squares method for spectrum analysis. We used this method for substance identification and extraction of concentration data. We compare the results with more common methods of absorption spectroscopy. Eventually, we prove the feasibility of using this simple method for quantitative and qualitative analysis of experimental data acquired with QCL.

1. Introduction

Nowadays spectroscopy is one of the most used research methods in different areas [1-5]. Infrared spectroscopy plays a crucial role in identification and qualitative analysis of organic and inorganic substances.

Quantum Cascade Laser (QCL) [6] is a semiconductor laser that emits due to electron transitions between multiple quantum well heterostructures. This technology makes it possible to achieve a wide range of radiation wavelengths. It was first introduced in 1994 and is used today in many research areas [8-10].

Distinctive features of QCLs are:

- radiation frequency manipulation by setting the appropriate active layer thickness in the heterostructure;
- frequency manipulation by changing the voltage applied to the superlattice.

The main goal of this paper was to apply QCL IR-spectroscopy in the range of 5.14 to 12.8 μm for qualitative and quantitative analysis of gases and thin films of test substances. To achieve this goal, the research team had to perform the following tasks:

- setting up a QCL-based experiment;
- registering absorption spectra of test substances;

IOP Conf. Series: Journal of Physics: Conf. Series **918** (2017) 012034 doi:10.1088/1742-6596/918/1/012034

- analysing the data acquired by means of the least squares method (LSM) in matrix form;
- estimating the efficiency of substance identification and concentration determination.

2. Experimental set-up

Figure 1 demonstrates the experimental set-up used in the investigation.

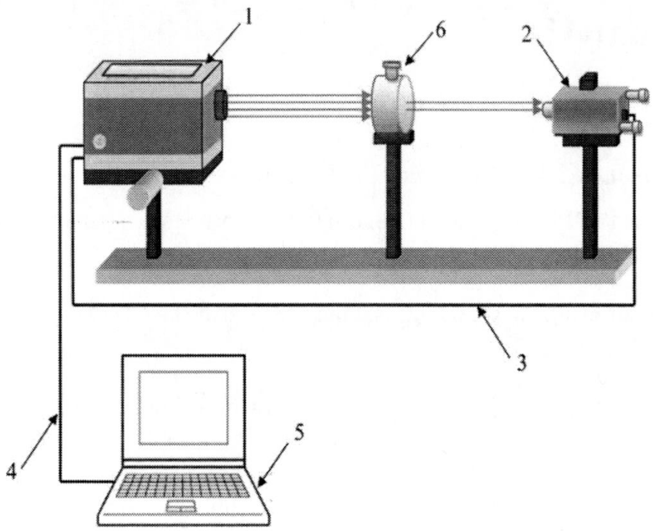

Figure 1. Experimental set-up

1 - QCL, 2 - external photodetector, 3 - QCL-detector channel, 4 - QCL-PC channel, 5 - PC with specialised software, 6 - gas container.

During the experiments, the distance between the QCL and the photodetector was 50 cm. It is possible to use a greater distance with corner reflectors [10].

This experimental set-up can acquire absorption spectra of substances in the gas phase as well as those of films that are up to several dozen μm thick.

3. Spectrum analysis

For spectrum analysis we chose one of the simplest methods for quantitative and qualitative spectroscopy, the least squares method in matrix form (LSM). Works [11, 12] describe its theoretical foundation.

We used Python 2.7 for all the computations and visualisation. The main idea of the method consists of solving a matrix equation for C_j:

$$B_{ij}C_j = E_i\,, \tag{1}$$

where C_j is the resulting vector of concentrations, B_{ij} is the matrix of base substance optical densities and E_i is the experimental spectrum of optical density. Index i is the wavelength (0..1023 cm^{-1}) and j is the number of the substance in the base substance matrix B_{ij}.

The data which comes from the QCL is a normed experimental spectrum of absorbance I_i. The base W_{ij} is a matrix of normed base spectra. Transforming absorbance spectra into optical density, we get the B_{ij} and E_i quantities that we use in the expression (1).

$$D_i = -ln(I_i)\,, \tag{2}$$

where D_i is the corresponding optical density at the i-th wavelength.

To solve (1) we need to get ${B_{ij}}^{-1}$. Generally $i \neq j$ so ${B_{ij}}^{-1}$ is a pseudoinverse of the matrix B_{ij}, which has the required property ${B_{ij}}^{-1}B_{ij} = I$, where I is the identity matrix. So the solution of the equation (1) becomes:

$${B_{ij}}^{-1}E_i = C_j \tag{3}$$

4. Results

We used the Pearson correlation coefficient (r) [14] to perform our qualitative analysis. While quantitative description was provided by the LSM (expression (3)), where the j-th place in the $C_j$vector shows the corresponding concentration of the substance j in the base. This method measures the concentration in the units of concentration of the corresponding base spectrum.

Figures 2, 3 and 4 present examples of the spectra acquired and their analysis.

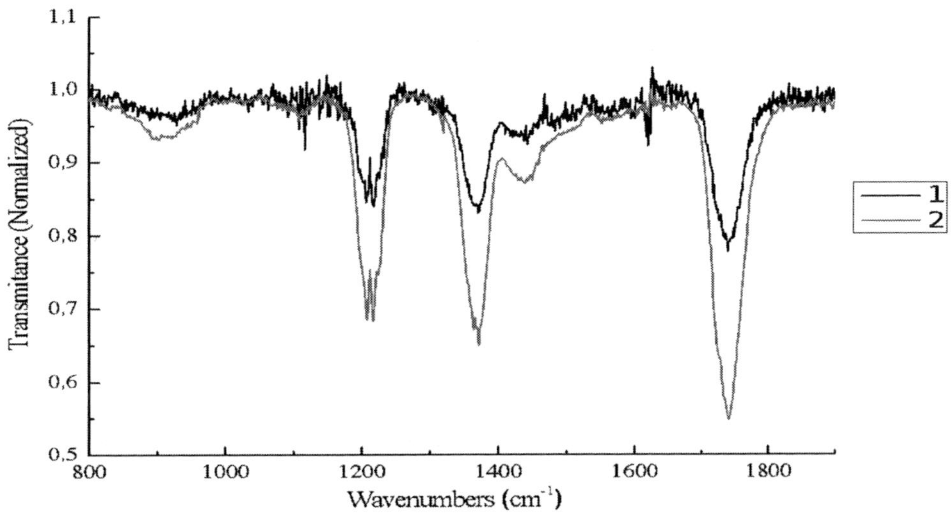

Figure 2. Aceton detection

1 - experimental spectrum, 2 - base spectrum. $r = 0.97$, $c = 64\ \frac{g}{cm^3}$, SNR = 7.4.

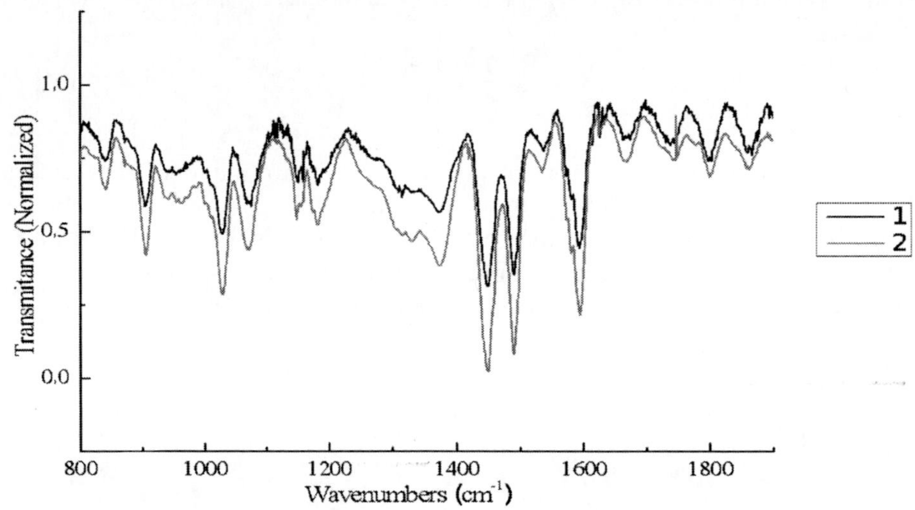

Figure 3. Polystyrene detection

1 - experimental spectrum, 2 - base spectrum. $r = 0.91$, $c = 5.6\mu m$, SNR = 28.7.

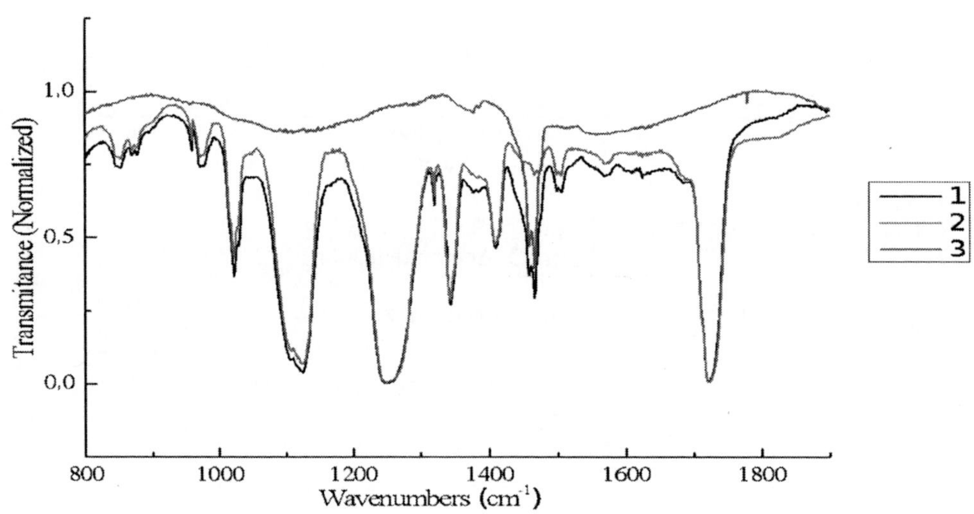

Figure 4. Lavsan (PET) and polyethylene mixture detection

1 - experimental spectrum, 2 - base polyethylene spectrum, 3 - base lavsan spectrum. $r2 = 0.96$, $r3 = 0.27$, $c2 = 10.4\mu m$, $c3 = 5.7\mu m$, SNR2 = 33.2, SNR3 = 19.5.

We performed mixture analysis as shown in figure 4 the following way. We analysed the most relevant spectrum using the correlation coefficient r, then we subtracted it from the experimental spectrum with the corresponding concentration value, then the procedure was repeated and all the remaining substances were identified.

5. Concentration error estimation

We conducted the following experiment to estimate errors. We prepared three identical polyethylene films of a known thickness (8 μm). Then we obtained absorption spectra of one, two and three films in sequence. Adding a film means that absorption lines on a normed spectrum should become deeper. This effect is evident in figure 5, and table 1 shows the results of estimating the concentration error.

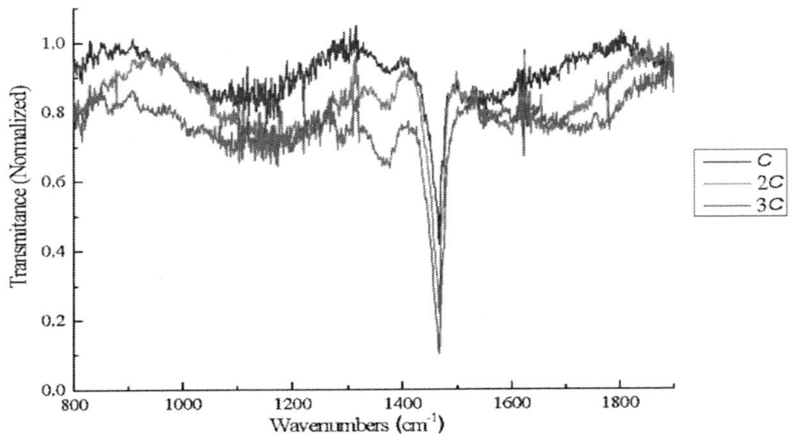

Figure 5. Concentration error estimation

C - 1 polyethylene film, 2C - 2 polyethylene films, 3C - 3 polyethylene films.

Table 1. LSM concentration error.

Number of films	LSM width, μm	Real width, μm	Relative error, %
1	7,870	8	1,62
2	13,251	16	17,18
3	20,233	24	15,67

6. Conclusion

This paper presented mid-infrared absorption spectra acquired with QCL. We showed the possibility to successfully conduct quantitative and qualitative spectrum analysis with the LSM. Moreover, we were able to analyse mixtures. This method gives a relative concentration error under 30% and may be a reliable method for substance identification. In real-world conditions, which are far from laboratory conditions, noise could have a critical effect. To this end, future research may be aimed at QCL noise analysis and ways of identifying substances in the natural environment.

References

[1] Morozov A.N and Svetlichniy S I 2016 *Osnovi furie-spectroradiometrii* [Fundamentals of Fourier spectroradiometry] (Moscow: Nauka) 456 p

[2] Golyak Ig S, Golyak Il S, Karfidov A O, Korolev P A, Morozov A N, Mironov A I, Strokov M A, Tabalin S E and Fufurin I L 2014 Panoramniy furie-spekrtroradiometr PHRDD-4 [Panoramic Fourier spectroradiometer PHRDD-4] *Pribory i tekhnika* eksperimenta Instruments *and Experimental Techniques* **6** 119-20

[3] Morozov A N, Svetlichniy S I, Tabalin S E and Fufurin I L 2013 Fizicheskie osnovi rascheta interferometra s vrashchayuscheysya plastinkoy [Principal physics for design calculations of an interferometer with a rotating plate]. *Opticheskiy zhurnal - Journal of Optical Technology* **80**(8) 37-41
[4] Zaytsev K I, Gavdush A A, Karasik V E and Yurchenko S O 2014 Highly accurate reconstruction of spectral optical characteristics of a medium using terahertz pulsed spectroscopy. *Vestn. Mosk. Gos. Tekh. Univ. im. N.E. Baumana, Estestv. Nauki - Herald of the Bauman Moscow State Tech. Univ., Nat. Sci.* **3** 69-92
[5] Bashkin S V et al 2016 Experimental results of investigating a panoramic Fourier transform infrared spectrometer. *Vestn. Mosk. Gos. Tekh. Univ. im. N.E. Baumana, Estestv. Nauki -Herald of the Bauman Moscow State Tech. Univ*., Nat. Sci. **2** 51–64 DOI: 10.18698/1812-3368-2016-2-51-64.
[6] Kazarinov R F and Suris R A 1971 O elektromagnitnihsvoystvahpoluprovodnikov so sverhreshetkoy [On electromagnetic properties of semiconductors with superlattice]. *Fizika i tehnika poluprovodnikov - Semiconductors* **5**(4)
[7] Herman M A 1986 *Semiconductor Superlattices* (Berlin: Akademic-Verlag)
[8] Schwaighofer A, Alcaráz M R, Araman C, Goicoechea H and Lendl B 2016 External cavity-quantum cascade laser infrared spectroscopy for secondary structure analysis of proteins at low concentrations. *Scientific Reports* **6**
[9] Deutsch E R, Kotidis P, Zhu N, Goyal A K, Ye J, Mazurenko A, Norman M, Zafiriou K, Baier M and Connors R 2014 Active and passive infrared spectroscopy for the detection of environmental threats *Advanced Environmental, Chemical, and Biological Sensing Technologies XI*
[10] Volkov V G 2016 Quantum Cascade Lasers and their Application in Safety and Communication Systems *Systems of Control, Communication and Security* **1**
[11] Kozintsev V I, Belov M L, Gorodnichev V A and Fedorov Yu V 2006 *Lazernyy optiko-akusticheskiy analiz mnogokomponentnyh gazovyh smesey* [Optoacoustic laser analysis of multicomponent gas mixtures] (Moscow: Bauman Moscow State Technical University) 387 p
[12] Tsej R and Shumafov M M 2011 Matrix condition number as a stability characteristic in solving applied problems *Trudy FORA - Works of the Adygheya Republic Physical Society* **16**
[13] Pearson K 1895 Notes on regression and inheritance in the case of two parents. *Proceedings of the Royal Society of London*
[14] Morozov A N, Kochikov I V, Novgorodskaya A V, Sologub A A and Fufurin I L 2015 Statistical estimation of the probability of the correct substance detection in FTIR spectroscopy *Computer Optics* **39**(4) 614-21 DOI: 10.18287/0134-2452-2015-39-4-614-621.
[15] Fufurin I L 2010 *Research on method and algorithm of substance recognition in the atmosphere for one measurement of fourier-spectroradiometer* Cand. phys. sc. diss. (Moscow)

IOP Conf. Series: Journal of Physics: Conf. Series **918** (2017) 012035 doi:10.1088/1742-6596/918/1/012035

Raman spectra of lithium compounds

V S Gorelik[1, 2], Dongxue Bi[2], Y P Voinov[1], A I Vodchits[3], B P Gorshunov[4, 5], N I Yurasov[2] and I I Yurasova[2]

[1]Lebedev Physical Institute of the Russian Academy of Sciences, Moscow, Russian Federation
[2]Bauman Moscow State Technical University, Moscow, Russian Federation
[3]Stepanov Institute of Physics of National Academy of Sciences of Belarus, Minsk, Belarus
[4]Moscow Institute of Physics and Technology, Dolgoprudny, Russian Federation
[5]Prokhorov General Physics Institute of the Russian Academy of Sciences, Moscow, Russian Federation

E-mail: gorelik@sci.lebedev.ru

Abstract. The paper is devoted to the results of investigating the spontaneous Raman scattering spectra in the lithium compounds crystals in a wide spectral range by the fibre-optic spectroscopy method. We also present the stimulated Raman scattering spectra in the lithium hydroxide and lithium deuteride crystals obtained with the use of powerful laser source. The symmetry properties of the lithium hydroxide, lithium hydroxide monohydrate and lithium deuteride crystals optical modes were analyzed by means of the irreducible representations of the point symmetry groups. We have established the selection rules in the Raman and infrared absorption spectra of LiOH, $LiOH \cdot H_2O$ and LiD crystals.

1. Introduction and background

The studies of the Raman spectra of lithium compounds with hydrogen and hydroxyl groups [1-3] are of great current interest due to the wide use of these crystals in the area of hydrogen energy and in other applications [4, 5]. In addition, the study of the lithium structures crystalline lattices dynamics is important thanks to the high mobility of lithium and the respective crystal lattices instability [6]. To date, the Raman spectrum of these crystals has not been well studied [7-9] and needs further clarification.

In this paper we present the results of the experimental studies of the spontaneous Raman scattering spectra of lithium hydroxide (LiOH), lithium hydroxide monohydrate ($LiOH \cdot H_2O$) and lithium deuteride (LiD) as well as the stimulated Raman scattering spectra of lithium hydroxide (LiOH) and lithium deuteride (LiOD).

2. Experimental technique

The schematic diagram of the experimental setup for the excitation and detection of the spontaneous Raman scattering spectra is shown in Figure 1a. The excitation and registration of the Raman spectra was carried out with the use of the fiber-optical recording technique [10]. As a source of excitation radiation we used the laser with the wavelength λ=785nm and the power of 200 mW. The laser (1) light was directed into the optical fiber (2) and the probe (3) (see Figure 1a). The radiation was focused by the lens (5) on the sample (6) which has the form of polycrystalline powder. The Raman signal arising in the sample was sent to the second optical fiber (8) and filtered in the focusator (9) by means of the photonic crystal (11) and then it entered the input of the minispectrometer (13) with the multielement receiver. The computer (14) has been designed for digital processing and storage of Raman spectra. The exposure time was 1-10 s. The Raman signal was recorded by a small amount (less than 1mg) of the polycrystalline sample.

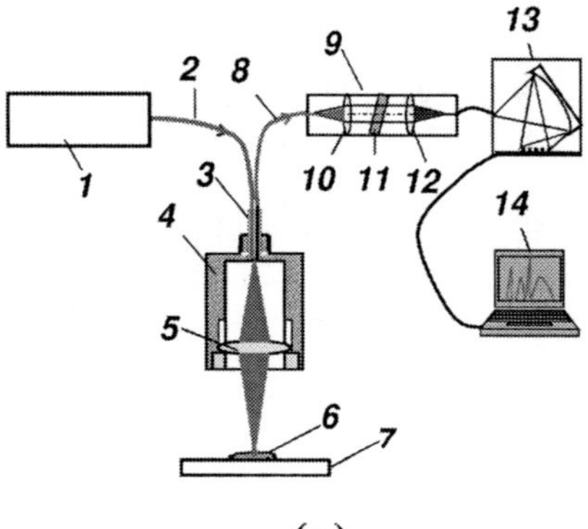

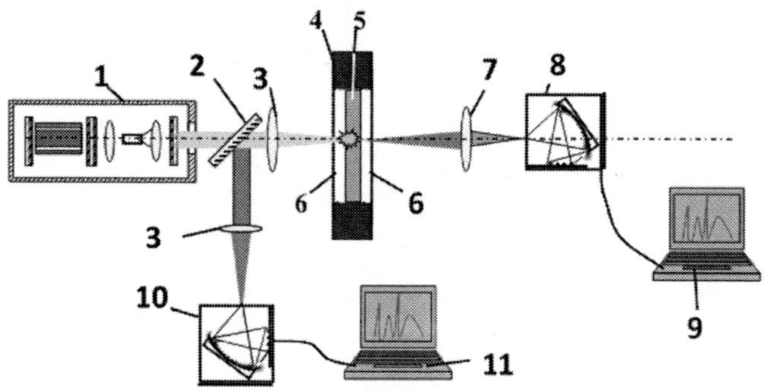

Figure 1. Schematics of the experimental setup. a) for excitation and detection of Spontaneous Raman Scattering spectra:1 – laser; 2 – fiber; 3 – probe; 4 – shell; 5,10,12 – lens; 6 – sample; 7 – substrate; 8 – fiber; 9 – focusator; 11 – photonic crystal; 13 – minispectrometer; 14 – computer; b) for excitation and detection of Stimulated Raman Scattering spectra:1 - laser; 2 - mirror; 3,7 - lens; 4 - holder; 5–sample; 6–windows; 8, 10–minispectrometer; 9,11 – computer.

The stimulated Raman scattering spectra of lithium hydroxide and lithium deuteride were studied using the experimental setup [11], the schematic of which is given in Figure 1b. The stimulated Raman scattering spectra were excited by the giant pulses of the second optical harmonic (532 nm) of a Nd^{3+}:YAG laser (1). Through the semitransparent mirror (2) the laser beam was focused by the lens (3) onto the sample (5) and then was brought to the minispectrometer (8). The stimulated Raman scattering in the "backward" geometry was brought to the entrance slit of the minispectrometer (10) after reflecting at the mirror (2).

3. Experimental results and discussion

The spontaneous Raman scattering spectra of lithium hydroxide, lithium hydroxide monohydrate and lithium deuteride obtained at room temperature are shown in Figure 2a-c.

IOP Conf. Series: Journal of Physics: Conf. Series **918** (2017) 012035 doi:10.1088/1742-6596/918/1/012035

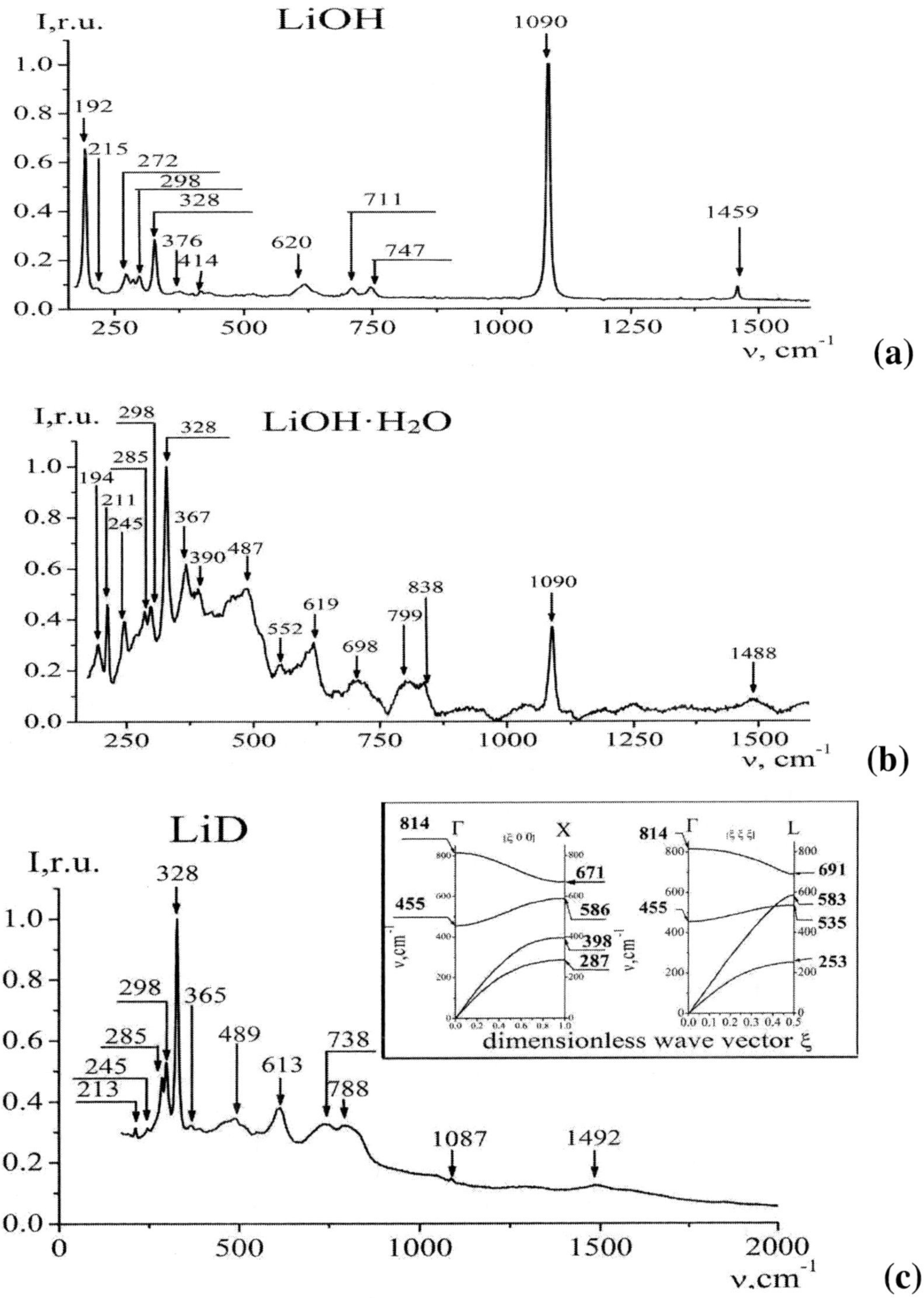

Figure 2. The spontaneous Raman scattering spectra of lithium compounds containing the hydroxyl groups, recorded at room temperature: a) lithium hydroxide; b) lithium hydroxide monohydrate; c) lithium deuteride. The data about the dispersion branches from the work [12] were presented at the inset in the upper right corner of Figure 2c.

IOP Conf. Series: Journal of Physics: Conf. Series **918** (2017) 012035 doi:10.1088/1742-6596/918/1/012035

As we can see from Figure 2, in the low-frequency Raman spectra of these crystals there are a large number of lines, the most intense of which has a frequency of 328 cm^{-1}.

The spectra of the stimulated Raman scattering in lithium hydroxide and lithium deuteride were shown in Figure 3a, b.

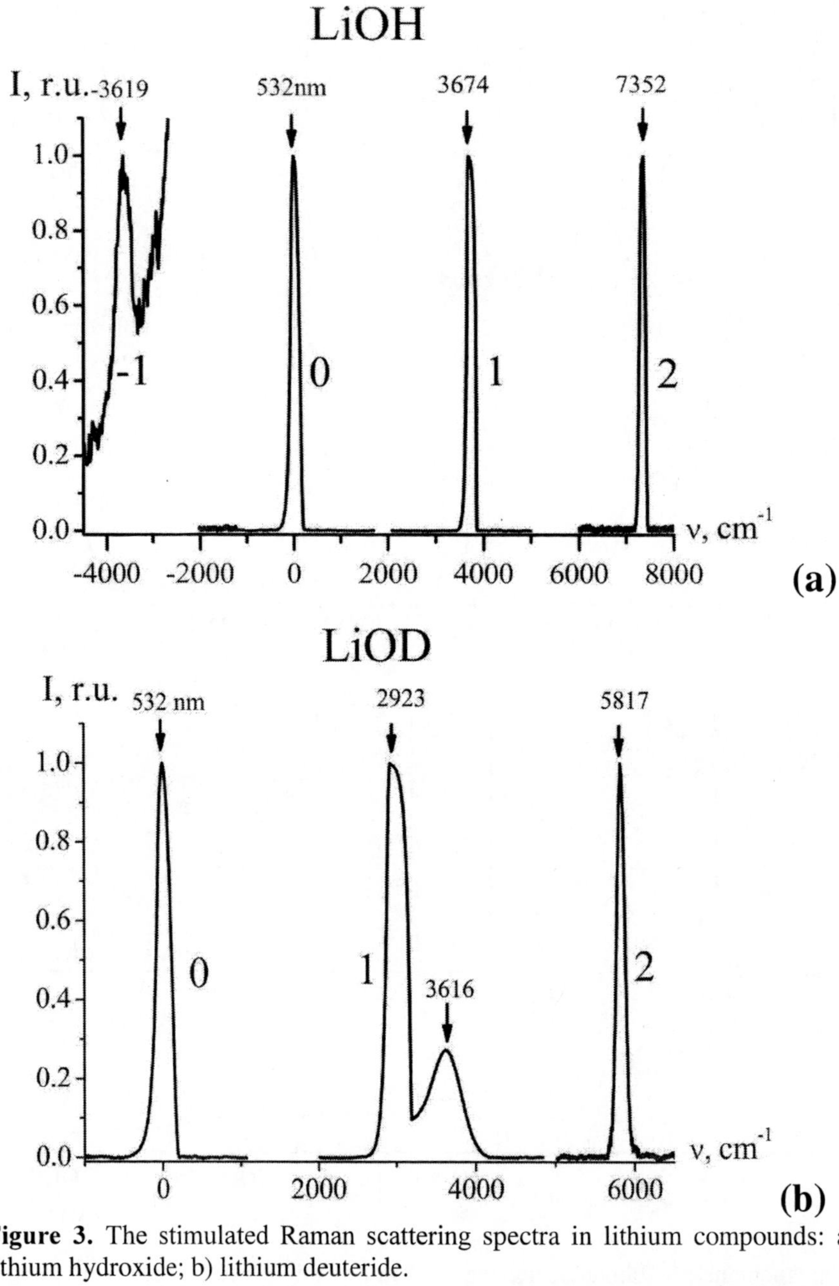

Figure 3. The stimulated Raman scattering spectra in lithium compounds: a) lithium hydroxide; b) lithium deuteride.

IOP Conf. Series: Journal of Physics: Conf. Series **918** (2017) 012035 doi:10.1088/1742-6596/918/1/012035

As we can see from Figure 3b, the first and the second order of the Stokes lines are manifested in the stimulated Raman scattering spectra of lithium hydroxide and lithium deuteride.

The types of the lithium compound primitive cells are presented in Figure 4.

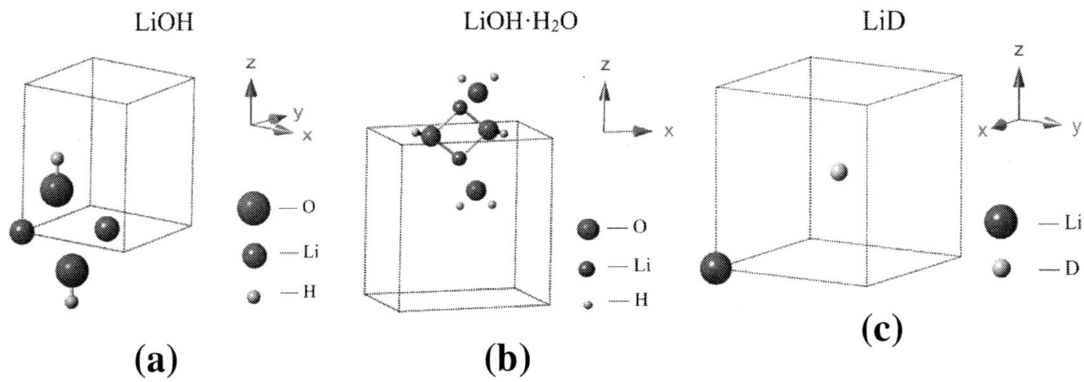

Figure 4. The types of the lithium compound primitive cells: a) lithium hydroxide; b) lithium hydroxide monohydrate; c) lithium deuteride.

LiOH crystals have a tetragonal structure (see Figure 4a) which is characterized by the space group symmetry D^7_{4h} ($P\frac{4}{n}mm$) and contains two formula units. The OH^- ions are located in positions with the symmetry C_{4v} [1].

We have analyzed the symmetry properties of the LiOH optical modes, classified these modes with the aid of the irreducible representations of the group D^7_{4h} and set the selection rules in the Raman and infrared absorption spectra. The results of the group-theoretic analysis are presented in Table 1.

Table 1. The results of the group-theoretical analysis of optical modes in lithium hydroxide.

The space group symmetry of the crystal LiOH – D^7_{4h}										
	E	$C_4^{\pm1}$ (z)	C_2(z)	$2C_2$	$2C_2'$	σ_h	$2\sigma_v$	$2\sigma_d$	$S_4^{\pm1}$	I
$\vec{\alpha}$ [a]	0;0;0	½;½;0	0;0;0	½;½;0	0;0;0	½;½;0	0;0;0	½;½;0	0;0;0	½;½;0
n_0 [b]	6	4	6	0	0	0	6	4	2	0
n_0-1	5	3	5	−1	−1	−1	5	3	1	−1
χ_V [c]	3	1	−1	−1	1	1	1	1	−1	−3
χ_{opt} [d]	15	3	−5	1	−1	−1	5	3	−1	3

[a] $\vec{\alpha}$ – translation vector.

[b] n_0 – the number of atoms remaining in place under the transformations of the symmetry.

[c] χ_V – the character of the vector representation.

[d] χ_{opt} – the character of the optical representation.

As a result of the analysis it is established that the vibrational representation contains the following types of optical modes:

$$T_{opt} = 2A_{1g}(Raman) + B_{1g}(Raman) + 2A_{2u}(Infrared) + 3E_g(Raman) + 2E_u(Infrared) \quad (1)$$

Lithium hydroxide monohydrate crystallizes in the monoclinic space group C_{2h}^3 with two formula units in the primitive cell at room temperature (see Figure 4b) [13]. The lithium ions and water molecules have a symmetry $C_2(Z)$, and the hydroxyl ions are lying in a mirror plane [14].

The results of the group-theoretical analysis of optical modes in lithium hydroxide monohydrate are presented in Table 2.

Table 2. The results of group-theoretical analysis of optical modes in lithium hydroxide monohydrate.

Space group symmetry of the crystal $LiOH \cdot H_2O - C_{2h}^3$				
	E	$C_2(z)$	σ_h	I
$\vec{\alpha}$	0;0;0	0;0;0	0;0;0	0;0;0
n_0	12	4	4	0
n_{0-1}	11	3	3	–1
χ_V	3	–1	1	–3
$\chi_{\tilde{V}}$ [a]	3	–1	–1	3
$\chi_{[V]^2}$ [b]	6	2	2	6
χ_{opt}	33	–3	3	3
χ_{tr} [c]	15	–3	1	3
χ_{lib} [d]	10	–2	0	0

[a] $\chi_{\tilde{V}}$ –the character of pseudovector representation.
[b] $\chi_{[V]^2}$ –the character of the second rank symmetrical tensor.
[c] χ_{tr} –the character of translational representation.
[d] χ_{lib} –the character of the librational representation.

The following types of vibrations are present in the spectrum of lithium hydroxide monohydrate optical vibrations:

$$T_{opt} = 9A_g(Raman) + 9B_g(Raman) + 6A_u(Infrared) + 9B_u(Infrared) \quad (2)$$

The structure of irreducible representations corresponding to the librational T_{lib}, translational T_{tr}, internal $T_{out} = T_{lib} + T_{tr}$ and external $T_{in} = T_{opt} - T_{out}$ types of lithium hydroxide monohydrate vibrations is shown in Table 3.

Table 3. The structure of irreducible representations in the lithium hydroxide monohydrate crystal.

	Li^+	OH^-	H_2O
T_{tr}	$T_{tr}(Li^+) = A_g + 2B_g + A_u + 2B_u$	$T_{tr}(OH^-) = 2A_g + B_g$	$T_{tr}(H_2O) = A_g + 2B_g + A_u + 2B_u$
		$T_{tr} = 4A_g + 5B_g + 2A_u + 4B_u$	
T_{lib}	-	$T_{lib}(OH^-) = A_g + B_g + A_u + B_u$	$T_{lib}(H_2O) = A_g + 2B_g + A_u + 2B_u$
		$T_{lib} = 2A_g + 3B_g + 2A_u + 3B_u$	
T_{out}		$T_{out} = 6A_g + 8B_g + 4A_u + 7B_u$	
T_{in}	-	$T_{in} = A_g + B_u$	$T_{in} = 2A_g + B_g + 2A_u + B_u$
		$T_{in} = 3A_g + B_g + 2A_u + 2B_u$	

The lattice modes corresponding to the lithium ions of the hydroxyl group OH^- and water molecules translations as well as to the water molecules and hydroxyl groups librations are detected in the spontaneous Raman scattering spectra of $LiOH{\cdot}H_2O$.

The lithium deuteride crystals have a sodium chloride type structure (see Figure 4c) characterized by the space group symmetry O_h^1 [15]. In the elementary cubic cell there is one formula unit, and the phonon dispersion curve has three optical branches:

$$T_{opt} = F_{1u}(\,Infrared\,) \tag{3}$$

According to the selection rule, for a perfect crystal of lithium deuteride the optical modes are forbidden in the Raman spectrum of LiD at Γ–point. However, as can be seen from the experimental spectrum (see Figure 2c), there are several Raman peaks with spectral positions close to the phonon frequencies at Brillouin zone edge (see upper inset in Figure 2c). Such effect may be explained as the result of the lattice disturbance due to disordering of lithium and deuterium ions.

4. Conclusions

We have detected the sharp lines of Raman scattering for all discussed crystals. The results allow characterizing the samples of lithium compounds with hydrogen and hydroxyl groups. It is necessary for their effective use. We have presented the results of the stimulated Raman scattering of lithium hydroxide and lithium deuteride. The isotopes influence on the Raman spectrum in the peak positions has been established.

Acknowledgments

This work was supported in part by the Russian Foundation for Basic Research grants №15-02-02882, 16-52-00026 (BRFFR- F16R-063), Ministry of Education and Science of the Russian Federation Projects 5top100 and N3.9896.2017/BY, China Scholarship Council.

References

[1] Wilkinson G R 1973 Raman Spectra of Ionic, Covalent, and Metallic Crystals *The Raman*

effect (Applications Vol 2*)* ed A Anderson (New York: Marcel Dekker) chapter 5 p 408
[2] Jaswal S S, Sharma T P and Wolfram G 1972 *Solid State Communications* **11** 1151
[3] Okazaki S, Ohtori N and Okada I 1989 *Journal of Chemical Physics* **91** 5587
[4] Megahed S and Ebner W 1995 *Journal of Power Sources* **54** 155
[5] Laude T, Kobayashi T and Sato Y 2010 *International Journal of Hydrogen Energy* **35** 585
[6] Sebastian L and Gopalakrishnan J 2003 *Journal of Materials Chemistry* **13** 433
[7] Sifuentes A, Stowe A C and Smyrl N 2013 *Journal of Alloys & Compounds* **580** 271
[8] Hase Y and Yoshid I V P 1979 *Chemical Physics Letters* **65** 46
[9] Anderson A and Lüty F 1983 *Phys. Rev.* B **26** 3415
[10] Gorelik V S, Litvinova A O and Umarov M F 2014 *Bulletin of the Lebedev Physics Institute* **41** 305
[11] Gorelik V S, Vodchits A I and Orlovich V A 2015 *Journal of Russian Laser Research* **36** 562
[12] Pepinsky B R 1940 *Journal of Crystallography – Crystalline Materials* **102** 119
[13] Verble J L, Warren J L and Yarnell J L 1967 *Physical Review* **168** 980
[14] Krishnamurti D 1959 *Proceedings – Mathematical Sciences* **50** 247
[15] Zintl E and Harder 1935 *Journal of Physical Chemistry* B*: Chemistry of Elementary Processes Structure of Matter* **28** 478

IOP Conf. Series: Journal of Physics: Conf. Series **918** (2017) 012036 doi:10.1088/1742-6596/918/1/012036

Reflection spectra and their angular dependences of one-dimensional photonic crystals based on aluminium oxide

V S Gorelik[1], M M Yashin[2], A V Pudovkin[1] and A I Vodchits[3]

[1]Lebedev Physical Institute of the Russian Academy of Sciences, Moscow, 119991, Russia
[2]Bauman Moscow State Technical University, Moscow, 105005, Russia
[3]Stepanov Institute of Physics of the National Academy of Sciences of Belarus, Minsk, 220072, Belarus

E-mail: gorelik@sci.lebedev.ru

Abstract. The article considers optical properties (transmission and reflection) of one-dimensional photonic crystals based on mesoporous anodic aluminum oxide, with periods of crystal lattices 188 and 194 nm. A comparison of the experimentally measured reflection spectrum in the spectral region of the first stop-zone with the theoretical dependence obtained from the dispersion relation for one-dimensional photonic crystal is carried out. The angular dependence of the first stop-zone spectral positions of one-dimensional photonic crystal is established. The authors analyze the possibility of applications of mesoporous one-dimensional photonic crystals based on aluminum oxide as the selective narrowband filters and mirrors.

1. Introduction and background

Currently the creation of new types of optical devices with controlled optical properties is extremely important. In this connection there is a great interest to study the so-called photonic crystals [1-4] with dielectric constant periodically varying in the space. One-*dimensional* photonic crystals are built of two alternating layers with two different refractive indices: n_1 and n_2. In the reflection spectrum of such a crystal, the regions with a jump-in reflection of electromagnetic radiation are observed. Such regions are called forbidden bands (band gaps) or stop-zones, the spectral positions of which depend on the parameters of the photonic crystal.

2. Experimental technique

Today, there are many different ways [5-8] to obtain porous photonic crystals. This paper is to study the stop-zones spectra of two different one-dimensional photonic crystals and to establish the possibility of their optical properties control [5,6,9-11]. Thin aluminum oxide films obtained by the electrochemical etching are used as samples for the research. The experimental installation diagram for the analysis of reflection spectra at normal incidence on the surface of the photonic crystal is shown in figure 1a. The light from the radiation source (6) is propagated along the light-guide (5) and falls on the surface of the photonic crystal (1) at incidence angle close to zero. The reflected radiation falls to the second light-guide and is analyzed with the help of spectrometer (7) and computer (8). In

IOP Conf. Series: Journal of Physics: Conf. Series **918** (2017) 012036 doi:10.1088/1742-6596/918/1/012036

this case the incidence angle of electromagnetic radiation on the surface of the mesoporous photonic crystal is close to zero. This experimental scheme (see figure 1a) permits to record the reflectance spectra from the surface of photonics crystal at spectral range 400-800 nm and find the spectral position of the first photonic band gap.

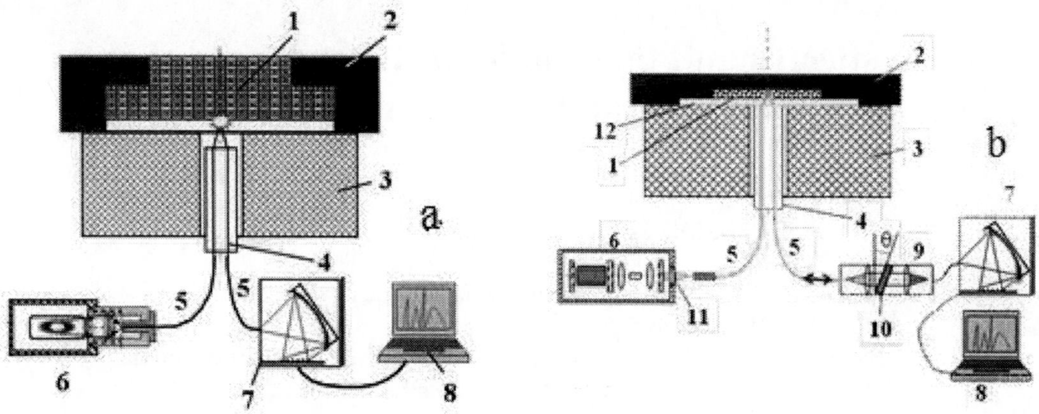

Figure 1. Installation diagram for registering of the reflection spectra from the surface (a) of photonic crystal and (b) from the analyzed substance; (a) 1 – photonic crystal, 2 – aluminum plate, 3 – teflon holder, 4- fiber-optics probe, 5 – quartz light-guides, 6 – radiation source, 7 - minispectrometer, 8 – computer; (b) 1- the sample under investigation, 2- aluminum plate, 3- teflon holder, 4- fiber- optical probe, 5- light guides, 6- laser, 7- spectrometer, 8- computer, 9- refocusator, 10- photonic crystal, 11 – fiber optic connector , 12 – quartz substrate

The installation diagram where we use the porous one dimensional photonic crystals based on aluminum oxide as selective narrowband filters when registering Raman scattering (RS)spectra from the molecular environment (1) is shown in figure 1b. The turn of the photonic crystal at a certain angle in the refocusator(9) provides the fine adjustment of the spectral position of the photonic crystal stop-zone to the frequency value of the laser (6) exciting radiation. The fiber optic minispectrometer(7) with multi-element light receiver and the computer (8) provide RS spectra express-registration in a wide spectral range.

We obtain the experimental reflection spectra of radiation from the surface of two photonic crystals given in figure 2a, b.

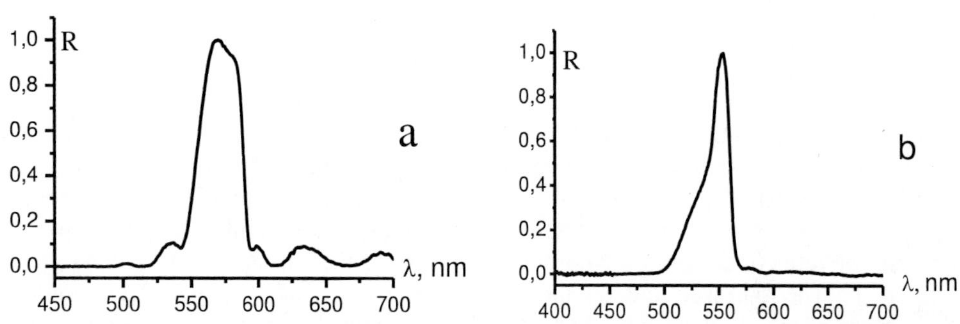

Figure 2. Reflection spectrum from the surface of two photonic crystals: №1(a) and №2(b).

We use the known Bragg relation (1), which specifies the relationship between the wavelength λ_B of the maximum reflection, the period a of one-dimensional lattice of photonic crystal and the effective refractive index n_{ef}: $\lambda_B = 2a\,n_{ef}$. Where $n_{ef}^2 = \frac{a_1}{a_1+a_2}n_1^2 + \frac{a_2}{a_1+a_2}n_2^2$; n_1 and n_2 - the refractive indices of the layers; a_1 and a_2 - their thickness [12].

$$\lambda_B = 2a\sqrt{n_{ef}^2 - \sin^2\theta} \tag{1}$$

Next, we obtain the linear dependence of the square of the photonic crystaltilt angle $\sin^2\theta$from the square of the wavelength l^2 at different angles of rotation: 8^0, 15^0, 30^0, 45^0, 60^0(see figure 3).

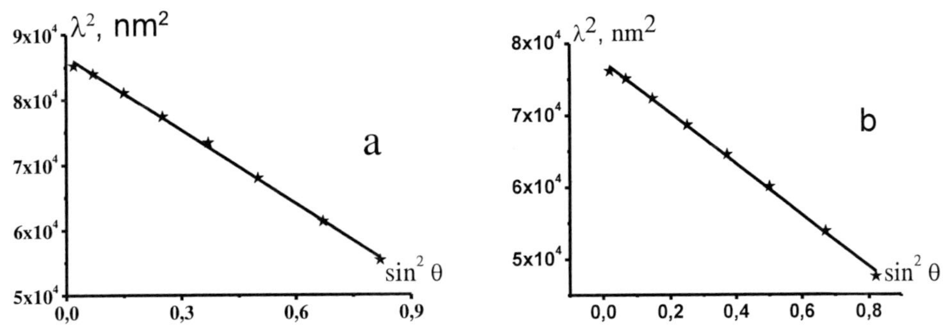

Figure 3. The dependence $(\lambda_B)^2$from $\sin^2\theta$ for two investigated samples:№1 (a) and №2 (b).

3. Experimental results and discussion

On the basis of the obtained data we compute the period values of studied photonic crystals and their effective values of refractive indices n_{ef}. In this case we use the known Bragg relation (1).

Thus, we obtain the following parameters of mesoporous one-dimensional photonic crystal based on aluminum oxide. For the sample №1, the effective refractive index n_{ef} = 1.52; the period of the structure a = 194 nm; the refractive indices of the median_1 = 1.36andn_2=1.57. For the sample №2 we have: the effective refractive index n_{ef}= 1.47; the period of the structure a = 188 nm; the refractive indices of the media n_1 =1.57; n_2 = 1.36. To analyze the observed optical properties of one-dimensional photonic crystal, we use the Kronig-Penny model. In this case the dispersion law of electromagnetic waves in one-dimensional photonic crystal with infinite number of layers is determined by the expression [13, 14]:

$$\cos ka = \cos k_1a_1 \times \cos k_2a_2 - \frac{1}{2}\left(\frac{n_1}{n_2} + \frac{n_2}{n_1}\right)\sin k_1a_1 \times \sin k_2a_2 \tag{2}$$

Here n_1 and n_2 are the refractive indices of composite layers with thicknessesa_1 and a_2

correspondingly; $k_1 = \frac{\omega}{c} n_1; k_2 = \frac{\omega}{c} n_2$; $c = 3\times108$ m/s - the speed of light in vacuum; $a = a_1 + a_2$ - the lattice constant of photonic crystal.

The reflection coefficient of the broadband radiation from the surface of one-dimensional photonic crystal at normal incidence of radiation on a surface is determined by Fresnel equation:

$$R(\omega) = \left| \frac{\frac{ck(\omega)}{\omega} - 1}{\frac{ck(\omega)}{\omega} + 1} \right|^2 . \qquad (3)$$

Figure 4 illustrates the comparison between theory results (curves 2, figure 4a,b) according (3) and experimental data of reflection spectra, obtained at experimental setup in figure 1.

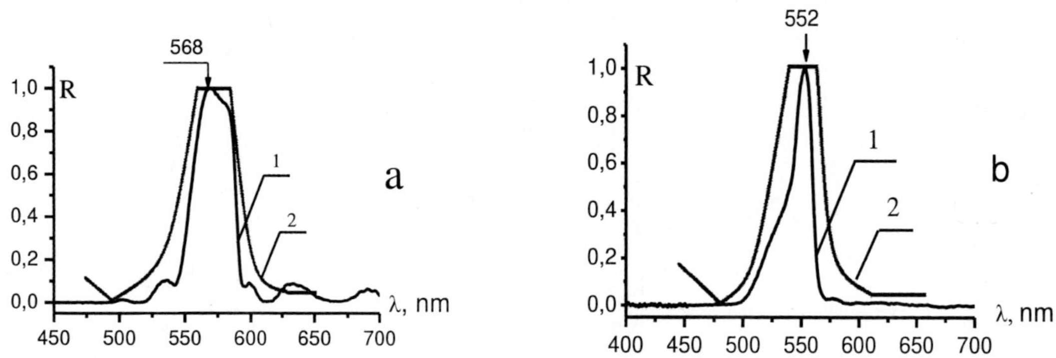

Figure 4. Comparison of theoretical (curve 2) and experimental (curve 1) data for sample № 1 (a), and for sample № 2 –(b).

As we can see, the experimentally determined reflection spectra (curves 1, figure 4a,b)are close to calculated dependences (curves 2). The observed differences between curves 1 and 2 are caused by the final number of layers in the analyzed photonic-crystal films.

4. Conclusions

Thus, we recorded the reflection spectra of broadband radiation from the surfaces of mesoporous one-dimensional photonic crystals based on aluminum oxide. A theoretical model confirming the presence of the stop-zones in these samples is built. We fulfilled the comparison of the experimentally and theoretically obtained data. There is a strong similarity between the mapped results, suggesting the possibility of this model using for further studies of photonic crystals. We found the parameters of these crystals, thereby it becomes possible to control their optical properties. Change in the photonic crystal parameters, such as: introduction of other substances into the crystal pores, and the rotation of the sample relative to the incidence plane of the electromagnetic radiation , allows to control the width and position of forbidden band. Thus, it is shown the possibility of creation of narrow band optical filters and selective mirrors, based on porous one-dimensional photonic crystals [15-18]. The rotation angle of the photonic crystal to suppress the excitation radiation of the optical laser at the wavelength of 532nm may be calculated. For sample № 1 we have θ=41^0 and for sample № 2 θ=2^0.Thus for these conditions the corresponding photonic crystals may be used for spontaneous and stimulated Raman spectra recording with YAG:Nd^{3+} laser (wave length 532 nm) as a source of exciting light at experimental setup, shown in figure 1a.

Acknowledgments

This work is partially supported by Russian Foundation for Basic Research (15-02-02882, 16-52-00026) and Belarusian Republican Foundation for Fundamental Research (F16R-063).

References

[1] Bykov V P 1972 Spontaneous emission in a periodic structure *JETP* **62** 505-512
[2] Yablonovitch E 1987 Inhibited spontaneous emission in solid-state physics and electronics *Phys.Rev. Lett.* **58** 2059-62
[3] John S1987 Strong localization of photons in certain disordered dielectric superlattices *Phys. Rev. Lett.* **58**2 486-89
[4] Gorelik V S 2007 Optics of globular photonic crystals *Quantum electronics* **37** 409-432
[5] Liu Y, Chang Y, Ling Z, Xing H and Li Y2011 Structural coloring of aluminum *Electrochemistry communications* **13** 1336-39
[6] Svyakhovskiy S E, Maydykovsky A I and Murzina T V2012 Mesoporous silicon photonic structures with thousands of periods *J. Appl. Phys.* **112** 013106
[7] Gorelik V S, Ivicheva S N, Kargin Yu Fand Filatov V V 2013 Optical Dispersion in Synthetic Opal Matrices Filled with Europium Oxide_Doped Silica Sols *Inorganic Materials* **49** 685–691
[8] Gorelik V S, Voinov Yu P, Zlobina L I and Sverbil P P 2013 Three-Dimensional Quantum Photonic Crystals and Quantum Photonic Glasses *Russian Journal of General Chemistry* **83** 2125–31
[9] Gorelik V S, Klimonskii S O, Filatov V V and Napolskii K S 2016 Optical Properties of One-Dimensional Photonic Crystals Based on Porous Films of Anodic Aluminum Oxide *Optics and Spectroscopy* **120** 534–539
[10] Gorelik V S and Shchavlev V V 2016 Optical properties of mesoporous one-dimensional photonic crystals*Bull. Lebedev Physics Institute* **43** 189-194
[11]Gorelik V S, Pudovkin A V and Filatov V V 2016 Mesoporous Photonic-crystal Films for Amplification and Filtering of Electromagnetic Radiation *J. Russ. Laser Research* **37** 604-610
[12] Gorelik V S, Lepnev L S and Litvinova A O 2014 The conversion of electromagnetic radiation in photonic crystals with SiO2 shortwave radiation *Inorganic materials* **50** 1086-90
[13] Yariv A and Yeh P 1984 *Optical waves in crystals. Propagation and control of laser radiation*(New York: Wiley) p 589
[14] Shirkov D V 1957 About the energy of the electromagnetic field in dispersive media *Optics and spectroscopy* **3** 308-312
[15] Asakawa K, Sugimoto Y, Watanabe Y, Ozaki N, Mizutani A, TakataY, Kitagawa Y, Ishikawa N, Ikeda N, Awazu K., Wang X, Watanabe A, Nakamura S, Ohkouchi S, Inoue K, Christensen M, Sigmund O, Borel P I and Baets R 2006 Photonic crystals and quantum dots technologies for all-optical switch and logic device *New J. Phys.* **8** 208
[16] Lodahl P, van Driel A F, Nikolaev I S, Irman A, Overgaag K, Vanmaekelbergh D and Vos W L 2004 Controlling the dynamics of spontaneous emission from quantum dots photonic crystals *Nature* **430** 654
[17] Raymond Ooi C H, Au Yeung T C, Kam C H and Lim T K 2000 Photonic band gap in a superconductor-dielectric superlattice *Phys. Rev. B***61** 5920 — 23
[18] Wu S-J, Chen M S and Yang T J 2005 Photonic band structure for a superconductor-dielectric superlattice*Physics C: Superconductivity* **432** 133-139

Simulation of droplet impact onto a deep pool for large Froude numbers in different open-source codes

V N Korchagova[1,2], M V Kraposhin[1], I K Marchevsky[1,2], E V Smirnova[1,2]

[1]Institute for System Programming of Russian Academy of Sciences, Alexander Solzhenitsyn st. 25, Moscow, 109004 Russian Federation
[2]Bauman Moscow State Technical University, 2-ya Baumanskaya st. 5, Moscow, 105005 Russian Federation

E-mail: v.korchagova@ispras.ru, m.kraposhin@ispras.ru, iliamarchevsky@mail.ru, alena.davidova@ispras.ru

Abstract. A droplet impact on a deep pool can induce macro-scale or micro-scale effects like a crown splash, a high-speed jet, formation of secondary droplets or thin liquid films, *etc*. It depends on the diameter and velocity of the droplet, liquid properties, effects of external forces and other factors that a ratio of dimensionless criteria can account for. In the present research, we considered the droplet and the pool consist of the same viscous incompressible liquid. We took surface tension into account but neglected gravity forces. We used two open-source codes (OpenFOAM and Gerris) for our computations. We review the possibility of using these codes for simulation of processes in free-surface flows that may take place after a droplet impact on the pool. Both codes simulated several modes of droplet impact. We estimated the effect of liquid properties with respect to the Reynolds number and Weber number. Numerical simulation enabled us to find boundaries between different modes of droplet impact on a deep pool and to plot corresponding mode maps. The ratio of liquid density to that of the surrounding gas induces several changes in mode maps. Increasing this density ratio suppresses the crown splash.

1. Introduction

Simulation of droplet impact hydrodynamics is a key element in a number of technological studies. Droplet impact simulation is important in such areas as ink-jet printing, internal combustion engine design and optimization, microchip manufacturing, solving meteorological problems and many others. However, numerical simulation of droplet impacts (considering flows with a free surface) is a complicated problem because high accuracy is critical for correct simulation of different phenomena arising after droplet impact. A suitable numerical algorithm should satisfy general requirements like numerical stability, convergence, low dissipation rate in the proximity of the liquid free surface, etc.

Worthington published the first studies of droplet impact onto a liquid layer in the late 19th century. Now there exist many papers on experimental and numerical studies in this area. The most popular ones deal with droplet impact on a deep pool [1, 2, 3], on a thin liquid film [1, 4, 5] or on a liquid layer whose depth is of the same order as the droplet diameter [6], and with droplet impact onto a dry solid surface, both hydrophilic and hydrophobic [1, 4, 7]. There are also some experimental results regarding thermal convection and the effect of other factors (for example, a magnetic field) on the behaviour of the liquid after the impact [1].

Different post-impact phenomena may appear, such as secondary droplet formation, crown splashing, high-speed jets. It appears reasonable to introduce a ratio of dimensionless criteria to correlate the differences between impact modes. This approach allows us to point out the degree of effect that different physical processes may have on the behaviour of the liquid after the impact in each case. Therefore, the main question is how the values of dimensionless criteria are related to appearance or suppression of different phenomena.

There are several known papers featuring in-depth research on this problem. They present mode maps for the phenomena mentioned, plotted depending on the ratio of dimensionless criteria [2, 5]. The authors of these works considered various ranges of dimensionless criteria for droplet impact on a deep pool or on a thin liquid layer, but they did not focus on droplet behaviour in a number of 'extreme' modes, i.e. when the values of dimensionless criteria are very large. For example, from the practical point of view it is interesting to consider droplet behaviour if the gravity force effect vanishes (when the Froude number value is over 105). Therefore, the main aim of our research is to plot mode maps for droplet impact on a deep pool under this assumption.

There exist two ways to investigate this problem: experimentally and by means of numerical simulation. However, for the 'extreme' modes mentioned it is sometimes impossible (or hardly possible) to perform precise experiments, so numerical simulation is preferable. Moreover, numerical studies allow researchers to vary problem parameters independently, which is difficult in an actual experiment.

Since code development is a very labour-intensive and time-consuming procedure even in the case of well-known problems, usage of open-source numerical codes is widespread. While some open-source codes may be based on nearly the same numerical methods, specific algorithms implemented in them may differ significantly, and as a result, it leads to differences in approximation properties, dissipation rate, numerical stability, etc. Therefore, it is interesting to compare numerical simulation results of several open-source codes and find the area where the results are in good agreement. We used two codes in this study, OpenFOAM and Gerris, which fully implement numerical models used to simulate the physical phenomena that interest us.

2. Governing equations

Consider a droplet of viscous incompressible liquid, the droplet diameter being d. The droplet falls vertically onto a deep pool (figure 1); the liquid in the pool is the same as in the droplet. We assume the pool depth H and length L to be infinitely large (with respect to the droplet diameter). U is impact velocity. The densities and dynamic viscosity coefficients of the liquid and the gas above the pool surface are ρ_f, μ_f and ρ_a, μ_a, respectively. The gravitational acceleration is $\vec{g}$; the surface tension coefficient is σ.

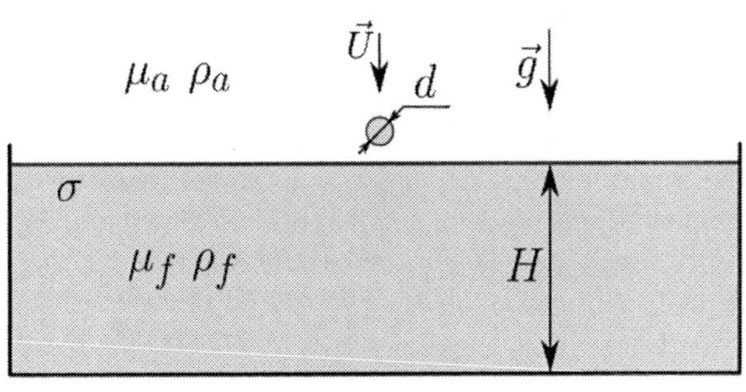

Figure 1. Problem statement. A droplet falling vertically into a deep pool

The governing equations are the continuity equation and the Navier — Stokes equations (mass and momentum conservation for incompressible Newtonian fluids):

$$\nabla \cdot \vec{U} = 0;$$

$$\frac{\partial \rho \vec{U}}{\partial t} + \left(\rho \vec{U} \cdot \nabla\right)\vec{U} = -\nabla p + \nabla \cdot \hat{\tau} + \rho \vec{g},$$

where p is pressure, $\hat{\tau} = \mu\left(\nabla \vec{U} + \nabla \vec{U}^T\right)$ is the viscous stress tensor.

Boundary conditions on the free surface between the liquid and surrounding gas should satisfy the surface energy balance and velocity continuity condition [8]:

$$\left(p_f - p_a\right)\vec{n} + \left(\hat{\tau}_f - \hat{\tau}_a\right) \cdot \vec{n} = \sigma \kappa \vec{n};$$

$$\vec{U}_f = \vec{U}_a.$$

Here $\vec{n}$ is a unit normal vector directed from the liquid into the gas; κ is the mean curvature of the interface. Introducing an additional force term into the momentum equations can automatically satisfy the first boundary condition [10].

We can write the governing equations down in a dimensionless form. We can use three dimensionless criteria to describe general properties of the case considered:

- Reynolds number $\mathrm{Re} = \rho U d / \mu$ defines the ratio of inertial forces to viscous forces;
- Weber number $\mathrm{We} = \rho U^2 d / \sigma$ is the ratio of inertial forces to surface tension;
- Froude number $\mathrm{Fr} = U^2 / (gd)$ shows the ratio of inertial forces to gravity.

3. Published results regarding droplet impacts

One way to classify the phenomena that appear after a droplet impact onto a deep pool is to divide them in two groups: large-scale and small-scale effects. Large-scale ones are crater and high-speed jet formation in the liquid after a droplet falls, secondary drops tearing off the top of the jet, and the crown splash [4]. All these phenomena have a common property: their characteristic scales are of the same order as the initial droplet diameter. Let us estimate the size of the mesh that is required to represent them correctly. For simplicity, we consider the mesh to be uniform; the resolution should be at least 20 cpd (cells per diameter). Our computational domain size should be on the order of $20d$ along each dimension; the initial position of the droplet is in the centre. Therefore, for 2D and axisymmetric simulations, the total number of cells is on the order of 10^5; 10^7 for a 3D case. Such meshes are acceptable for numerical simulation (in parallel mode), so simulating large-scale phenomena is feasible in principle. Mesh refinement procedures can significantly reduce the number of cells further. In this study, we investigate large-scale phenomena: the crown splash and high-speed jet formation.

We should note that small-scale phenomena (bubble entrapment, microdroplet splash and ejecta sheet at the moment when the droplet touches the pool surface [4, 5, 9]) are more difficult for numerical simulation. The mesh resolution required here is at least 200 cpd, so the only way to perform a simulation is to use adaptive mesh refinement.

A number of papers [2, 4, 5] attempted to predict appearance of large-scale effects as depending on physical properties of the liquid and gas as well as on droplet diameter and impact velocity. Such studies are mostly experimental and cover a small range of dimensionless criteria. Certain authors attempted to generalize previously published results and supplement them with experimental and numerical results of their own. Ultimately, they endeavoured to represent their results by plotting so-called mode maps.

An experimental study [5] considers a droplet impact on a thin liquid layer (of an approximately $0.2d$ depth). The authors fully investigated different types of crown effects, such as crown splash, microdroplet splash, ejecta sheet, Peregrine sheet, and obtained the diagram shown in figure 2. The boundaries between different modes are clear. The authors assumed that it is possible to neglect the effect of gravity. For all cases, Froude numbers [5] are larger than 10^2.

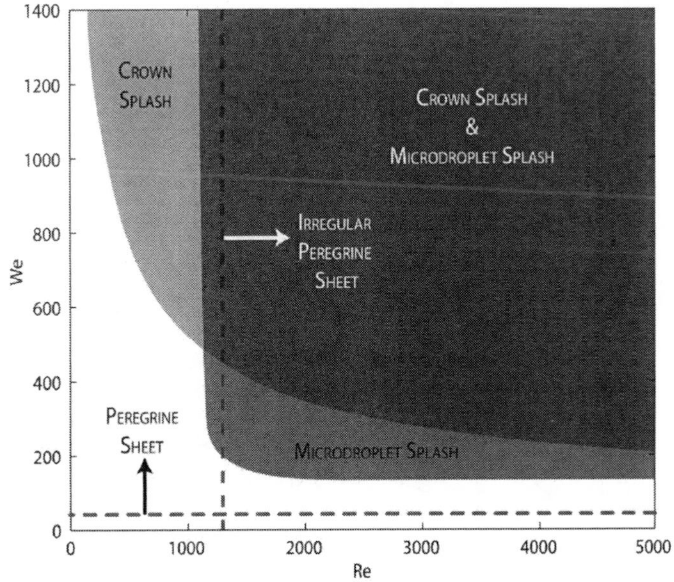

Figure 2. Mode map for crown effects after a droplet impact on a thin liquid layer [5]

Another large-scale phenomenon is high-speed jet formation. Authors of the paper [2] investigated shapes of high-speed jets being formed after a droplet impacts a deep pool (its depth was $7.2d$). Figure 3 shows a mode map from [2] for jet shapes depending on surface tension and gravity, with a Reynolds number of approximately 10^3. This mode map presents joint results of experiments and OpenFOAM axisymmetric numerical simulations. The paper [2] states the results are in good agreement with experimental data. The authors compared empirical 'boundaries' for the Weber and Froude numbers corresponding to different modes from previously published experimental studies with the numerical results and noted that gravity forces and surface tension have a significant effect on the form of high-speed jets and the behaviour of liquid in general.

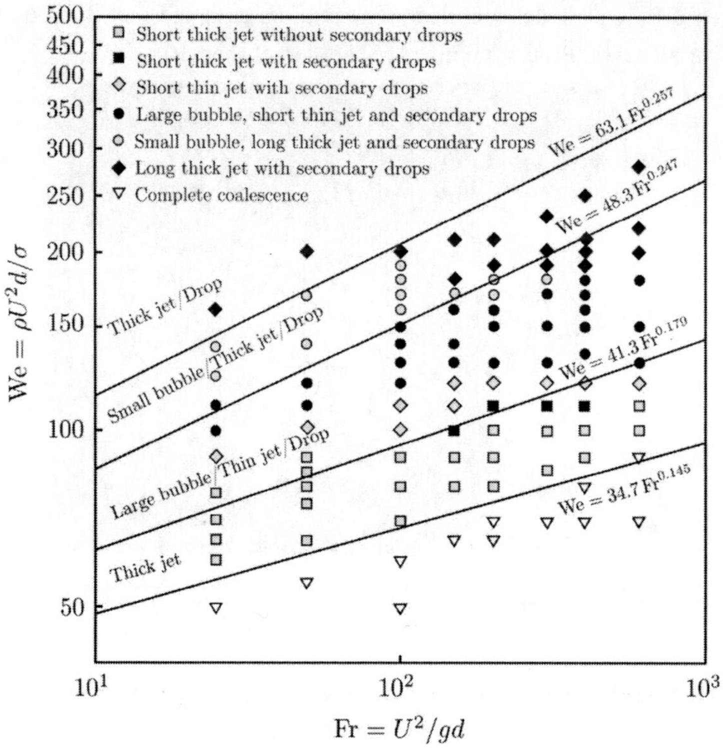

Figure 3. Mode map for high-speed jets after a droplet impact onto a deep pool, $\mathrm{Re} \approx 10^3$ [2]

Power functions can approximate plots of mode 'boundary' values. They do not have any asymptotes, but it is obvious that they exist because large values of dimensionless criteria mean that it becomes possible to neglect the corresponding forces as compared to inertial forces. For example, a large Froude number means that gravity is negligible. However, such modes are interesting in practice because they appear in the cases of very small droplets (with diameters of about a few micrometers) when impact velocity is rather high (over 100 m/s). These cases are common in ink-jet printing, microchip manufacturing, *etc*.

We consider gravity forces to be negligible if the Froude number exceeds a certain critical value, close to 10^5. We found this value through a series of simulations in which we gradually decreased gravity values.

4. Tools and numerical schemes

It is necessary to choose a suitable tool for numerical simulation. Different open-source codes can be used for scientific research, so we reviewed the degree to which these codes are applicable for the problem under consideration. According to the main aims of our investigation and our problem statement, correct surface tension models must be implemented in selected packages since surface tension forces play a crucial role in fluid behaviour.

As a result of comparing different codes, we chose two of them: OpenFOAM and Gerris. We also looked at some other codes, which are based on other numerical methods, like meshfree methods (for example, Smoothed Particle Hydrodynamics), hybrid methods (Particle Finite Element Method), immersed boundary methods. However, either these codes implement no surface tension models, or they need to be validated and/or modified significantly.

Both OpenFOAM and Gerris are based on the Finite Volume Method [10, 11]. The main idea is that, for the case of two media in the computational domain, governing equations are solved over the whole domain for a 'mixture' with corresponding physical properties, *i.e.*, according to Volume of Fluid (VOF) approach [10, 11].

An additional variable is added to the model, a volume fraction function α_i of the i-th phase. It shows the ratio between the liquid and gas in the cells (usually the liquid is the 1st phase, gas is the 2nd phase):

$$\alpha_1 = \begin{cases} 1, & \text{in the liquid subdomain;} \\ 0, & \text{in the gas subdomain;} \\ (0, 1), & \text{in the proximity of the interface.} \end{cases}$$

It is obvious that for two phases

$$\alpha_1 + \alpha_2 = 1.$$

The transport equation for the volume fraction should be solved together with the continuity and momentum equations:

$$\frac{\partial \alpha_1}{\partial t} + \nabla \cdot (\vec{U}\alpha_1) + \nabla \cdot (\vec{U}_R \alpha_1 \alpha_2) = 0.$$

Here $\vec{U} = \alpha_1 \vec{U}_1 + \alpha_2 \vec{U}_2$ is the velocity of the mixture, $\vec{U}_R = \vec{U}_1 - \vec{U}_2$ is the relative velocity of the phases; there exist special discretization approaches to approximate it [10], making it possible to introduce a certain 'artificial' velocity which 'compresses' a subregion of the interface.

The volume fraction gradient approximates surface tension forces $\vec{f}_\sigma$ [12] considered as an additional term in the Navier — Stokes equations:

$$\vec{f}_\sigma \approx \sigma \kappa \nabla \alpha_1.$$

OpenFOAM and Gerris calculate mean interface curvature κ differently. OpenFOAM uses the gradient of the volume fraction function for curvature approximation:

$$\kappa \approx \nabla \cdot \left(\frac{\nabla \alpha_1}{\|\nabla \alpha_1\|} \right).$$

Gerris uses a level-set function to identify the interface [13]. For 2D simulations, an efficient algorithm exists to estimate curvature for a Cartesian mesh and a 7×3 stencil.

5. Discretization

The flow domain is axisymmetric, so in OpenFOAM we consider a 'wedge' with a single cell in the azimuthal direction with the corresponding boundary conditions on the sides of the wedge, and a flat region in Gerris. Artificial influence of the outer boundary conditions on the flow inside the domain should be as small as possible. Figure 4(a) shows the schematic of the flow domain and boundary conditions.

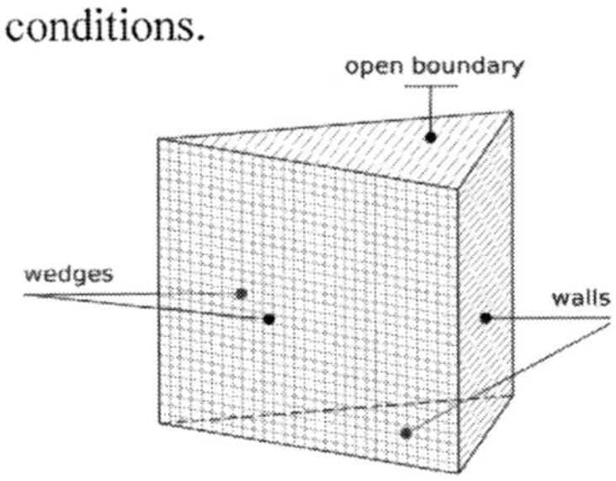

(a)

(b)

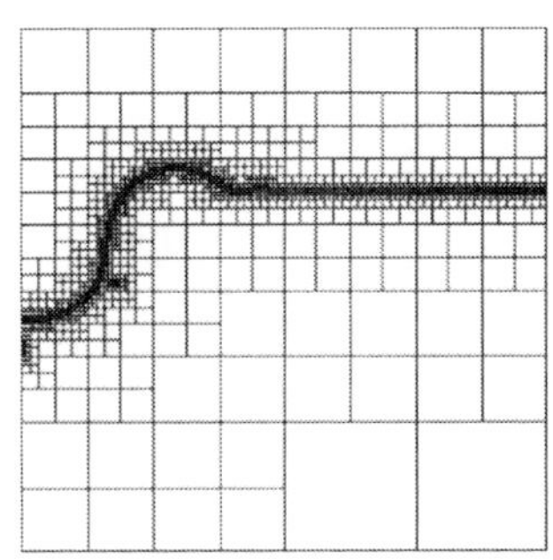

(c)

Figure 4. Discretization domain schematic: (a) schematic of boundary conditions; (b) sketch of a typical OpenFOAM mesh generated by the snappyHexMesh utility; (c) sketch of a dynamically refined mesh in Gerris

Mesh refinement technologies in OpenFOAM and Gerris differ significantly. In OpenFOAM, the mesh is static and the refinement is performed prior to calculations in the subdomains where the free surface can move during the solution process. In Gerris, the mesh is adaptively and dynamically refinable in the regions in the proximity of the interface (or in some other regions, if it is necessary).

Time step selection is automatic according to the value of the Courant-Friedrichs-Lewy (CFL) number. This approach makes it possible to obtain a numerically stable solution and provide a low level of non-physical disturbances.

6. Validation

We reviewed the information we found in previous publications dealing with validation of OpenFOAM and Gerris codes. Here we briefly describe the most interesting validation cases.

Validation test #1: Droplet impact and crown formation.
Paper [14] describes simulating a 3D droplet impact problem in OpenFOAM, using a mesh with 48 cells per diameter (48 cpd). We compared the results with experiments from [15] (figure 5).

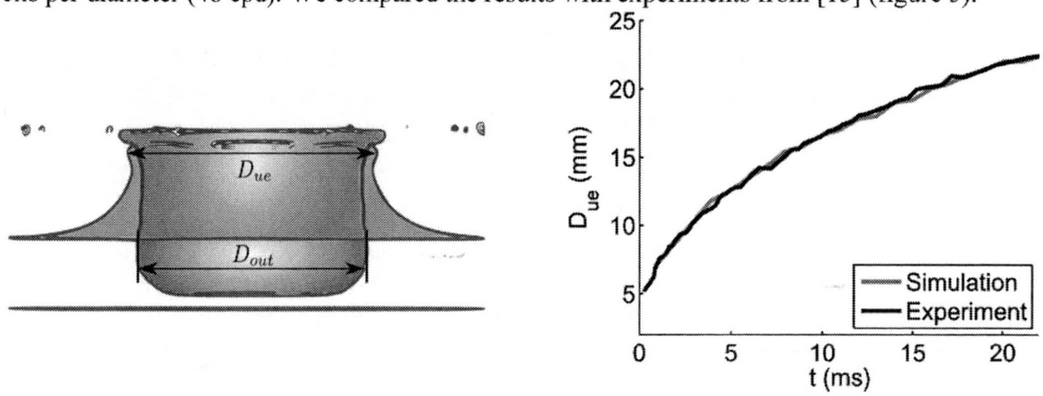

Figure 5. Comparison of droplet impact simulation from [14] with experimental data from [15]

Validation test #2: 2D standing capillary wave.
In the paper [14], the authors considered the problem of damping interface perturbation under surface tension and compared the kinetic energy from their OpenFOAM simulation with Lamb's analytical solution found in [16]. They also investigated mesh convergence and obtained good agreement between numerical and analytical solutions.

Validation test #3: Rayleigh breakup of a liquid jet.
An important feature of a Computational Fluid Dynamics code is the possibility of simulating instabilities. A well-known type of hydrodynamic instability is the Rayleigh — Plateau instability (breakup of a liquid jet). Paper [14] compares results of OpenFOAM simulation with a non-linear theory [17] that permits predicting size of droplets formed after jet breakup (figure 6 (a)). Results of simulation in Gerris were compared with linear Weber [18] and Rayleigh [19] theories that predict maximum and minimum jet radii (figure 6 (b)).

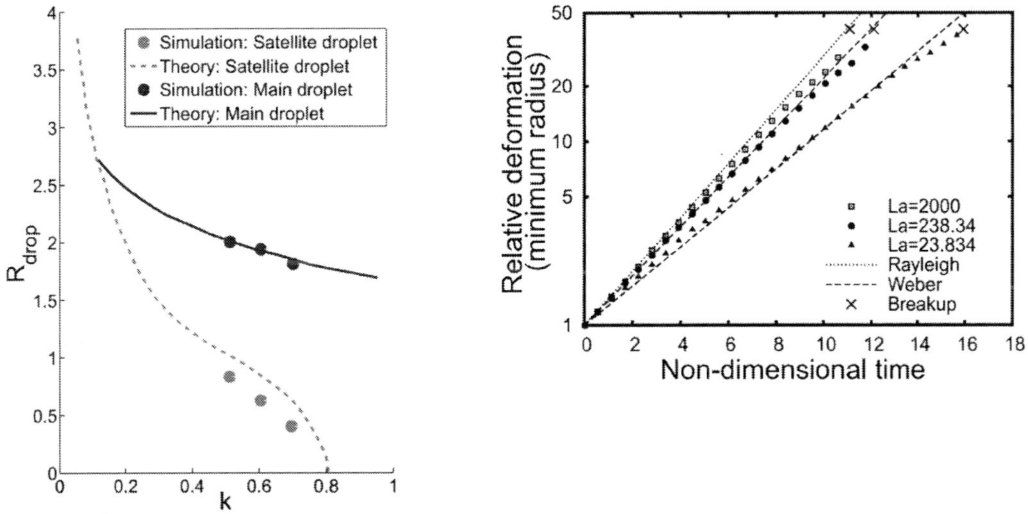

(a) (b)

Figure 6. Results of simulation of Rayleigh — Plateau instability: (a) OpenFOAM, comparison with non-linear theory; (b) Gerris, comparison with linear theory

Validation test #4: Retraction of a liquid column.
An initially stationary liquid column deforms under the action of surface tension. Results of 3D simulation in OpenFOAM (retraction of a cylindrical liquid column) [14] are compared with VoF results [20]. Mesh resolution is 32 cpd with respect to the column diameter. Paper [21] describes a similar case for Gerris, comparing the computed value of the so-called Taylor — Culick velocity with an analytical expression from [22]. There were 6×115 cells in the cross-section of the liquid region in the initial state (230 cpd).

Validation test #5: Kelvin — Helmholtz instability.
Paper [23] uses OpenFOAM to simulate the problem of Kelvin — Helmholtz instability, comparing the outcome with the numerical results of Rogers & Moser [24]. The authors noted that the results are in good agreement, but the paper does not present any quantified comparison results.

Validation test #6. Droplet deformation.
A droplet is placed in a gas flow and deforms with time. Results of OpenFOAM simulation [26] were compared with the theory of Stokes and Oseen [27], with the results of numerical simulations from [28] and the experiments of Temkin & Mehta [29].

Two numerical tests were performed to validate large-scale and certain small-scale phenomena simulation in OpenFOAM and Gerris.

Test #1. Large-scale phenomena.
We consider crater and high-speed jet formation in the axisymmetric case according to the paper of Berberović et al. [6]. In this paper, there are both experimental data and the necessary information about computational domain, mesh refinement and other key points for numerical simulation. Table 1 presents our input parameters.

Table 1. Input data for test 1 (for large-scale phenomenon validation).

Property	Value
Droplet diameter, d	2.67 mm

Impact velocity, U	2.56 m/s
Depth of the liquid layer, H	5.34 mm
Density of liquid, ρ_f	1179 kg/m^3
Dynamic viscosity of liquid, μ_f	$18.57 \cdot 10^{-3}$ Ns/m^2
Density of air, ρ_a	1.204 kg/m^3
Dynamic viscosity of air, μ_a	$1.9 \cdot 10^{-3}$ Ns/m^2
Surface tension coefficient, σ	0.063 N/m
Characteristic flow domain size, L	45 mm
Mesh refinement	30 cpd (cells per diameter)
CFL	0.2

Figure 5 shows the results of numerical simulation. To enable quantitative comparison of the results obtained, we measured two parameters at different time moments: crater depth (penetration depth) and crater diameter, measured halfway up the initial height of the liquid layer.

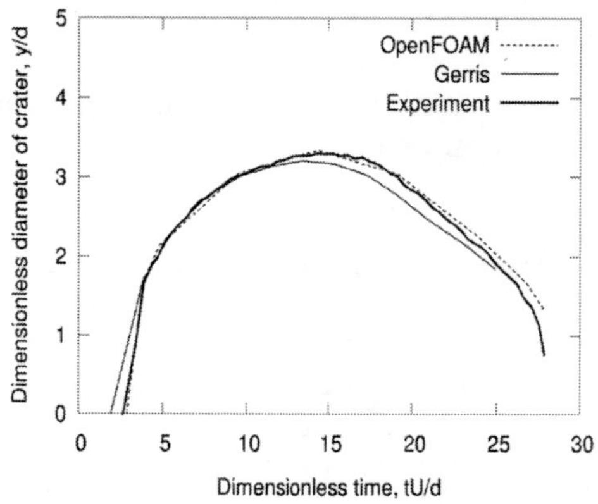

(a)

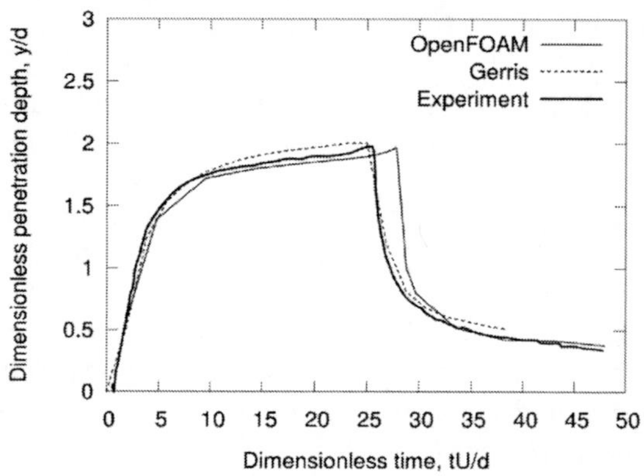

(b)

Figure 7. Comparison of experimental data with a numerical solution, test #1, Berberović et al. [6]: (a) crater diameter; (b) penetration depth

There is good agreement between experimental and numerical data. Therefore, mesh refinement over 30 cpd is enough to solve for large-scale effects correctly.

Test #2. Small-scale phenomena.

In paper [3], Tran et al. described an interesting small-scale phenomenon. When the falling droplet touches the surface of liquid, the bottom part of the droplet captures a thin film of air. This film keeps its 'continuous' shape for a short time and then it ruptures into a sequence of bubbles. There are experimental data for the penetration depth of the air film in this paper; therefore, this test makes for a good validation case. Table 2 shows our input data.

Table 2. Input data for test 2 (for small-scale effect validation).

Property	Value
Droplet diameter, d	1.9 mm
Impact velocity, U	0.55 m/s
Depth of the liquid layer, H	10 mm
Density of liquid, ρ_f	953 kg/m^3
Dynamic viscosity of liquid, μ_f	0.1906 Ns/m^2

Density of air, ρ_a	1.204 kg/m^3
Dynamic viscosity of air, μ_a	$1.9\cdot10^{-3}$ Ns/m^2
Surface tension coefficient, σ	0.0197 N/m
Characteristic flow domain size, L	45 mm
Mesh refinement	200 cpd (cells per diameter)
CFL	0.2

Figure 6 shows numerical simulation results for an air film penetrating a deep liquid pool. The $t = 0$ moment corresponds to the moment when the air film ruptures into bubbles. The quantitative results are in good agreement with the experimental data but the air film in the numerical solution turns into a bubble sequence shortly after the moment when the interaction between the droplet and the pool surface begins.

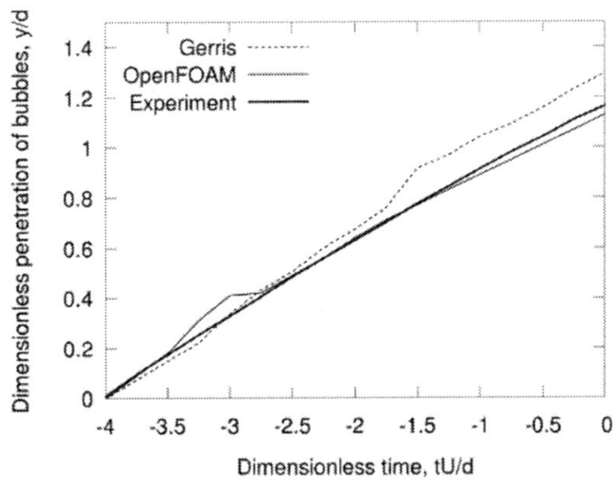

Figure 8. Penetration depth of the air film

7. Results and discussion

We note once again that we made the following assumptions for simulations in validation cases:

- the problems are considered to be axisymmetric because it is enough for correct simulation of crater and high-speed jets, and also can be useful for prediction of crown splash formation;
- mesh refinement was at least 20 (normally about 30) cells per diameter of initial droplet.

Post-impact behaviour of the liquid depends on the ratio between viscous forces, surface tension, inertial forces and external air pressure [30]. As for incompressible fluids, the main parameter is the pressure gradient, and during computations the pressure gradient can be defined by the liquid/gas density ratio δp.

In this study, we performed a series of computations in OpenFOAM and Gerris and obtained mode maps for two different values of the liquid/gas density ratio (figure 7–9). It is clear that increasing δp leads to suppression of modes featuring crown splashes and large drop formation on tops of high-speed jets. As for comparing computation results for these two open-source codes, the axisymmetric model in Gerris gives non-physical results in modes with the crown splash (when the Re and We numbers are large enough) — the secondary 'drops' pull to the centre of the flow region. It is the reason why we do not show these points on the Gerris mode maps.

IOP Conf. Series: Journal of Physics: Conf. Series **918** (2017) 012037 doi:10.1088/1742-6596/918/1/012037

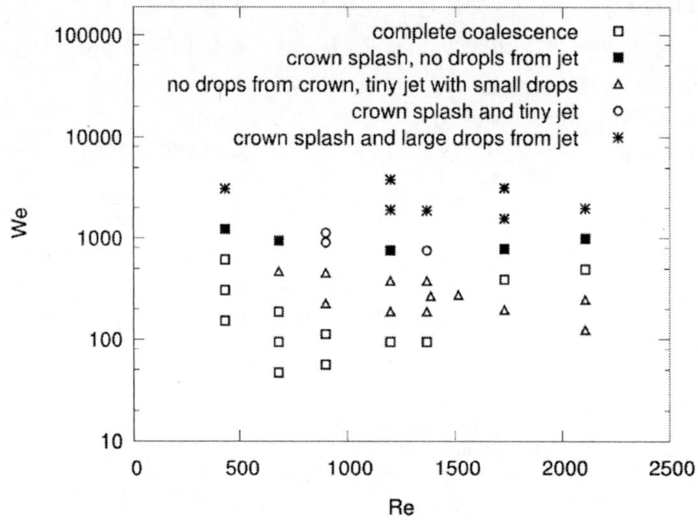

Figure 9. Mode map for droplet impact, generated from OpenFOAM results; the liquid/gas density ratio is 10^3

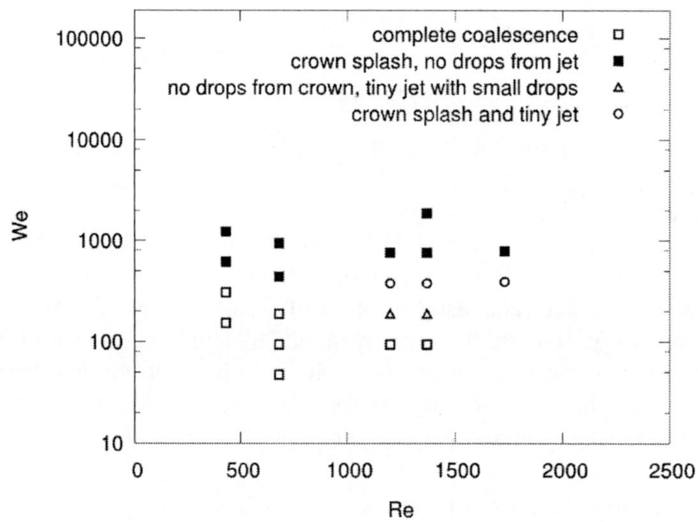

Figure 10. Mode map for droplet impact, generated from Gerris results; the liquid/gas density ratio is 10^3

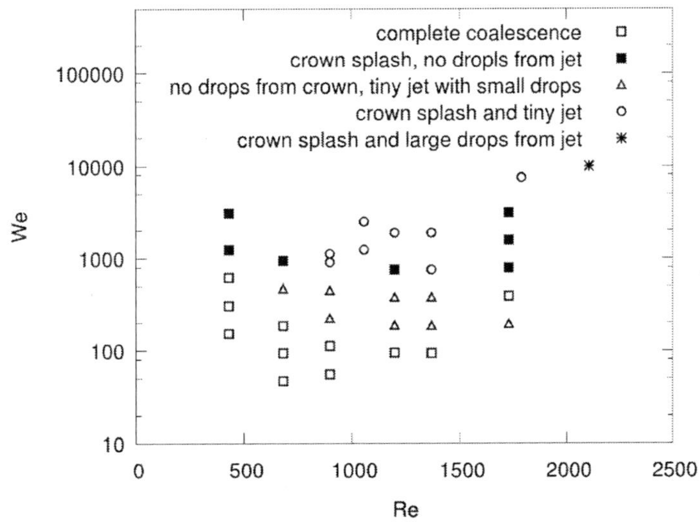

Figure 11. Mode map for droplet impact, generated from OpenFOAM results; the liquid/gas density ratio is 10^4

8. Conclusions

We investigated the problem of simulating a droplet impact onto a deep pool. First, we reviewed previously published results and analysed both experimental data and numerical results. Dimensionless criteria (Re, We, and Fr numbers) indicate ratios between various forces (inertia, surface tension, viscosity, gravity) and make it possible to predict droplet behaviour. Maps of various droplet impact modes can represent the results obtained.

Previous studies featured several maps for a number of specific modes in a small range of dimensionless criteria. Empirical functions for mode boundaries are power functions, but it is clear that certain 'saturation' values should exist. If the value of a dimensionless criterion exceeds a critical value, it means that it is possible to disregard the influence of the corresponding force, and the current mode position on the map will not change.

For example, low gravity conditions are of practical interest. Numerical investigation is the most convenient way to obtain mode maps because a full-scale experiment in these conditions is too difficult. It is important to use 'good' tools for simulation. We reviewed various open-source packages and chose two codes, OpenFOAM and Gerris. Various test cases helped us to validate these codes. We also found a lot of information relevant for code validation in published papers.

Numerical results of using these codes enabled us to plot mode maps. We considered the effect of the liquid-to-gas density ratio. Increasing it suppressed the crown splash. We obtained good agreement between maps for a high-speed jet. While we were able to predict results for a crown splash, it is necessary to run 3D simulations to obtain a correct solution.

Acknowledgements

The work of M.V. Kraposhin and I.K. Marchevsky is supported by the Russian Science Foundation (grant 17-79-20445).

References

[1] 2002 *Drop-Surface Interactions* (Berlin: Springer-Verlag) ed M Rein
[2] Ray B, Biswas G and Sharma A 2015 Modes during liquid drop impact on a liquid pool *J. Fluid Mech.* **768** 495–522
[3] Tran T, de Maleprade H, Sun C and Lohse D 2013 Air entrainment during impact of droplets on

liquid surfaces *JFM Rapids* **726** R3

[4] Yarin A I 2006 Drop Impact Dynamics: Splashing, Spreading, Receding, Bouncing... *Annu Rev. Fluid Mech.* **38** 159–92

[5] Deegan R D, Brunet P and Eggers J 2008 Complexities of splashing *Nonlinearity* **21**(1) C1

[6] Berberović E, van Hinsberg N P, Jakirlić S, Roisman I V and Tropea C 2009 Drop impact onto a liquid layer of finite thickness: Dynamics of the cavity evolution *Phys. Rev.* **79**

[7] Romanov A S and Semikolenov A V 2015 Film shape of partially wetting liquid while flowing down wetted surface *Vestn. Mosk. Gos. Tekh. Univ. im. N.E. Baumana, Estestv. Nauki [Herald of the Bauman Moscow State Tech. Univ., Nat. Sci.]* **5** 88–99

[8] Landau L D and Lifshitz E M 1959 *Fluid Mechanics, Volume 6 of A Course of Theoretical Physics* (New York: Pergamon)

[9] Zhang L V, Toole J, Fezzaa K and Deegan R D 2012 Evolution of the ejecta sheet from the impact of a drop with a deep pool *J. Fluid Mech.* **690** 5–15

[10] Rusche H 2002 *Computational Fluid Dynamics of Dispersed Two-Phase Flows at High Phase Fractions: Ph.D. thesis* (University of London, Imperial College)

[11] Ferziger J H and Perić M 2002 *Computational Methods for Fluid Dynamics* (Springer-Verlag)

[12] Brackbill J U, Kothe D B and Zemach C 1992 A Continuum Method for Modelling Surface Tension *J. Comput. Phys.* 100(2) 335–54

[13] Popinet S 2009 An accurate adaptive solver for surface-tension-driven interfacial flows *J. Comput. Phys.* **228** (16) 5838–66

[14] Deshpande S S, Anumolu L and Trujillo M F 2012 Evaluating the performance of the two-phase flow solver interFoam. *Comput. sci. disc.*, 2012 **5** (1)

[15] Cossali G E, Marengo M, Coghe A and Zhdanov S 2004 The role of time in single drop splash on thin film *Exp. Fluids* **36** 888–900

[16] Lamb H 1932 *Hydrodynamics* (New York: Dover)

[17] Lafrance P 1975 Nonlinear breakup of a laminar liquid jet *Phys.* Fluids **18** 428–32

[18] Weber C 1931 Disintegration of liquid jets *Zeitschrift fur Angewandte Mathematik und Mechanik* **11**(2) 136–59

[19] Rayleigh L 1892 On the instability of a cylinder of viscous liquid under capillary force *Phil. Mag.* **34**(207) 145–54

[20] Umemura A 2011 Self-destabilizing mechanism of a laminar inviscid liquid jet issuing from a circular nozzle *Phys. Rev.* **83**

[21] Fuster D, Agbaglah G, Josserand C, Popinet S and Zaleski S 2009 Numerical simulation of droplets, bubbles and waves: state of the art *Fluid Dyn. Res.* **41**(6)

[22] Savva N and Bush J W M 2009 Viscous sheet retraction *J. Fluid Mech.* **626** 211–40

[23] Wedolowski K, Bajer K and Kwiatkowski K 2011 Analysis and modelling of the effective reaction rate in a developing mixing layer *J. Phys.* **318**(9) 2688–97

[24] Rogers M M and Moser R D 1992 The three-dimensional evolution of a plane mixing layer: the Kelvin — Helmholtz rollup *J. Fluid. Mech.* **243** 183–226

[25] Orazzo A and Hoepffner J 2012 The evolution of a localized nonlinear wave of the Kelvin — Helmholtz instability with gravity. *Phys. Fluids*, **24**

[26] Jalaal M and Mehravara K 2014 Transient growth of droplet instabilities in a stream *Phys. Fluids* **26**

[27] Clift R, Grace J R and Weber M E 1978 *Bubbles, Drops and Particles* (Mineola: Dover Publ.)

[28] Quan S and Schmidt D P 2006 Direct numerical study of a liquid droplet impulsively accelerated by gaseous flow *Phys. Fluids* **18**, 102103

[29] Temkin S and Mehta H K 1982 Droplet drag in an accelerating and decelerating flow *J. Fluid Mech.* **116** (1) 297–313

[30] Xu L, Zhang W W and Nagel S R 2005 Drop Splashing on a Dry Smooth Surface. *Phys. Rev. Lett.* **94**, 184505

IOP Conf. Series: Journal of Physics: Conf. Series **918** (2017) 012038 doi:10.1088/1742-6596/918/1/012038

Simulation of forming a flat forging

K Solomonov[1], L Tishchuk[2], N Fedorinin[3]
[1]Professor, Moscow State University of Railway Engineering (Voronezh branch), Russia
[2]Head of the Laboratory, Moscow State University of Railway Engineering (Voronezh branch), Russia
[3]Associate Professor, Moscow State University of Railway Engineering (Voronezh branch), Russia

E-mail: konssol@list.ru

Abstract. The metal flow in some of the metal shaping processes (rolling, pressing, die forging) is subjected to the regularities which determine the scheme of deformation in the metal samples upsetting. The object of the study was the research of the metal flow picture including the contour of the part, the demarcation lines of the metal flow and the flow lines. We have created an algorithm for constructing the metal flow picture, which is based on the representation of the metal flow demarcation line as an equidistant. Computer and physical simulation of the metal flow picture with the help of various software systems confirms the suggested hypothesis.

1. Introduction

The blank parts with thin blades are produced at the special production facilities of iron and steel plants equipped with powerful pressing machines. Such products often have flaws which worsen their macrostructure. Therefore it is of great current interest to find new techniques of simulating the forming processes for forgings with stiffening fins.

It is hard to make a mathematical model of metal forming processes, which would describe the mode of deformation of metal plastic forming. One way to solve the problem of simulating the metal flow pattern and the spatial pressure profile is a so-called «theory of thin-layer flow» [1]. It is based on the assumptions which simplify the original system of differential equations. Therefore the problem comes to purely geometric tasks and can be solved in terms of the so-called «sand analogy» using the suggested technique [2].

2. The basic provisions of the developed method

This method is based on the following provisions [3, 4]:

- The principle of the shortest normal current line determines the current line direction orthogonal to the contour of the forging, which represents a sudden change in the layer thickness (including stiffening ribs or elevations on the forging blade). Hence, at the initial stage of deformation, when the terminal pressure remains the same along the boundary contour, the metal flows move orthogonally to the contour and the flowed metal amount at each point on the boundary is defined by the length of the current lines.
- In the process of deformation the boundary conditions change and along the contour the contact pressures become unequal. In this regard, the current lines will be directed at an acute angle to the contour of the forging. However, given the fact that the spatial diagram of contact pressure is a ruled surface, the fall lines (and so the current lines) are directed orthogonally to the level curves of this surface. Projecting the three-dimensional picture on the plane of the forging blade it is possible to add a conditional path along which the contact pressures are equal. Then the flow lines are orthogonal to this conditional loop.
- In general, the conditional contour is a rather complicated curve. In accordance with the principle of the least perimeter slab tends to take the form of a circle in the plan. Therefore, we can assume that

the flow lines are directed along the radii of a circular arc. Then the conditional contour is a circle, and the flow pattern of metal on the forging web will be radial.

- Note that the radial flow pattern of the metal is more universal than normal. It is also applicable in the initial stage of deformation for the forging when its circuit consists of the curved lines sections. Approximating the contour of forgings with circular arcs you can also apply a radial circuit at the initial moment of deformation when the current lines are perpendicular to the contour.
- In view of the above, the spatial diagram of contact pressure is a combination of conical surfaces at any stage of the workpiece deformation except the initial one. Concurrently, the boundary contact pressure is lying in the vertical planes intersecting these surfaces.
- The value of the boundary contact pressure at any time point for any deformation in the arbitrary point on the contour depends on several parameters: the thickness of the forging blade, the sizes of the die cavity, the gutter width and the amount of the metal flowed into the cavities. Consideration of all these parameters entails using rather complex formulae to calculate the boundary contact pressure.
- Since the spatial diagram of contact pressure is the surface of the same slope, the line section of metal flow is the locus of points equidistant from the contour of forging. The contour of forging can be approximated by straight lines and circular arcs. Therefore, the problem of constructing a line section of metal flow is reduced to finding the locus of points equidistant from the circles and straight lines.
- Because any multiply connected circuit with the sufficient degree of accuracy can be approximated by straight lines and circular arcs, we can assume that the surface of the contact pressure spatial diagram consists of flat and conical sections. Their intersection lines form the edges (so-called ridges).
- Frontal and profile projections of these edges allow determining the volume of the contact pressure diagram and therefore the force, required for deforming the metal. The horizontal projection (plan view) is a dividing line of the metal flow, which characterizes the metal flow distribution on the contact surface.

3. A new algorithm for constructing the equidistant

Most of the parts used in mechanical engineering, in view of technological design requirements, consist of the revolution surfaces and polyhedra. Therefore, in the practice of die forging a lot of parts produced from flat blanks have planar elements. In this article we shall confine ourselves to the simplest case: we shall consider the problem of constructing an equidistant for the contour which represents a piecewise-linear closed line, i.e. polygon (Fig. 1, a).

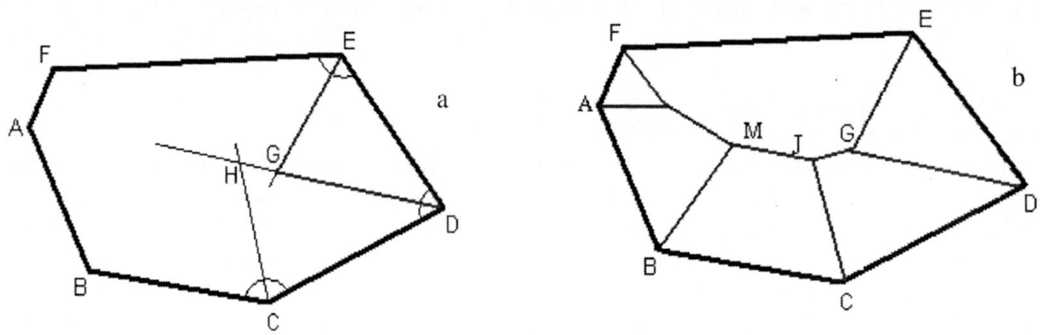

Figure 1. The scheme of constructing an equidistant

The equidistant of two lines is the bisector of the angle. Therefore, the very first equidistant contour line is the bisector of the smallest angle at the apex D. Then hold the bisector of two adjacent angles (C and E) to the intersection with the bisector of the smallest angle (points G and H). The first equidistant lines finish at the nearest point G of the intersection with the bisectors of the adjacent angles.

Next, we eliminate from consideration the direction DE. The bisectors of the angles adjacent to the direction DE have formed the equidistant lines. Continue to the intersection side FE and CD adjacent to the dropped line.

The dimension of the contour has decreased per unit: instead of the hexagon we shall now consider a pentagon. It is evident that now the smallest angle in the polygon of the contour is a newly obtained angle. The procedure is repeated, but the new equidistant line is built from the end point of the last equidistant line (point G) rather than from the angle of the contour. Then again look for the smallest of the remaining angles of contour (i.e. the angle at the vertex F) and repeat the above-described algorithm until the polygon is reduced to the triangle. As we know from geometry, the bisectors of the triangle always intersect at one point, so to finish the construction of equidistant lines it is sufficient to connect the points in which the stop of the sequential actions took place. The construction is completed (the result is shown in Fig. 1, b).

In a similar way, one can construct an equidistant of any polygon. Currently, this algorithm is implemented in the visual programming environment DELPHI.

A similar algorithm was developed for the multivariable piecewise-nonlinear contour [5].

4. Computer-generated simulation

The following example of simulating the forming of closed die forging with contour finning shows the potentialities of the developed method and software package (Fig. 2).

Figure 2. Forging with a technological cut-out

For express analysis of a lug (or a notch) application expediency in this forging, a metal flow pattern with different circle center positions and radii has been simulated by means of the developed software complex [6]. The end-point analysis shows that the metal outflow to the lug (or the notch) reduces the unevenness of metal flowing into the die cavity. Therefore its application is advisable.

A program based on the developed method has been used for simulating the forging.

When simulating the following requirements should be taken into consideration. We can change only those geometric elements which do not influence the product design, for example, the width and height of the gutter, the lug radius, the initial thickness of the primary blank or upsetting stage. The lug radius belongs to these variables because it should be replaced by a 240 mm hole in the fine product and the lug can be removed during machining. In the calculating process the lug size is the controlling factor which allows getting different metal flow patterns on the die impression surface and different stiffening rib profiles.

Fig. 3, a shows the pattern of the metal flow along the forging cloth.

Similarly, we can get the metal flow picture for any complexity of the contour (Fig. 3, b).

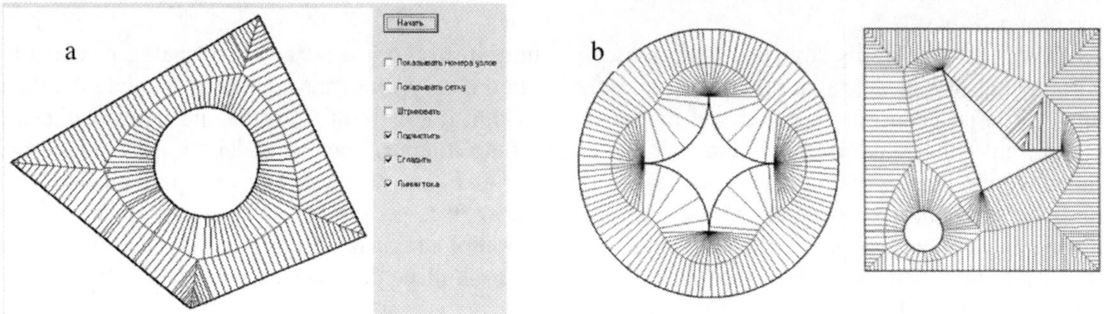

Figure 3. Metal flow picture model

5. Physical simulation

To test the simulation results of forming the closed die forging with the contour finning we have conducted an industrial experiment (Fig. 4) on stepwise upsetting of the flat forged primary blanks made of AK6 alloy by means of a hydraulic press of 150 MN deforming force.

The full forging was not achieved because of the low power of the hydraulic press. At the last step

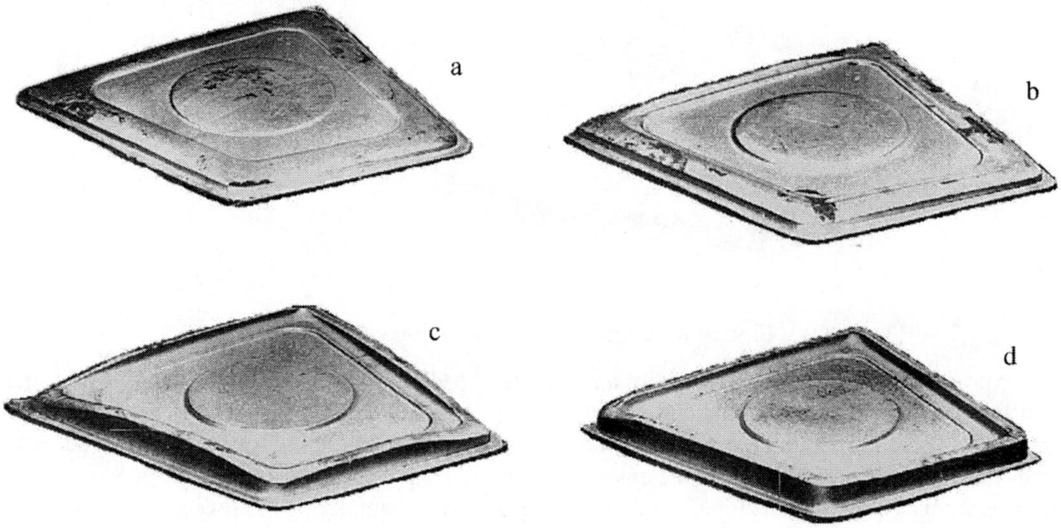

Figure 4. Forming of stamped forging

of upsetting the lug was entirely formed whereas one of the stiffening fins angular areas didn't reach the estimated height. (Fig. 4, d).

The central areas of stiffening fins are formed much quicker than the angular ones, which results in metal flowing above the die cavities under the stiffening fins in the central areas and worsens the product macrostructure (Fig. 5, a). The figure shows that the fin was cut from the side of the groove.

The application of the software complex for simulating different technologies in order to obtain the given serial forging served as a proof. As it was stated above, the forging could not be formed in one step with the technique suggested by the factory.

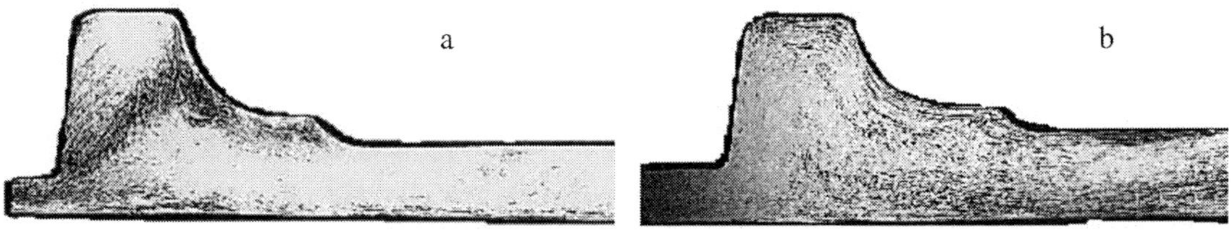

Fig. 5 The macrostructure of forging

The calculation analysis provided recommendations for the die design and mass production technology. As the use of the lug of a big radius does not eliminate nonuniformity of certain stiffening fins, which may result in flaws forming, it is suggested that die forming should be performed in two steps in one final die, with cutting a hole in the centre of forging after the first step.

Industrial die stamping with account of the recommendations has proved their appropriateness: the force of the hydraulic press of 100 MN (Fig. 2) power was enough to produce a quality product; whereby the die forging macrostructure was significantly improved (Fig. 5, b).

Summary

The software complex may perform different functions: express analysis of the metal flow picture, the calculation of primary blank deformation at each stage, which makes it possible, while examining the values of die impression geometric parameters, to obtain various metal flow patterns and stiffening fin profiles, and to select those providing the most uniform metal flow into the die cavity under the stiffening fins, which ensures flawless product manufacturing.

References

[1] Ilyushin A A 1948 *Plasticity* (Moscow: Gostekhizdat) p 376.
[2] Solomonov K 2011 *Simulation of forming forgings* (Saarbrűcken: LAMBERT Academic Publishing) p 165.
[3] Lisunets N L, Solomonov K N and Tsepin M A 2009 *Bulk forming of aluminum workpieces* (Moscow: Mechanical Engineering) p 172.
[4] Solomonov K N, Kostarev I V and Abashkin V P 2008 *Simulation of processes bulk forming and forging of flat workpieces* (Moscow: Pub. house MISA) p 128.
[5] Solomonov K N, Fedorinin N I and Tishchuk L I 2016 The method of constructing the dividing line of metal flow in the processes of upsetting slabs *Bulletin of scientific and technological development* **No** 2 pp 36-55.
[6] Solomonov K 2014 Development of software for simulation of forming forgings *Procedia Engineering* **No** 81 pp 437-443.

Soliton sustainable socio-economic distribution

M A Dresvyannikov[1], M V Petrova[2] and A M Tshovrebov[3]

[1]Savings Bank, Moscow, Russia, [2]Moscow State University. University, Moscow, Russia, [3]P.N.Lebedev Physical Institute of the Russian Academy of Sciences, Moscow, Russia

Email : dresvyannikovm@gmail.com

Abstract. In the work presented, from close positions, we consider: 1) the question of the stability of socio-economic distributions; 2) the question of the possible mechanism for the formation of fractional power-law dependences in the Cobb /Douglas production function; 3) the introduction of a fractional order derivative for a general analysis of a fractional power function; 4) bringing in a state of mutual matching of the interest rate and the production function of Cobb/Douglas.

1. Introduction

According to the common view of the middle class, it is the social category in which people can make savings.

Figure 1. Man at the Crossroads (1934) was a fresco by Diego Rivera in New York City's Rockefeller Center

At the empirical level, it is well known that sustainable social state of society can be characterized with a "single hump" distribution of the population' welfare. And the hump corresponds precisely to the very middle class. In the language of the simplest functions hump can be described by the formula depending on the frequency of the oscillation amplitude in the conditions close to resonance (Lorentz peak). However, it requires huge imagination to apply resonance model to the social phenomena ... Another, seemingly more natural concept of an isolated social hump can be described with a Gaussian distribution that is follows from the Central Limit Theorem. However, CLT doesn't take into account transitions from one social group to another. Such "jumps" provide a dynamic balance and long-term stationarity of single-humped distribution.

2. Stability mechanism socioeconomic distributions

A well-known dynamic generalization (that allows a transition from one level of living to another) of a Gaussian distribution is the Wiener process. Consider discrete equidistant segments of the welfare, there it is assumed that all people with the level of life, whom are in the ± half step unit discrete scale "sit" in this segment. Also consider discrete time with the step τ (and variable of the scale of living standards - ℓ).

Suppose that at some point time t in the i-th node (corresponding to a standard of living $(i-½)\ \ell < L < (i-½)\ell$) there are $\aleph_i$ people. After time interval τ: firstly, when the probability of transition ω from $(i - 1)$-th node to the i-th number $\aleph_i$ increase by $+ \aleph_{i-1} \times \omega$, and in view of the probability of transition ω from the $(i +1)$-th node in the i-th $\aleph_i$ also will increase and $+ \aleph_{i+1} \times \omega$; second, when the transition probability ω with i-th node to the $(i - 1)$-th number $\aleph_i$ should decrease by $- \aleph_i \times \omega$ respectively, and i-th node on the $(i+1)$-th also decrease by $- \aleph_i \times \omega$. As a result, after the time step τ, the number of people with the standard of living in the vicinity of the i-th node will be $\aleph_i + \delta\tau\aleph_i = \aleph_i + \aleph_{i-1}\omega + \aleph_{i+1}\omega - \aleph_i\omega - \aleph_i\omega = \aleph_i + (\aleph_{i-1} - 2\aleph_i + \aleph_{i+1})\omega$. The growth $\delta_\tau\aleph_i = (\aleph_{i-1} - 2\aleph_i + \aleph_{i+1})\omega$ is proportional to the numerator of the finite-difference representation of the second derivative of the standard of living $\frac{\aleph_{i+1} - 2\aleph_i + \aleph_{i-1}}{\ell^2} \underset{\ell \to 0}{=} \frac{\partial^2 \aleph}{\partial \ell^2}$. Similarly, the instantaneous rate of change of the number of people with any specific standard of living as a result of unconscious "Brownian motion" to "better life" in the finite-difference form is $\frac{\delta\aleph_i}{\tau} \underset{\tau \to 0}{=} \frac{\partial \aleph}{\partial \tau}$. Finally, both finite-difference formula can be reduced to the overall equation $\omega \ell^2 \times \frac{\aleph_{i+1} - 2\aleph_i + \aleph_{i-1}}{\ell^2} = \tau \times \frac{\delta\aleph_i}{\tau} \underset{\ell, \tau \to 0}{\Rightarrow} \frac{\partial^2 \aleph}{\partial L^2} = \frac{\tau}{\omega \ell^2} \frac{\partial \aleph}{\partial t}$, which under setting grid steps close to zero turns to the old/good Fokker/Planck. Fokker/Planck equation is a statistical form of an even older (Fourier work of the Great French Revolution) the heat equation (and then also the diffusion, viscosity and so on up to the Schrodinger equation for the amplitude of the non-Kolmogorov probability of non-relativistic particle localization). The solution in non-normalized form is presented as a model solution $\aleph(L,t) \sim \frac{1}{\sqrt{t}} e^{-\frac{\tau L^2}{\omega \ell^2 t}} = \frac{1}{\sqrt{t}} e^{-\frac{L^2}{\sigma^2(t)}}$, where the standard deviation $\sigma = \sigma(t)$ (in this case the deviation from average level of the welfare) increases with time as the $\sigma(t) = \sqrt{\frac{\omega \ell^2 t}{\tau}}$ (here and below the standard of living of the middle class is regarded as the zero point of reference throughout the life of the scale levels, thus "poor" corresponds to the $L<0$, "average» - $L \approx 0$, « the rich» - $L>0$). The increase $\sigma(t) = \sqrt{\frac{\omega \ell^2 t}{\tau}}$ means that over time, the Wiener process a random walk through the various levels of life will lead to the expansion of distribution's hump full scale and asymptotically form a "flat" social distribution from the poorest to the richest. In practice, this is not observed, and hence the simplest Wiener process is not useful as a mechanism of sustainable humped distribution of living standards.

Complicate this scheme by entering the term $-\frac{\omega}{\tau}\aleph$ that is responsible for the rate of drawdown on the L-th standard of living $\frac{\partial\aleph(L)}{\partial t}=\frac{\omega\ell^2}{\tau}\frac{\partial^2\aleph}{\partial L^2}\to\frac{\partial\aleph(L)}{\partial t}=\frac{\omega\ell^2}{\tau}\frac{\partial^2\aleph}{\partial L^2}-\frac{\omega}{\tau}\aleph$. In fact, this term corresponds to the individual consumption of the excess of income over expenditure, which leads to a reduction in the number of L-th level. However, considering the scheme in such form, people consume all savings and roll down to the low standard of life. Thus, along with the consumption it should be added the factor responsible for welfare growth on L-th level, respectively, in the equation with the opposite sign (ie, with "+"). The principle of the expansion of smooth functions in power series suggests that if the previous term $\left|-\frac{\omega}{\tau}\aleph\right|\sim\aleph$, the following must be of type $\aleph^2$. However, as is well known, the algebraic sum of a linear and quadratic contribution can always be reduced to a full square by a change in the given case of the origin $\aleph$. Thus result will not differ from the case without consumption. The element responsible for the growth of wealth in the L-th level, should be sought in the form of an entire degree $\aleph$ higher than 2, such as $2\aleph^3$,which corresponds to the non-linear partial differential equation of parabolic type $\frac{\partial\aleph}{\partial t}=\frac{\omega\ell^2}{\tau}\frac{\partial^2\aleph}{\partial L^2}+\frac{\omega}{\tau}\left(2\aleph^3-\aleph\right)$. The equation is close to the famous soliton equations such as "nonlinear Schrödinger".

Since we are interested in the case of a stationary dynamic equilibrium $\frac{\partial\aleph}{\partial t}=0\Rightarrow\ell^2\frac{\partial^2\aleph}{\partial L^2}+2\aleph^3-\aleph=0$ is not normalized and the solution is a steady soliton $\aleph(L)\sim\frac{1}{ch\left((L-L_0)/\ell\right)}$ providing humped distribution of the standard of living L, measured from the level of the middle class L_0. Meaning ℓ is already slightly different than it was in the consideration of discrete components on the scale L. Now ℓ is not the step between the nodes, which close to zero in transition from the finite-difference equations to the differential now ℓ is rather made dimensionless scale of the range of possible levels of life values.

General view of the equation $\frac{\partial\aleph}{\partial t}=\frac{\omega\ell^2}{\tau}\frac{\partial^2\aleph}{\partial L^2}+\frac{\omega}{\tau}\left(2\aleph^3-\aleph\right)$ allows, apparently, also make inference on the structure and formation of income/consumption, ensuring the stability of the distribution as a whole. According to the expression in the bracket income arises in the interaction of the three representatives of the social group ($+2\aleph^3$), while the spending made individually ($-\aleph^1$).

3. Mikrosemantichesky approach to the Cobb/Douglas model

Use the same approach for the semantic study of the form of the production function of the Cobb-Douglas $Y=aC^{\alpha}L^{\beta}$ obtained by the authors in 1926 [1] empirically. In this formula, Y - output, C - capital L - labor invested in the production of goods, $0<\alpha<1$, $0<\beta<1$ (or even a stronger restriction $0<\alpha+\beta<1$). Thus, due to the smaller exponents unit with increasing C and L partial derivatives of $Y = Y(C, L)$ for C and L are reduced. In particular the increase in production should decrease productivity $\frac{\partial Y}{\partial L}\xrightarrow[Y,L\to\infty]{}0$ ("fatigue effect"), which was useful to explain without resorting to empiricism.

Assume, that for production of Y is required to involve the amount of manpower L broken (by the professionalism, educational and social level, etc.) on m groups of performers. The probability of full utilization of labour resources (respectively perform manufacturing tasks) is the product of the probabilities of development work, specified for each selected group $w_i=\omega_i L_i$. As a result,

$W(L) \sim \prod_{i=1}^{m} w_i = \prod_{i=1}^{m} \omega_i L_i \sim \prod_{i=1}^{m} \omega_i \prod_{i=1}^{m} L_i \sim \prod_{i=1}^{m} L_i \approx \langle L_i \rangle^m \sim L^m$ that $W(L) \sim L^m$ where $<L_i>$ - the labour force, specified for each of the selected group and the average for all groups of performers.
On the other hand for the production of Y good it is required to master the volume of the labour force L in the sequential chain of operations of the n links. It is obvious that the probability of performing manufacturing jobs is the product of the probability of its implementation in each area of the chain $p_i=\rho_i t_i$, which in turn is proportional to the time spent on this operation. As a result $P(L) \sim \prod_{i=1}^{n} p_i = \prod_{i=1}^{n} \rho_i t_i = \sim \prod_{i=1}^{n} \rho_i \prod_{i=1}^{n} t_i \sim \prod_{i=1}^{n} t_i \approx \langle t \rangle^n$, where $<t>$ - time average for all transactions.
At the same time $W(L)=P(L)$ therefore, $L^m \sim <t>^n$ and $<t> \sim L^{m/n}$. Thus, the average time spent time on the individual transaction in proportion to the amount of reclaimed labour force L in degree of m/n which in the case of low qualification (the number of professional groups m is less than the number of links of the production chain) would be less one $\left(\left(\frac{m}{n} < 1 \right) \& (\beta < 1) \longrightarrow \frac{m}{n} = \beta .... \right)$. The total time T spent on the production of the product obviously will be proportional to the average time spent on the individual transaction $T \sim <t>$.
According to Marx, the cost of production Y is determined by the T, averaged over all commodity producers. Then $Y \sim T \sim <t> \sim L^{m/n}$ ie, $Y \sim L^{m/n}$, that in fact in the case of $m < n$ is an integral part of the Cobb-Douglas formula $Y=aC^{\alpha}L^{\beta}...$

4. Analysis of fractional powers dependencies such as the Cobb-Douglas and the concept of non-integer order derivatives

The presence in the formula of fractional powers of the Cobb-Douglas demonstrates a simple functional dependence. However, there may be more complicated in principle functional relationships $Y \sim \lambda L^{\beta} \rightarrow Y \sim \left(\lambda_1 L^{\beta_1} + \lambda_2 L^{\beta_2} + \lambda_3 L^{\beta_3} + ... \right)$. By analogy with the "integer" power series $f(x) = \sum_{n=0}^{\infty} a_n (x - x_0)^n$, where the coefficients of the respective powers of the integer argument can be defined in terms of derivatives of integer order $a_n = \frac{1}{n!} \left(\frac{d}{dx} \right)^n f(x) \Big|_{x \rightarrow x_0}$, should introduce the concept of non-integer order derivatives that allow to find the coefficients of the corresponding non-integer powers of argument. To develop the analogy to non-integer degree first consider an alternative procedure for calculating the derivatives of the whole order $a_m = \lim_{x \to x_0} \frac{f(x) - \sum_{n=0}^{m-1} a_n (x - x_0)^n}{(x - x_0)^m}$. It is seen that, for example, for the derivative the first order, this procedure is reduced to the normal "asymptotically difference" scheme $f'(x_0) = a_1 = \lim_{x \to x_0} \frac{f(x) - a_0}{(x - x_0)}$.
Further, similar to the passage to the limit of the Fourier series for a periodic function to the integral of the Fourier integral for non-periodic introduce replacement power series $\phi(x) = \int_0^{\infty} \alpha(\upsilon) (x - x_0)^{\upsilon} d\upsilon$. Continuing this analogy will lead to the definition of ν-th coefficient α

IOP Conf. Series: Journal of Physics: Conf. Series **918** (2017) 012039 doi:10.1088/1742-6596/918/1/012039

in an integrated power series expansion $\alpha(\upsilon)=\lim\limits_{\substack{x\to x_0\\ \delta\to 0}}\dfrac{\phi(x)-\int_0^{\upsilon-\delta}\alpha(\varsigma)(x-x_0)^{\varsigma}d\varsigma}{(x-x_0)^{\upsilon}\delta}$, which is proportional to the derivative of the v-th order of the function $\phi(x)$ ie $\left\{\dfrac{d}{dx}\right\}^{\upsilon}\phi(x)=\alpha(\upsilon)\Gamma(\upsilon)$.

5. The modified law of compound interest and "right relationship" depreciation

Next, we consider the simplest model of expanded reproduction (approximately such that described in discrete form Marx in his Capital) $\frac{d}{dt}Y=\gamma Y$ with the aim of comparing the law of monetary growth (here - γ macroeconomic parameter corresponding to the percentage of contributions from the produced by the moment of time t value Y for the modernization, which stimulating labour productivity Y'_t). The obvious solution $Y=Y_0e^{\gamma t}$ to the recorded above ordinary differential equation is comparable to the usual law of compound interest $F_V(n)=P_V(1+r)^n$, where - n time measured by the total number of treatment cycles, and the logarithmic growth factor γ production corresponds to the logarithm $ln(1+r)$, expressed through the interest rate r. It is necessary to calculate a geometric progression, corresponding to the law of compound interest at the time forced the Europeans to move from the Roman nonpositional numbering system to the decimal positional, "brought" from India, a young mathematician Fibonacci (to XI ÷ XII century the Catholic Church strongly condemns usury, and in fact this only positional system used in Europe for scientific astronomical calculations were cumbersome ancient sumerian six decimal of calculation, which marks "and to this day" felt in the measures of angles and times). Comparison of $Y=Y_0e^{\gamma t}$ with $F_V(n)=P_V(1+r)^n$ shows, that the law of compound interest is not a manifestation of voluntarism, but is a reflection of the objective (in accordance with the law of value Marx) dynamic harmonization macroeconomic growth in the total value of manufactured goods and the money supply (the geometric progression in the law of compound interest is the expression of a discrete exponential functional $F_V(n)=P_V(1+r)^n=P_Ve^{t\ln(1+r)}$, $\gamma=ln(1+r)=const.$ However, the exponential growth in the value of goods in the course of expanded reproduction of production does not take into account "the effect of fatigue" in Cobb/Douglas model. For taking into account for this effect we should formally agree on money supply growth is not an exponentially growing number of products and dependence on wasted man/hours (ie, time t) replace the discrete L on continuous t formula: $Y=\alpha C^{\phi}L^{\beta}=\lambda L^{\beta}\to Y=\lambda t^{\beta}$. Further, in the law of compound interest if should be made $r=const\to r=r(t)$ (at which this law would have been modified formula Cobb/ Douglas $r=r(t)=(t^{\beta/t}-1)\to F_V(t)=P_V(1+r)^t=P_V(1+(t^{\beta/t}-1))^t=P_Vt^{\beta}\leftrightarrow Y=\lambda t^{\beta}$. This fixed formally negotiate a dynamic growth of money supply and the total value of manufactured goods, but not according to the laws of expanded reproduction, and now, in accordance with the model Cobb/Douglas. Thus, taking into account the effect of fatigue balanced and, apparently, more just becomes a modified law of compound interest with a constantly decreasing rate $r=r(t)=(t^{\beta/t}-1)$.

Apparently, not useless to compare the differential equation "normal" expanded reproduction model Cobb / Douglas, using the concept of elasticity (ie actually logarithmic derivative):

$$\begin{cases}\dfrac{d}{dt}Y=\gamma Y\\ \dfrac{d}{dL}Y=\dfrac{d}{dL}(aC^{\alpha}L^{\beta})=\dfrac{\beta Y}{L}\end{cases}\to\begin{cases}\dfrac{dY}{Y}=\gamma dt\\ \dfrac{dY}{Y}=\dfrac{\beta dL}{L}\end{cases}\to\ \gamma\Leftrightarrow\frac{\beta}{L}$$

IOP Conf. Series: Journal of Physics: Conf. Series **918** (2017) 012039 doi:10.1088/1742-6596/918/1/012039

We see that in the model of Cobb/Douglas effective rate of deductions for the modernization of production is instead of a constant (expanded reproduction $\gamma = const$) decreases in inverse proportion to the time worked $\gamma=\beta/t$. Perhaps this decline corresponds to the amortization (over time, the same money invested in productivity growth "is getting worse and worse working"). At least if it is, then the amortization (depreciation of equipment, as aging) should be calculated not according to the law of simple interest with a negative rate (arithmetic progression or linearly decreasing function of time) as an inverse function of time.

References

[1] Cobb C and Douglas P 1928 A theory of production American Economic Review **18**, 139-250

Solution of the *k*-space diffusion equation for the problem of aLangmuir plasmons motion in an inhomogeneous turbulent plasma

V V Gorev[1,2]**and D E Strekalov**[1]
[1]Bauman Moscow State Technical University, National Research University, Moscow, 105005, Russia
[2] National Research Centre "Kurchatov Institute", Moscow, 123182, Russia

Abstract. This article considers the interaction ofLangmuir plasmons with areas of nonsteady density inhomogeneity in the longwave approximationinion-acoustic weak turbulenceplasma. The interaction process is described by the kinetic equation for the distribution function of the plasmonsat the wave numbers obtained on the basis of Liouville's theorem. According to the basic equation,the evolution of the distribution function representsthe sequential processes of spherical symmetrization, diffusionover ion-acoustic turbulence and "runaway" along the decreasing density profile under the action of regularinhomogeneity. The article focuses on finding the exactself-similar solutions of the initial-boundary value problems to thediffusion and "runaway" equations. It is found that in the interaction of Langmuirplasmons withion-acoustic turbulennce,there occurs the pumping of the ion-acoustic oscillation energy as well asrelatively short Langmuir wave energy into the longwave part of the Langmuir wave spectrum.

1. Formulation of the problem and derivation of the *k*-space diffusion equation

We will examine thedistribution functionof L-plasmonsat thewave numbers represented as a sum of weakly changing and rapidly oscillating component:

$$N_{\vec{k}} = N_0 + \delta N_{\vec{k}}, \quad \frac{\partial N_0}{\partial t} = 0, \quad \frac{\partial N_0}{\partial \vec{r}} = 0, \langle \delta N_{\vec{k}} \rangle = 0. \tag{1.1}$$

We apply the Liouville's theorem, $\frac{dN_{\vec{k}}}{dt} = 0$:

$$\frac{\partial N_{\vec{k}}}{\partial t} + \frac{\partial \omega}{\partial \vec{k}} \frac{\partial N_{\vec{k}}}{\partial \vec{r}} - \nabla\omega \frac{\partial N_{\vec{k}}}{\partial \vec{k}} = 0. \tag{1.2}$$

Plasma will be further considered non-isothermal ($T_i \ll T_e$), and we will considerdensity fluctuationspresented in the form of a superposition of ion-acoustic harmonic oscillations:

$$\delta n \sim \sum n_{|\vec{q}|} exp(-i\Omega t + i\vec{q}\vec{r}), \ \Omega = \frac{qC_s}{\sqrt{1 + q^2 r_D^2}} \cong qC_s, \ qr_D \ll 1. \tag{1.3}$$

At the same time, we will work in geometrical optics approximation, $\lambda / L_0 \ll 1$, where L_0 –a characteristic length of inhomogeneity.

Rapidly oscillating fraction can also be represented as a superposition:

$$\delta N_{\vec{k}} \sim \sum N_{|\vec{q}|} exp(-i\Omega t + i\vec{q}\vec{r}). \tag{1.4}$$

It is assumed, that the dispersion relation for *L*-plasmons has the form:

$$\omega = \omega_p \sqrt{1 + 3k^2 r_D^2 + \frac{\delta n}{n}} \approx \omega_p \left(1 + \frac{3}{2} k^2 r_D^2 + \frac{1}{2} \frac{\delta n}{n}\right), \qquad kr_D \ll 1. \tag{1.5}$$

IOP Conf. Series: Journal of Physics: Conf. Series **918** (2017) 012040 doi:10.1088/1742-6596/918/1/012040

Thus, it is necessary to consider the interaction of L-plasmons with ion-acoustic oscillations (S-waves). The process, leading to the nonlinear absorption of S-waves, i.e. $L + S \leftrightarrow L'$, is the only possible.

Substituting (1.1) into the Liouville's equation (1.2), we obtain:

$$\delta N_{\vec{k}} = -\frac{\vec{q}\frac{\partial N_0}{\partial \vec{k}}}{\Omega - \vec{q}\frac{\partial \omega}{\partial \vec{k}}} \frac{\delta n}{2n} \omega_p. \tag{1.6}$$

Using this value, it is possible to average the Liouville'sequation (1.2) over the statistical ensemble and proceed to the diffusion equation for weaklychanging component of distribution function:

$$\frac{\partial N_0}{\partial t} + \frac{\partial \omega}{\partial \vec{k}}\frac{\partial N_0}{\partial \vec{r}} - \nabla\omega \frac{\partial N_0}{\partial \vec{k}} = \frac{\partial}{\partial k_\alpha} D_{\alpha\beta} \frac{\partial N_0}{\partial k_\beta}, \tag{1.7}$$

where:

$$D_{\alpha\beta} = \pi \frac{\omega_p^2}{4} \sum_{\alpha,\beta} q_\alpha q_\beta \left(\frac{\delta n}{n}\right)^2 \delta\left(\Omega - \vec{q}\frac{\partial \omega}{\partial \vec{k}}\right). \tag{1.8}$$

The right part of equation (1.7) should be understood as divergence of vector quantity$\vec{A} = \{D_{\alpha\beta}\}\frac{\partial N_0}{\partial k_\beta}$. A turbulence level is assumed to be set and included in $D_{\alpha\beta}$ as a parameter.

Physically, the process described by diffusion equation (1.7) can be divided into two stages: the first component is the short in time ion-acoustic turbulence diffusion process, and the second component is a relatively longer "runaway" process under the action of regular inhomogeneity.The diffusion process is also divided into two stages. The first one involves angle diffusion, leading to isotropization of the k-space distribution function.

In a spherical coordinate systemthe equation (1.7) will have the form:

$$\begin{aligned} &\frac{\partial N_0}{\partial t} - \nabla\omega \left(Cos\theta \frac{\partial N_0}{\partial k} - \frac{Sin\theta}{k}\frac{\partial N_0}{\partial \theta}\right) = \\ &= \frac{1}{k^3 Sin\theta}\frac{\partial}{\partial \theta} Sin\theta \left(A\frac{\partial N_0}{\partial \theta} - B\frac{\partial N_0}{\partial k}\right) + \frac{1}{k^2}\frac{\partial}{\partial k}\frac{1}{k}\left(D\frac{\partial N_0}{\partial k} - B\frac{\partial N_0}{\partial \theta}\right), \end{aligned} \tag{1.9}$$

where:

$$\{A, B, D\} = \pi \frac{\omega_p^2}{4} \int d\vec{k} \left(\frac{\delta n}{n}\right)^2 \delta\left(\Omega - \vec{q}\frac{\partial \omega}{\partial \vec{k}}\right)\{q_\theta^2, -\Omega q_\theta, \Omega^2\}. \tag{1.10}$$

After integration of the equation (1.9) over the solid angle, we obtain an equation for the symmetrized distribution function$N(k,t)$:

$$\frac{\partial N}{\partial t} = \frac{1}{k^2}\frac{\partial}{\partial k}\left(\frac{1}{2k}\int_0^\pi \left(D - \frac{B^2}{A}\right) Sin\theta d\theta + \frac{\omega_p^2}{32}(\nabla\omega)^2 k^5 \int_0^\pi \frac{Sin^3\theta}{A(\theta)} d\theta\right)\frac{\partial N}{\partial k}. \tag{1.11}$$

We introduce $a \sim (\nabla\omega)^2$, $b \sim w_s$, where w_s –the energy density of ion-acoustic oscillations, and further, introducing dimensionless variables $u = \frac{k}{k_D} = \frac{k r_D}{2\pi}$ and $\tau = t\omega_p$, we obtain an equation:

$$\frac{\partial N}{\partial \tau} = \frac{b}{u^2}\frac{\partial}{\partial u}\left(\frac{1}{u}\frac{\partial N}{\partial u}\right) + \frac{a}{u^2}\frac{\partial}{\partial u}\left(u^5 \frac{\partial N}{\partial u}\right). \tag{1.12}$$

Thus, the basic equation describing the interaction of L-plasmons with ion-acoustic oscillations is obtained, where the first term on the right-hand side of (1.12) corresponds to the diffusion process, and the second – to the "runaway" process. Solution $N(u,\tau)$of this equation is the angular symmetrized k-space distribution function.The quantity u is non-negative, and we seek a solution on the right half-axis. In this way:

$$u \in [0, +\infty), \qquad \tau \in [0, +\infty). \tag{1.13}$$

For equation (1.12) we will set an initial-boundary value problem:

$$N(u,\tau)=0,\quad \frac{\partial N(u,\tau)}{\partial u}=0,\quad u\to+\infty,\quad u\to 0, \tag{1.14}$$

andwill seek a self-similar solution with an initial condition in the form of delta function:

$$N(u,\tau\to 0)=\delta(u). \tag{1.15}$$

We must also consider the normalization condition:

$$4\pi\int_0^{+\infty}N(u,\tau)u^2du=1. \tag{1.16}$$

We will use an approximation:

$$\frac{b}{u^2}\frac{\partial}{\partial u}\left(\frac{1}{u}\frac{\partial N}{\partial u}\right)\sim\frac{bN}{u^5},\qquad \frac{a}{u^2}\frac{\partial}{\partial u}\left(u^5\frac{\partial N}{\partial u}\right)\sim auN. \tag{1.17}$$

Thus, the original complex process (1.12) can be divided into two simpler ones:

$$\frac{\partial N}{\partial\tau}=\frac{b}{u^2}\frac{\partial}{\partial u}\left(\frac{1}{u}\frac{\partial N}{\partial u}\right),\qquad au\ll\frac{b}{u^5}. \tag{1.18}$$

$$\frac{\partial N}{\partial\tau}=\frac{a}{u^2}\frac{\partial}{\partial u}\left(u^5\frac{\partial N}{\partial u}\right),\qquad au\gg\frac{b}{u^5}. \tag{1.19}$$

2. Constructing a self-similar solution of the diffusion equation

We will solve the initial-boundary value problem (1.13), (1.14), (1.15), (1.16) for the equation (1.18) and simplify the problem by $y=u^3$and $b'=9b$. Equation (1.18) will have the form:

$$\frac{\partial N(y,\tau)}{\partial\tau}=b'\frac{\partial}{\partial y}\left(y^{1/3}\frac{\partial N(y,\tau)}{\partial y}\right). \tag{2.1}$$

We will seek a solution of equation (2.1) in the form:

$$N=\frac{\Theta(\xi)}{\sqrt[5]{(b'\tau)^3}},\quad \xi=\frac{y}{\sqrt[5]{(b'\tau)^3}}, \tag{2.2}$$

Substituting (2.2) into equation (2.1), we had a result:

$$-\frac{3}{5}\left(\xi\cdot\Theta(\xi)\right)'_\xi=\left(\xi^{1/3}\cdot\Theta'_\xi\right)'_\xi. \tag{2.3}$$

Integration (2.3) over the ξ gives a result:

$$\Theta'_\xi+p(\xi)\cdot\Theta(\xi)=g(\xi),\qquad p(\xi)=\frac{3}{5}\xi^{2/3},\ g(\xi)=C'_1\xi^{-1/3}. \tag{2.4}$$

Equation (2.4) is the Bernoulli's equation, and its solution has the form:

$$\Theta(\xi)=I_2(\xi)\cdot exp\left(-I_1(\xi)\right) \tag{2.5}$$

Now, introducing an auxiliary variable:

$$z(\xi)=-\frac{9}{25}\xi^{5/3}=-\frac{9u^5}{25b'\tau}, \tag{2.6}$$

we write down a general solution of the ODE (2.5) in the form:

$$\Theta(\xi)=\left(C_1\int z^{-3/5}e^{-z}dz+C_2\right)e^z, \tag{2.7}$$

where integral is the incomplete gamma function, that has a connection with theKummer's confluent hypergeometric function:

$$\int z^{-3/5}e^{-z}dz\equiv\gamma(z,2/5)=\frac{5}{2}z^{2/5}e^{-z}M(1,s+1,z). \tag{2.8}$$

Boundary conditions (1.14) are satisfied simultaneously if $C_2=0$, and thus:

$$N(u,\tau)=\frac{C_1}{\sqrt[5]{(9b\tau)^3}}\gamma\left(-\frac{u^5}{25b\tau},\frac{2}{5}\right)exp\left(-\frac{u^5}{25b\tau}\right), \tag{2.9}$$

where$C_1=-\frac{1}{12\pi}\pm 10^{-4}$ for $b=1$.

IOP Conf. Series: Journal of Physics: Conf. Series **918** (2017) 012040 doi:10.1088/1742-6596/918/1/012040

Now, for$b = 1, \tau_i = 10^{-4} i,\ \ i = 1 \dots 20$, we visualize a time evolution of the distribution function, satisfying the diffusion equation (1.18) (Figure 1).

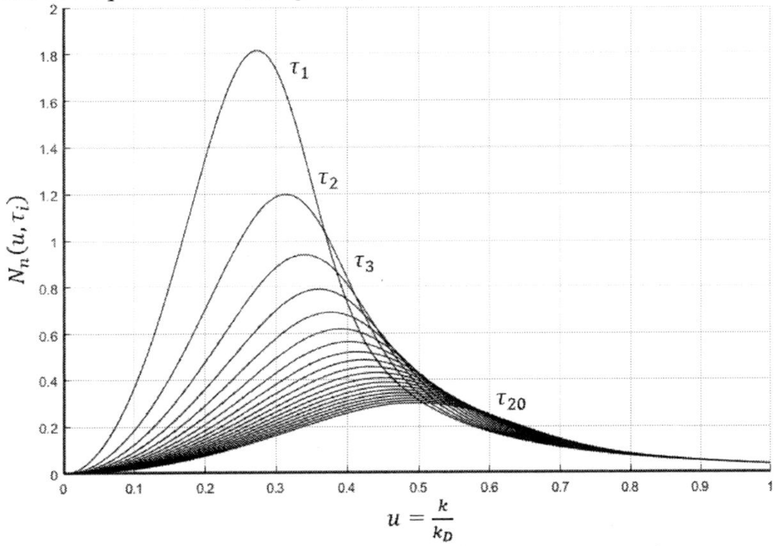

Figure 1. Self-similar solution of the diffusion equation

3. Constructing a self-similar solution of the "runaway" equation

We now proceed to construct a self-similar solution for the initial-boundary value problem (1.13), (1.14), (1.15), (1.16). An initial condition for this problem should be assumed asaninfinitesimal perturbance$\delta^0(u) \underset{\forall u}{\rightarrow} 0$.

We will use self-similar substitution:

$$N(u, \tau) = \frac{X(\xi)}{u^3},\ \ \xi = -\ln u\tau, \tag{3.1}$$

and obtain the equation (1.19) in the form of an ODE:

$$\frac{d^2 X(\xi)}{d\xi^2} + p(\xi)\frac{dX(\xi)}{d\xi} + q(\xi)X(\xi) = 0,\ \ p(\xi) = \left(\frac{e^{\xi}}{a} + 2\right),\ \ q(\xi) = -3. \tag{3.2}$$

Using transformation$X(\xi) = e^{-\frac{1}{2}\int p(\xi)d\xi} g(\xi)$,we obtain the equation (3.2) in the form:

$$g''_{\xi\xi} + \left(-4 - \frac{3}{2}\frac{e^{\xi}}{a} - \frac{1}{4}\left(\frac{e^{\xi}}{a}\right)^2\right)g(\xi) = 0. \tag{3.3}$$

Equation (3.3) can be reduced to the Whittaker's differential equation. To do this, we move to a new variable$z(\xi) = e^{\xi}$. Further, dividing equation (3.3) by z^2, we obtain:

$$g''_{zz} + \frac{1}{z}g'_z + \left(-\frac{1}{4a^2} - \frac{3}{2az} - \frac{4}{z^2}\right)g(z) = 0, \tag{3.4}$$

and then, passing to a new function $w(y) = \sqrt{z}\, g(z)$, $y = z/a$, and multiplying equation (3.4) by $a^2\sqrt{z}$, we obtain the Whittaker's equation:

$$w''_{yy} + \left(-\frac{1}{4} + \frac{\lambda}{y} + \frac{1 - 4\eta^2}{4y^2}\right)w(y) = 0. \tag{3.5}$$

Solution of the equation (3.5), has a connection with the Kummer's confluent hypergeometric function:

$$w(\lambda, \eta, y) = e^{-y/2} y^{\frac{1+2\eta}{2}} M\left(\eta - \lambda + \frac{1}{2}, 1 + 2\eta, y\right). \tag{3.6}$$

Thus, we have found a self-similar solution for the problem (1.13), (1.14), (1.15), (1.16):

$$N(u,\tau) = Ca^{-5/2}\frac{1}{u^4\tau}M\left(4,5,\frac{1}{au\tau}\right)e^{-\frac{1}{au\tau}}. \quad (3.7)$$

When $\frac{1}{au\tau} \ll 1$, we have an approximation$M\left(4,5,\frac{1}{au\tau}\right) \approx 1$, therefore:

$$N(u,\tau) = C\,a^{-5/2}\frac{1}{u^4\tau}e^{-\frac{1}{au\tau}},\quad \frac{1}{au\tau} \ll 1,\quad C = \frac{a^{3/2}}{4\pi}. \quad (3.8)$$

Function (3.8) is shown in Figure 2for $\tau_i = 0.5 + 0.1i$, $i = 1 \dots 10$, and $a = 1$.

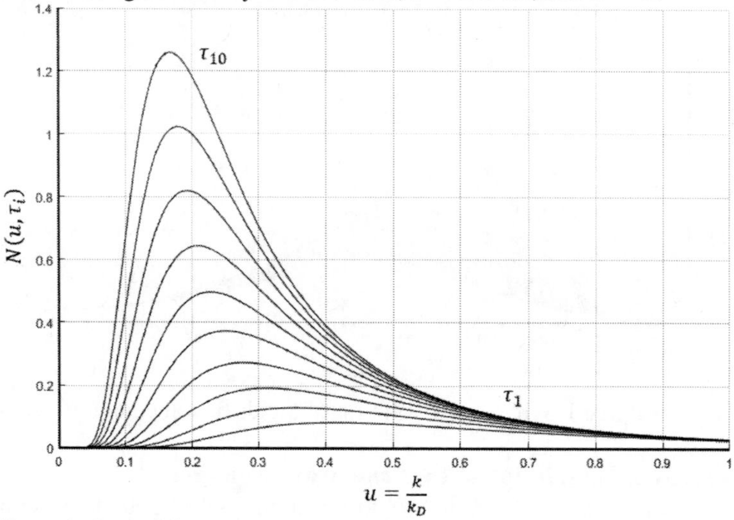

Figure2. Self-similar solution of the “runaway” equation

References

[1] ZeldovichYa and Raizer Yu 1966Self-similar heat wave from the instant flat source *Physics of Shock Waves and High-Temperature HydrodynamicPhenomena* (Moscow: Nauka) pp 506–519

[2] Gorev V and Kingsep A 1974 Interaction of Langmuir solitons with plasma particles *JETP*vol 66 issue 2 pp 2048–55

[3] Kadomtsev B 1988 *Collective Phenomena in Plasma* (Moscow: Nauka) 303 p

[4] Akhiezer A Akhiezer I and Polovin R 1974 *Plasma Electrodynamics*ed A Akhiezer(Moscow: Nauka) pp 489–491

Stimulated low-frequency Raman scattering in plant virus suspensions

E K Donchenko[1], O V Karpova[1], A D Kudryavtseva[2], S M Pershin[3], V I Savichev[2,4], M A Strokov[2], N V Tcherniega[2,4] and K I Zemskov[2]

[1]M.V. Lomonosov Moscow State University, Vorobevy Gory, 1, Moscow, 119991, Russia

[2]P.N. Lebedev Physical Institute of the RAS, Leninskii pr, 53, Moscow, 119991, Russia

[3]A.M. Prokhorov General Physics Institute of the RAS, Vavilova, 38, Moscow, 19991, Russia

[4]Bauman Moscow State Technical University, ul. 2 Baumanskaya, 5, Moscow, 105005, Russia

Abstract. The study deals with laser pulse interaction with plant viruses: we investigated tobacco mosaic virus (TMV) and two types of potato viruses (PVX and PVA) in Tris-HCl pH7.5 buffer and in water. We used 20 ns ruby laser pulses for excitation. We employed Fabry–Pérot interferometers to record spectra of the light passing through the sample and reflected from it. For TMV and PVX in Tris-HCl pH7.5 buffer, same as for PVA in water, we observed additional spectral lines corresponding to the stimulated low-frequency Raman scattering (SLFRS). We believe we were the first to measure SLFRS frequency shifts, conversion efficiency and threshold. High conversion efficiency of the scattered light is evidence of laser pulses efficiently exciting gigahertz vibrations in viruses. SLFRS can be used to identify and affect biological nanoparticles.

1. Introduction

Biological nanoparticles are now the subject of intense experimental and theoretical investigations [1]. It is important to develop new methods of identifying and affecting biological nanoparticles. One of the most effective ways to affect a virus system (same as any other biological system) is to stimulate its vibrational modes by resonant excitation. Vibration eigenfrequencies in viruses lie in the gigahertz range, which means that it possible to excite resonance using microwave electromagnetic radiation, hypersound or pulsed optical electromagnetic radiation. The first two approaches are difficult to implement in real-world biological systems due to strong absorption of microwave radiation by the water contained in all biosystems and a short hypersound propagation distance, which is on the order of several hundred nanometres in the systems under consideration. Using pulsed electromagnetic radiation in the optical range ensures the maximum possible penetration depth into the system under study and, consequently, the highest effect. As shown in [2, 3], short laser pulses can provide a powerful impact on a virus system by means of the impulsive stimulated Raman scattering mechanism [4, 5]. Another approach is to use a stimulated analogue of low frequency Raman scattering (LFRS) [6, 7], that is, stimulated low frequency Raman scattering (SLFRS) [8]. LFRS is the inelastic scattering of light by localized acoustic vibrations of nanoparticles. This type of scattering manifests itself by additional spectral components appearing in the spectrum of scattered light, these components displaying frequency shifts corresponding to nanoparticle vibration eigenfrequencies in the gigahertz or terahertz ranges. Each nanoparticle has a set of natural acoustic frequencies which are determined by its morphology. It is possible to calculate nanoparticle eigenfrequencies with the help of Lamb's theory, which deals with free vibrations of a spherically shaped homogeneous elastic body under stress-free boundary conditions [9, 10].

LFRS was first experimentally observed in nucleated cordierite glass [6], to be recorded later in many nanoparticle systems. Currently LFRS is commonly seen in systems containing metallic, insulator, or semiconductor nanoparticles. Analysing the LFRS spectrum provides a unique possibility to obtain information about morphological properties of a nanoparticle system.

Under certain conditions it is possible to excite SLFRS in systems of nanosized or submicron particles. SLFRS was observed in different materials: highly ordered samples such as opal matrices and nanocomposites based on them, nanostructured thin films, and disordered materials such as suspensions of nanoparticles (metal, semiconductor and dielectric) [11-13]. Frequency shifts of the SLFRS components are in the terahertz or gigahertz range and are determined by the shape, size and elastic properties of nanoparticles. Since SLFRS frequency shifts are small, to excite SLFRS efficiently, it is necessary to use a laser source with narrow spectral linewidth. Another important factor affecting the efficiency of SLFRS excitation is the monodispersity of the systems under study.

Cylindrical or spheroidal viruses vibrating at their eigenfrequencies are a good example of a highly monodisperse (uniform) nanoparticle system. SLFRS aids in studying such systems, having several substantial advantages as compared to spontaneous scattering. One of the difficulties in exciting spontaneous scattering is damping of particle vibrations due to the energy radiating into the environment. Stimulated scattering, excited by short and powerful laser pulses, solves this problem. The possibility to control the frequency shift in the giga- and terahertz range along with high conversion efficiency not only make SLFRS an efficient biharmonic excitation source for nanosized and submicron system spectroscopy, but also allow this phenomenon to induce a powerful resonance in such systems. The purpose of this article was to implement SLFRS excitation in certain types of plant viruses and to determine its spectral and energetic characteristics.

2. Samples

We recorded SLFRS in plant viruses for the first time. In [14] we presented our first experimental results of recording SLFRS in tobacco mosaic virus (TMV) in Tris-HCl pH7.5 buffer ($C_4H_{12}ClNO_3$). In this work we present results of investigating SLFRS in TMV and two types of potato viruses: PVA and PVX. Figure 1 shows a typical transmission electron microscopy image of a TMV sample.

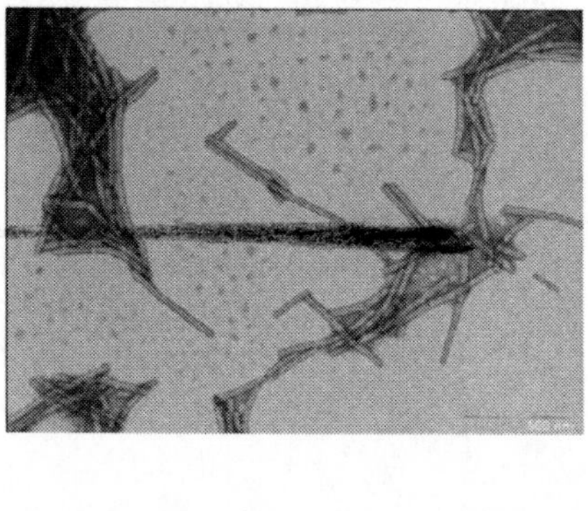

Figure 1. TMV sample in Tris-HCl buffer; picture obtained with a transmission electron microscope.

Tobacco mosaic virus (TMV) is 18 nm in diameter and 300 nm in modal length; it consists of 2130 identical 17.5 kDa protein molecules assembled into a rigid rod-like helical structure. The viral RNA is intercalated between the protein turns [15]. Potato viruses are thinner and longer than TMV and less rigid. PVA virions are filamentous, usually flexuous, 730 nm long and 15 nm in diameter, featuring helical symmetry with a pitch of 3.4 nm. Virions consist of 5% nucleic acid and 95% protein. Their genome RNA is single-stranded and contains 9700 nucleotides. The coat protein (CP) has a molecular weight of 29.8 kDa. PVX virions are also filamentous and flexuous, 500 nm long and 13 nm in diameter. Their symmetry is helical, with a pitch of 3.6 nm. Virions consist of 6% nucleic acid and 94% protein (6400 nucleotides, and CP with a molecular weight of 25 kDa). We isolated the TMV U1 strain from systemically infected Nicotiana tabacum L. cv. Samsun plants [16]. The virion concentration in the test sample was 50 μg/ml; we used Nanoparticle Tracking Analysis (NTA) according to [17, 18] to determine the number of particles in the sample, which turned out to be 0.5×10^{12} particles/cm^3.

3. Experimental

SLFRS was excited by single pulses of ruby laser (λ = 694.3 nm, τ = 20 ns, E_{max} = 0.3 J, $\Delta\nu$ = 0.015 cm^{-1}, divergence $3.5\cdot10^{-4}$ rad). We used a lens with a 5 cm focal length to focus the laser light at the centre of a 1 cm quartz cuvette containing our sample. We recorded SLFRS spectra by means of Fabry–Pérot interferometers with different baselines (and, correspondingly, different dispersion ranges). We used the same quartz cuvette filled only with Tris-HCl pH7.5 buffer or water as a reference sample for comparison. Figure 2 shows the experimental setup for SLFRS investigation in virus suspensions.

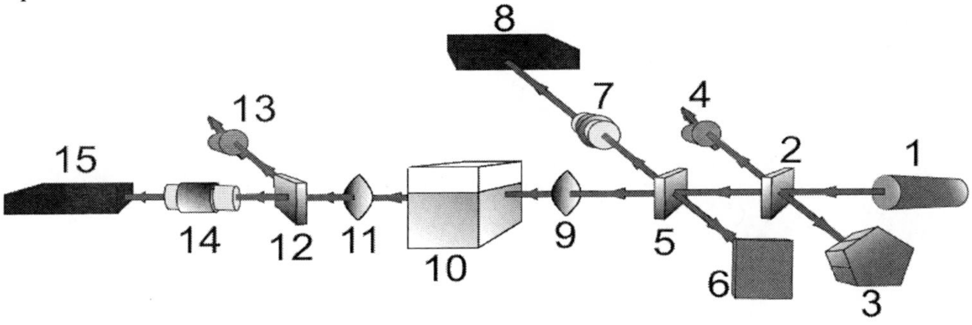

Figure 2. 1 – ruby laser; 2, 5, 12 – glass plates; 3 – system for measuring laser light parameters; 4, 13 – systems for measuring scattered light energy in backward and forward directions; 6 – mirror; 7, 14 – Fabry–Pérot interferometers; 8, 15 – photocameras recording SLFRS spectra, 9, 11 – lenses; 10 – quartz cuvette containing a sample.

We recorded the spectra of the light scattered by the virus suspension simultaneously in forward and backward directions. Figure 3 presents SLFRS interferograms in TMV in Tris-HCl pH7.5 buffer for various excitation energy values. A small excitation light intensity caused only a single ring system to be recorded, the system corresponding to the excitation light frequency (Fig. 3a). When the intensity exceeded a certain threshold, additional lines appeared in the spectra in both forward and backward directions. These lines corresponded to the SLFRS components (Fig.3b).

IOP Conf. Series: Journal of Physics: Conf. Series **918** (2017) 012041 doi:10.1088/1742-6596/918/1/012041

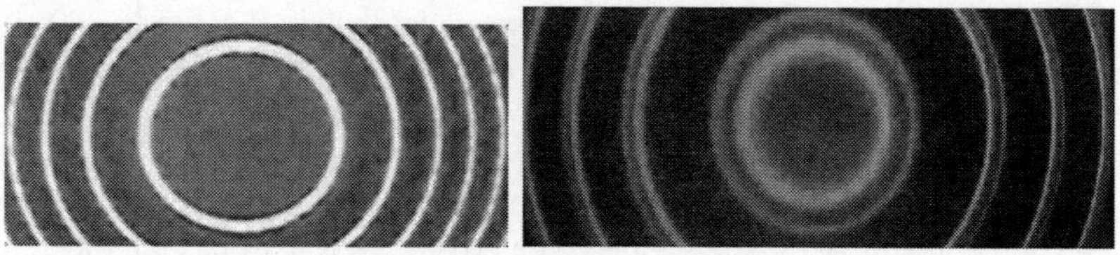

a) b)

Figure 3. Fabri-Perot interferograms corresponding to forward-scattered radiation in TMV suspension in Tris-HCl pH7.5 buffer for laser intensities of a) 0.02 GW/cm^2, b) 0.08 GW/cm^2.

Figure 4 presents the SLFRS spectrum corresponding to the interferogram 3b.

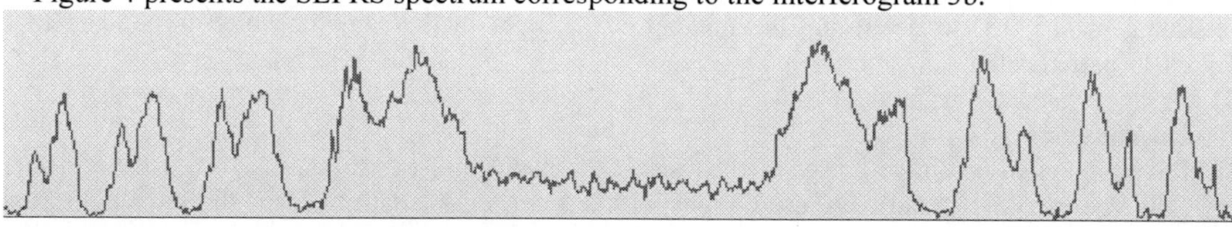

Figure 4. Spectrum of the SLFRS in TMV suspension in Tris-HCl pH7.5 buffer for a laser intensity of 0.08 GW/cm^2.

It is necessary to note that the same experimental conditions did not lead to recording SLFRS in the cell filled only with Tris-HCl pH7.5 buffer.

Figure 5 presents the interferograms and spectra of the SLFRS in potato viruses.

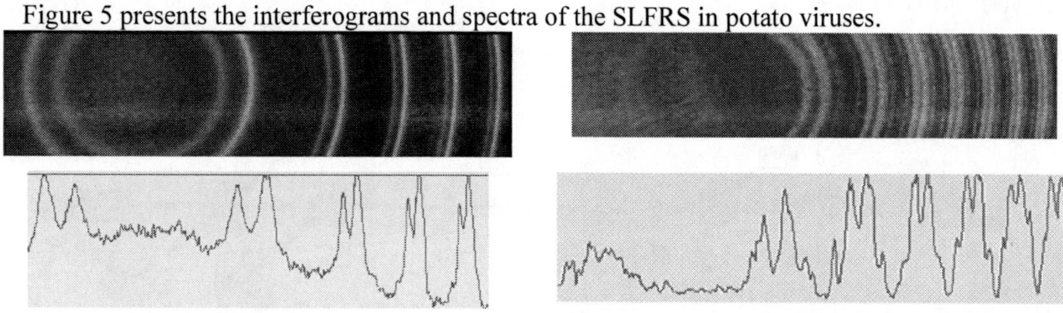

a) b)

Figure 5. Interferograms and spectra of the SLFRS in potato virus suspension: a) PVX in buffer, b) PVA in water.

The SLFRS line width and divergence are close to the corresponding values of the excitation light. Table 1 lists the following SLFRS characteristics: conversion efficiency (η), threshold (*P*), Stokes shifts (Δ*v*), virus size (*D* – diameter, *L* – length), range of dispersion ($\Delta\nu_0$).

Table 1. SLFRS characteristics

Virus type	η%	*P* GW/cm^2	Δν GHz	*D*x*L* nm	$\Delta\nu_0$ cm^{-1}
TMV	5	0.07	60	18x300	2.50
PVA	10	0.03	9, 18	15x730	1.25
PVX	10	0.035	6, 12	13.5x715	1.00

For rather rigid viruses of a simple shape it is possible to estimate vibration eigenfrequencies. For TMV, which is a rigid cylinder, the LFRS radial breathing mode frequency was calculated in [7]. According to these calculations, it equals 2.1 cm^{-1} (63 GHz), which is close to our experimental value of 60 GHz. The main difficulty concerning calculating eigenfrequencies of biological nanoobjects is that their complex composition prevents knowing the exact values of their acoustic characteristics (for instance, speed of sound). We assumed the speed of sound to match that of the main constituent protein. As for potato viruses, they are much more flexuous than tobacco mosaic virus; therefore it is difficult to calculate their vibrations.

4. Conclusions

Analysing the low frequency spectrum of inelastically scattered light in rigid biological structures, same as in any nanoparticle system, can provide very important information about their mechanical properties and can be used for their identification. Any spectral component of the SRWS corresponds to a vibrational mode of the virus. To estimate the theoretical value of a vibrational frequency, it is necessary to use a proper theoretical model (for a spherical virus one can use a liquid drop model or an elastic sphere model). In the simplest case of a spherical virus it is necessary to divide the sonic speed by the diameter of the virus. But in the case of a non-spherical shape and (or) poorly defined elastic constants, calculating the vibrational frequency exactly becomes rather complicated, and it becomes easier to obtain vibrational frequency information from experimental data. It is very important to know this value so as to be able to affect the virus through resonance, which can even lead to virus destruction. Matching the external source frequency to the virus eigenfrequency exactly makes this very important application possible. The external source may be ultrasound or electromagnetic radiation of a suitable frequency. These methods are not very effective because of environmental damping. Another way to implement this effect efficiently is biharmonic excitation, that is, electromagnetic radiation containing two spectral components separated by the frequency corresponding to the eigenfrequency of the virus vibration. SLFRS can be a very effective biharmonic excitation source. The scattered light contains two waves with close wavelengths and intensities, and the frequency shift matches the virus vibration eigenfrequency exactly.

References

[1] Tsen K, Dykeman E, Sankey O et al 2006 *Virology Journal.* **3** 79
[2] Tsen K-T, Tsen S W D, Chang C L, Hung C F, Wu T C and Kiang J G 2007 Inactivation of viruses by coherent excitations with a low power visible femtosecond laser *Virol. J.* **4** 50
[3] Tsen K-T, Tsen S-W D, Chang C-L, Hung C-F, Wu T-C and Kiang J G 2007 Inactivation of viruses with a very low power visible femtosecond laser *J. Phys.: Condens. Matter* **19** 322102
[4] Yan Y X, Gambel E B Jr and Nelson K A 1985 Impulsive stimulated scattering: general importance in femtosecond laser pulse interactions with matter, and spectroscopic applications *J. Chem. Phys.* **83** 5391–9
[5] Tsen K-T, Tsen S-W D, Sankey O F and Kiang J G 2007 Selective inactivation of micro-organisms with near-infrared femtosecond laser *J. Phys.: Condens. Matter* **19** 472201
[6] Duval E, Bukenter A and Champagnon B 1986 *Phys. Rev. Lett.* **56,** 2052
[7] Balandin A and Fonoberov V 2005 *Journal of Biomedical Nanotechnology.* **1** 90
[8] Tcherniega N and Kudryavtseva V 2009 Nonlinear-optical properties of photonic crystals. *Journal of Surface Investigation: X-ray, Synchrotron and Neutron Techniques.* **3** 513
[9] Lamb H 1882 *Proc.London Math.Soc.* **13** 189
[10] Saviot L, Murray D, Mermet E and Duval E 2004 *Phys. Rev.* E **69** 023901
[11] Tcherniega N, Samoylovich M, Kudryavtseva A, Belyanin A, Pashchenko P and Dzbanovski N 2010 *Optics Letters* **35** 300
[12] Kudryavtseva A, Tcherniega, Samoylovich M and Shevchuk A 2012 *International Journal of Thermophysics* **33** 2194
[13] Tcherniega N, Zemskov K, Savranskii V, Kudryavtseva A, Olenin A, and Lisichkin G 2013 *Optics Letters* **38** 824
[14] Karpova O V, Kudryavtseva A D, Lednev V N, Oshurko V B, Pershin S M, Petrova E K, Tcherniega N V and Zemskov K I 2016 Stimulated low-frequency Raman scattering in

tobacco mosaic virus suspension *Laser Phys. Lett.* 13, 085701
[15] Klug A 1999 The tobacco mosaic virus particle: structure and assembly *Philos Trans R Soc Lond B Biol Sci.* **354** 531–535
[16] Karpova O, Nikitin N, Chirkov S, Trifonova E, Sheveleva A, Lazareva E and Atabekov J 2012 Immunogenic compositions assembled from tobacco mosaic virus-generated spherical particle platforms and foregn antigens *Journal of General Virology* **93** 400–407
[17] Nikitin N, Trifonova E, Karpova O and Atabekov J 2013 Examination of Biologically Active Nanocomplexes by Nanoparticle Tracking Analysis *Microscopy and Microanalysis* **19** 808-813.
[18] Petrova E, Nikitin N, Trifonova E, Protopopova A, Karpova O and Atabekov J 2015 The 5'-proximal region of Potato virus X RNA involves the potential cap-dependent "conformational element" for encapsidation *Biochimie* **115** 116-119

Structural investigation of laser luminescent response in a rotating ruby crystal

V O Gladyshev[1], D S Prozorovskiy[1], D I Portnov[1], V L Kauts[1]

[1]Department of Physics, Bauman Moscow State Technical University, Moscow, Russia
E-mail: vgladyshev@mail.ru

Abstract. This paper presents the study of fluorescent trace in a rotating ruby crystal occurring upon irradiation of coherent emission with a wavelength λ = 532 nm. As a sample we used an oriented cylinder-shaped ruby single crystal of 5 mm in length and 10 mm in diameter.

The theoretical framework of moving media optics enables calculations of electromagnetic radiation propagation in moving media away from the absorption medium.
In active media near the resonance absorption frequency one can observe luminescence, and the time of atom relaxation from the excited state to the ground state, may take macroscopic values. Provided radiation passes through a moving medium, during relaxation the medium is able to shift, resulting in a spatial redistribution of luminescent glow intensity.
This shift of the fluorescent trace greatly exceeds the value of the wavefront shift caused by the medium motion as a result of the Fizeau effect.
Indeed, the effects of moving media optics, such as light entrainment, or a rotation of the polarization plane, appear quite weak. The effect size is proportional to the medium velocity, the optical path in the medium, as well as the Fresnel drag coefficient, which depends on the refractive index of the medium $k = 1 - \frac{1}{n^2}$.
Therefore, having increased the refractive index, one can influence the effect size, and that will open the ways for their practical use.
The purpose of this work is to select the optical material with a long relaxation time that can be regarded as an equivalent increase in the effective refractive index.
This paper presents the study of fluorescent trace in a rotating ruby crystal occurring upon irradiation of coherent emission with a wavelength λ = 532 nm. As a sample we used an oriented cylinder-shaped ruby single crystal of 5 mm in length and 10 mm in diameter. The crystallographic axis c coincided with the rotation z. During the experiment the crystal rotation frequency varied between 2 and 130 Hz. The design of the experiments is depicted in Figures 1, 2. Figure 1 shows the observation results of the laser emission passed through a rotating ruby, and Figure 2 gives observation data on the reflected radiation.

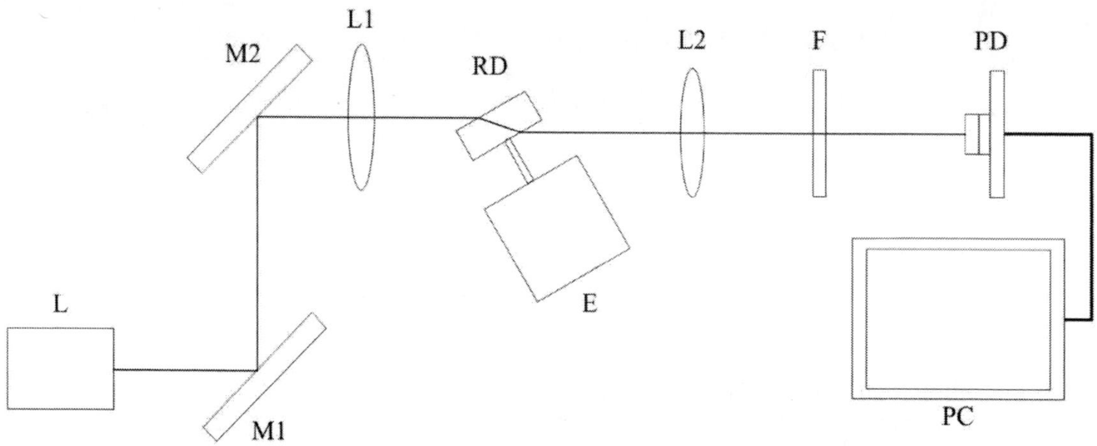

Figure 1. Schematic diagram of the first experiment

To ensure accurate alignment, radiation of the power-stabilized laser L passes through a system of mirrors M1, M2 and falls onto a collecting lens L1. The beam is focused on the side surface of the crystal RD, which is fixed to the engine axis E. The refracted beam exits the ruby, then it is re-focused by the second collecting lens L2, passes through the filter F and falls onto the photodetector PD. From the photodetector the data is transmitted to computer PC for processing the received optical images. When green laser is emitted through the ruby, red radiation propagation is observed, which is due to fluorescent properties of the ruby. To increase the number of produced polaritons, the beam inside the crystal was focused. The red light filter is needed, so that the obtained images record only the emission caused by ruby atoms excitation and not generated by the laser.

Since the fluorescent light propagates in all directions, in an attempt to detect the polaritons entrainment, we made the second experiment (Figure 2): the beam profile detector was placed in a plane perpendicular to the disk rotation. The image of the "ring" within the crystal as it rotated was almost without distortion, the illumination was significantly reduced due to the fact that the photodetector array was not on the path of the laser beam.

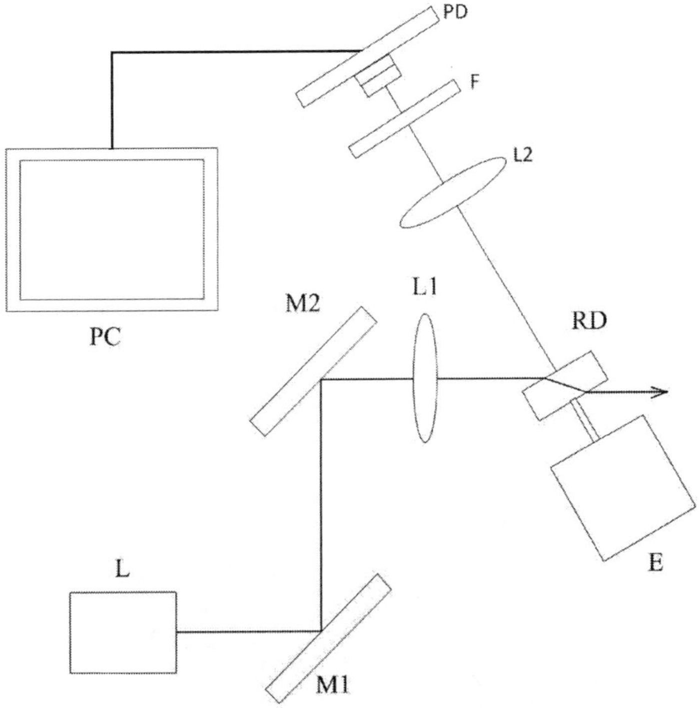

Figure 2. Schematic diagram of the second experiment

Experimental results are presented in Figure 3.

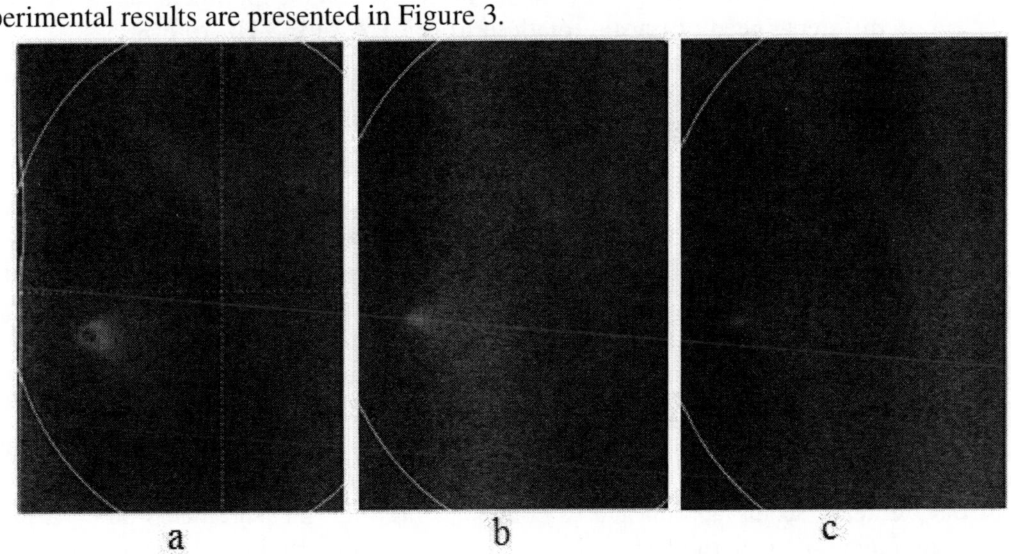

Figure 3. Images of the optical response obtained in the first experiment at different speed of crystal rotation: a) f=80 Hz, b) f=100 Hz, c) f=110 Hz

The observed effect of the arc-wise beam path formation on the front surface of the sample is due to the finite lifetime of the impurity electronic state of chromium ions in the ruby excited at green crystals irradiation by ruby laser radiation.

When the rotation frequency changes, the trace area is increased, but the image itself becomes weaker. The integrated intensity remains constant.

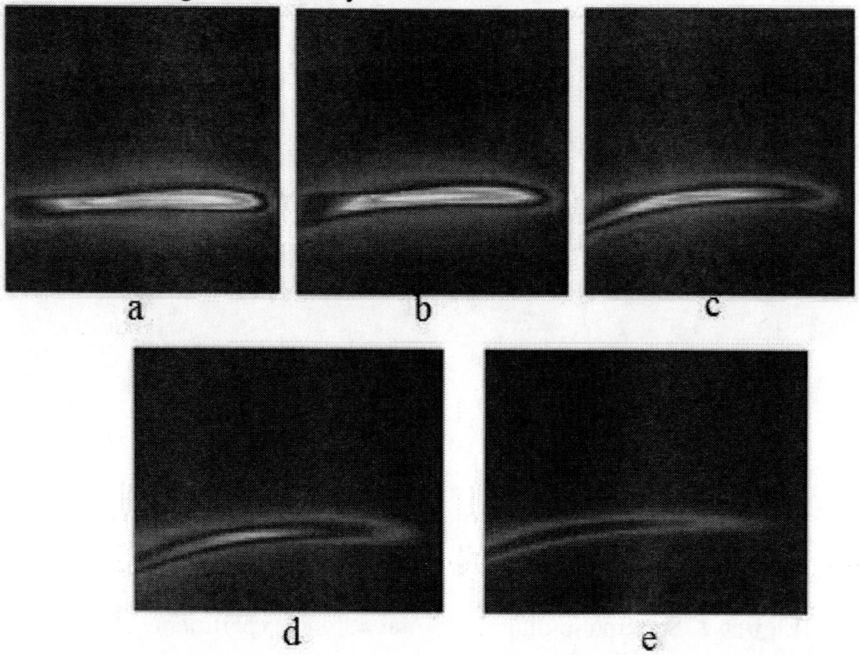

Figure 4. The images obtained in the second experiment at different speed of crystal rotation: a) f=5 Hz, b) f=8 Hz, c) f=80 Hz, d) f=100 Hz, e) f=110 Hz

In the second experiment, depending on the disc speed, the center-of-intensity shift was observed: in the case of the resting ruby the maximum is at the point of beam incidence on the crystal, but at rotation the maximum begins to shift along the beam path.

At low rotation frequencies, the "trace" expansion at the end of the beam was observed.

In the course of experimental studies of photoluminescence beam traces in a single ruby crystal rotating at speed of 2 to 123 Hz, we registered a type of the laser beam trajectory on the input crystal surface, as well as a form of fluorescent light trace inside the crystal. Depending on the disc rotation speed we observed trajectories as a finite length arc. When the rotational speed increased, the arc length on the crystal surface increased as well. At high rotation speeds the arc degenerated into a circle, and the intensity peak shifted.

In experiments at certain angles we observed the "comet trail", similar in shape to the expected polariton trail.

In the future, it seems appropriate to conduct amplitude measurements along a path of luminescent radiation in a ruby crystal. The research will provide the radiation intensity propagation caused by the conversion of polaritons, drifting to the edge of the rotating ruby disc in fluorescent light. The light response integrated intensity will also be measured at different rotation frequencies.

In prospect the research results can be used in systems determining the space trajectory, spatial orientation, to identify angular speeds and accelerations.

The work was supported by Russian Foundation for Basic Research grants №16-02-00488 A and №16-08-00618 A.

References

[1] Gladyshev V O, Gladysheva T M, Dashko M, Trofimov N, Sharandin E A 2007 The first measurement results of dependency of the spatial light entrainment in a rotating medium on the rotation speed. *Technical Physics Letters* vol. **33** no. 21 pp. 17-24.

[2] Gladyshev V O, Portnov D I, Kauts V L, Sharandin E A 2013 Experimental studies of laser radiation polarization in the rotating optical glass. *Optics and Spectroscopy* vol. **115** no. 3 pp. 398-405.

[3] Gladyshev V O, Gladysheva T M, Zubarev V E 2002 Registration of light entrainment effect in the disc optical interferometer. *Technical Physics Letters* no. **3** pp. 88-94.

IOP Conf. Series: Journal of Physics: Conf. Series **918** (2017) 012043 doi:10.1088/1742-6596/918/1/012043

Supplementary education program for school students "The measurements in nature and technology"

E Burkova[1], A Kravtsov[1] and G Pogodin[1]
Physics Department, Bauman Moscow Technical University, 5 Vtoraya (2nd) Baumanskaya ul, Moscow, 107055, Russia

E-mail: burkelen@gmail.com

Abstract. The continuity of training students in methods of experimental data processing is under consideration. The development of training programs for pupils at an early stage of learning is reported.

When studying natural phenomena in a mathematical circle for school students, the fundamental idea about the experimental basis of natural-sciences materializes in a training (demonstration and laboratory) experiment. Training in the methods of measurement error estimation is an essential component of this process.

In the paper by A. Kravtsov and G. Pogodin [1], it is noted that to form a single learning chain for the methods of experimental data processing is very important. Special attention is to be paid to measurement error estimation. It should be based on a "simple-to-complex" basis: starting with simple measurements in a school physics course, then doing the course on practical physics in the profession-oriented general education institutions, and proceeding with professional activities in academic institutions. An impressive arsenal of developed methodologies, techniques, and standards is supposed to serve this purpose. Unfortunately, nowadays it is not possible to achieve this continuity as the very first link, school physics course, is out of this chain. A. Zaydel [2] notes that the error theory is rather difficult for school students. Citing this fact, the authors of school textbooks and guidelines for laboratory work present the problem of measurement error in a deliberately simplified form. Based on their experience in the Specialized Scientific Training Center of Bauman Moscow State Technical University, the authors of this report, like many of their colleagues, believe that a school student is quite capable to learn the basics of a rigorous error theory at the level of quality explanations and ready-made formulae.

The result of a physical measurement makes sense if the following three values are specified:

- Best estimation of the actual value of measurement quantity.
- Limits (confidence interval) within which the actual value with possible probability (confidence probability) defined by an experimenter range.
- Confidence probability itself.

What kind of understandable information about the nature of error can be conveyed to a school student in the process of learning in order to provide him with a fairly perceived, though incomplete, conception about the measurement data processing?

In the modern mathematics course the fundamentals of probability theory are studied in the 8th grade of secondary school. School students learn a number of concepts and how to solve the simplest tasks for a probability of the single test outcome (the probability of taking a ball of the right colour out of a box with a set of coloured balls, etc.). Based upon this, while processing the experimental results during laboratory work in physics classes, we may talk about uniformly, normally, and Poisson distributed random variables.

The authors propose the following sequence of arguments, which allows us to show how the set of uniformly distributed quantities tends to a normally distributed quantity with increasing a number of quantities. A cube with numbers from 1 up to 6 on its faces is considered as a model object. The probability of occurrence of any number with a single throw of the cube is the same and equals to 1/6, i.e. we can talk about a uniform distribution of the probability of occurrence of each number. If you throw two identical cubes at the same time (or the same cube consistently two times), then the interval of the throws will expand from 2 up to 12, and the distribution of the probability of occurrence of these numbers will no longer be uniform since the same total sum can be obtained by several combinations of throws. The sum distribution of the throws will tend to normal when throwing more cubes.

Analyzing the results obtained, it can also be shown that the boundary method used in the school physical experiment for measurement error estimation gives a satisfactory result only for a very small interval of the confidence probability values and has extremely limited application for a rough estimation of errors.

The authors developed the supplementary education program "The Measurements in Nature and Technology". It was tried out in the Summer School conducted by the Center for Pre-University Training of Bauman Moscow State Technical University in 2016 at the International Children's Center "Artek". The module "Measurement Errors", in which the above idea is realized, is one of the main sections of this program. To conduct classes in this module, a set of methodological materials was developed; material and technical support was prepared. The work on this program is also carried out within the framework of the Center for Technological Support of Education and in the physical laboratory of the department "Fundamentals of Physics" of Bauman Moscow State Technical University.

The authors believe that this approach to the study of methods for processing experimental results in the school physics course will make it possible to lay a strong foundation for the experimental competencies of the future engineer.

References

[1] Kravtsov A and Pogodin G 2015 *Trudy Vosmoy Vserossijskoy Conf. Neobratimiye processy v prirode i tekhnike - Proc. 8th All-Russian Conf. Irreversible Processes in Nature and Technology (Moscow)* vol 3 (Moscow: BMSTU) pp 48-150

[2] Zaydel A 2005 *Oshibki fizicheskikh izmerenij - Errors in the Measurement of Physical Quantities* (St. Petersburg: Lan')

The behavior of nanosatellite body materials during electromagnetic launch

Yu V Gerasimov[1] , I A Konstantinova[2], L A Konstantinova[3], R G Rakhimov[2] and A B Selivanov[1]

[1] Associated professor, Bauman Moscow State Technical University, 5, 2-ya Baumanskaya str, Moscow, 105005 Russia

[2] Master Degree, Bauman Moscow State Technical University, 5, 2-ya Baumanskaya str, Moscow, 105005 Russia
[3] Student, Moscow Aviation Institute, 4, Volokolamskoe shosse, Moscow, 125993 Russia

E-mail: i.a.konstantinova@gmail.com

Abstract. Current development of aerospace technologyand demands for the economic feasibility have led to a reduction in weight and size characteristics of the on-board electronics and other on-board equipment.There is a tendency to use small-scale spacecraft: midi-satellites, mini-satellites, nanosatellites etc. Reducing the weight and size characteristics of the satellites makeselectromagnetic launching techniques more promising compared to traditional methods of sending payload into orbit. Electromagnetic launchdoes not require expensive space centers – it iscost- efficient, environmentally friendly and enables frequent low-cost launches.

1. Introduction

The aim of this work was to develop an effective approach to the calculation and analysisof the stress-strain behavioraccounting for the thermal effect on ananosatellite bodyunderconstant exposure to a concentrated particles stream.

The aim required solving the following objectives:

- to identify the nanosatellite bodyelements subjected to the greatest load and to determine the methods of improving the strength characteristics of these components;
- to determine the stress-strain behaviour of a rigidly fixed structure element from OT 4 alloy (the most frequently used material in aircraft with the highest density and strength characteristics), when exposed to a concentrated stream of high-energy particles and mechanical load;

The approachused in this paper is based on the finite element method that allows solving the heat transfer and the strengththeory equations to the best degree of accuracy, taking into account all conditions, such as non-linear temperature distribution and plastic deformation. The numerical calculation of the most loaded elements of the nanosatellite bodywas performed using the ANSYS software package. In the process of solving the thermal stressproblem, a specialized macros forthe ANSYS Mechanical APDL software package was created using the APDL language to determine theresulting temperature fields and the stress-strain parameters (at proton energies from 3 to 12 MeV with 1 MeVincrement).

2. Identifying the most loaded zones of a nanosatellite body

According to [1], the most effective nanosatellite design is a streamlined shape with 1:10 elongation. This design is shown in Fig. 1; its basic characteristics are given in Table 1. The following additional parameterswere used for the numerical modeling: 1:10 aerodynamic shape extension;titanium wrought alloys OT 4 as the body material; 5000 m altitude; 6000 m/s velocity.

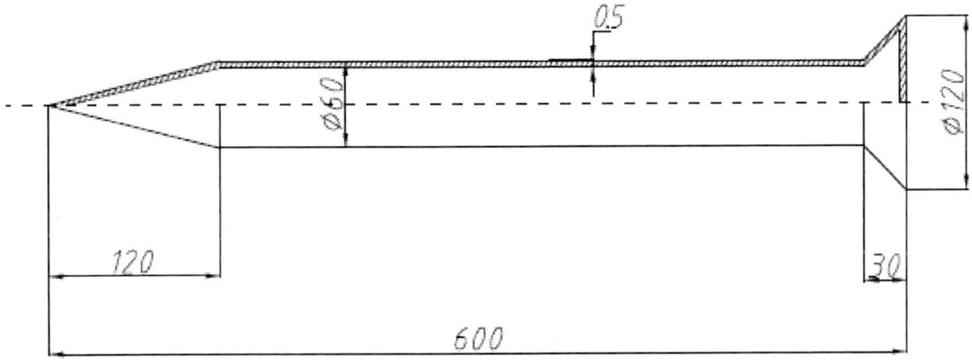

Figure 1. Nanosatellite design

Table. 1.Nanosatellite main characteristics

Midship section diameter, mm	60
Skirt diameter, mm	120
Total length, mm	600
Head cone length, mm	120
Cylinder length, mm	450
Wallthickness, mm	0.5
Total mass, kg	10

In case of the high velocity incident flow(6000 m/s) with compression shocks, the nanosatellite bodycan be divided into three zones with regard to itsstress-strain behavior, as shown in Fig. 2 and 3.

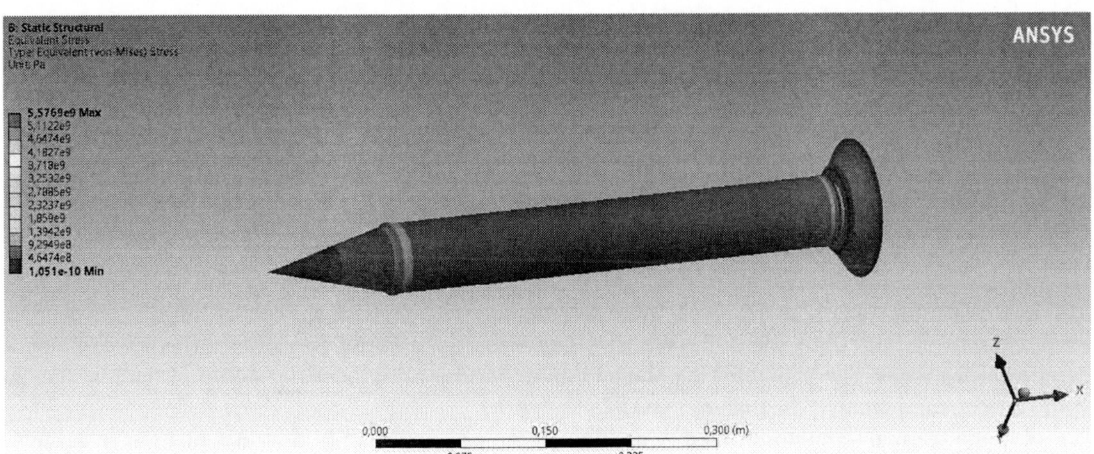

Figure 2. Stress distribution along the aircraft case in accordance with the Huber-Mises theory of plastic yield inhypersonic flight (6000 m/s)

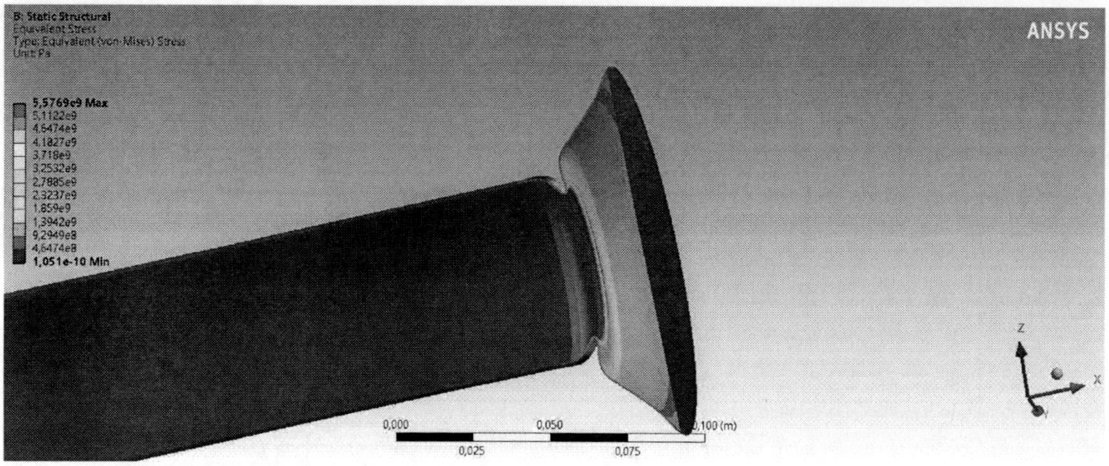

Figure 3. Stress distribution in the tail-end sectionthe aircraft case in accordance with the Huber-Mises theory of plastic yield in hypersonic flight (6000 m/s). Longitudinal cross-section

One of the most loaded zones (zone 1) is located in the nose part ofthenanosatellite hull. The form of this zoneis affected by the incident streamvelocity and depends on the position of the (compression) wave front determined by the incident flow velocity and the headcone divergence angle.According to Fig. 2,the compression stress values increase considerably following the first shock wave front.

Zone 2 is a very narrow area on the border of the nose cone and the cylindrical part of the body. Significant loads acting upon the fairing nose as well as the bodyinertia-loading component lead to stress-strain localization.

The zone subjected to the greatest loadis the aerodynamic skirt contour. This happens becausethe compression fronts overlap with the skirt contour.In the instance of a serious manufacturing error leading to the insufficient structural stiffness, elastic vibrations may occur in the nanosatellite. So, with the flow being unstable, the shock wave front will becomeasymmetricin the aerodynamic skirt area.That, in turn,causes a high gradient stress-strain on the skirt andadditional aerodynamic vibration, which leads to an unsteady flight and consequently a crash.

The designprocess should take into account these stress-strain areas. For instance, the fairing and the aerodynamic skirt of the aircraftcould be strengthened and a thermal protection shield added.

3. Estimating the protons stream effect

After determining the most loaded and, as a result, the weakest parts of the nanosatellite body during the electromagnetic launch, we need to consider these parts in detail and take into account the impact of the quasi-stationary concentrated stream of protons through aselected volumesurface. The volume selectionis dictated by the available computer operating memory and speed, as well as by the character of the stress-strain state in agiven area. With the memory and performance level limited, all available resource must be used to the full. We can conveniently assume the selected volumeto have a parallelepiped form. The following parameters appear to be most realistic: 140 mm length, 100 mm width and5mm thickness. The material isOT4alloy.

To solve the thermal-strength problem we obtained temperature distribution in a plate exposed to3 to 12 MeV protons stream,with 1 MeV increment. By way of illustration,Fig. 4shows the temperature distributions for 3 MeVprotons stream.

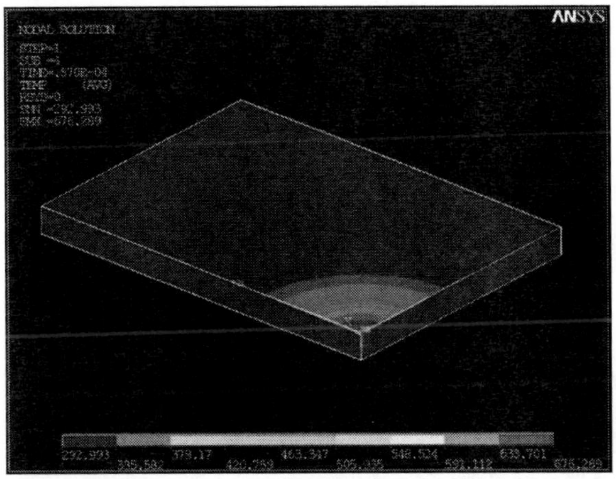

Figure 4. Temperature distribution for 3 MeV protons stream

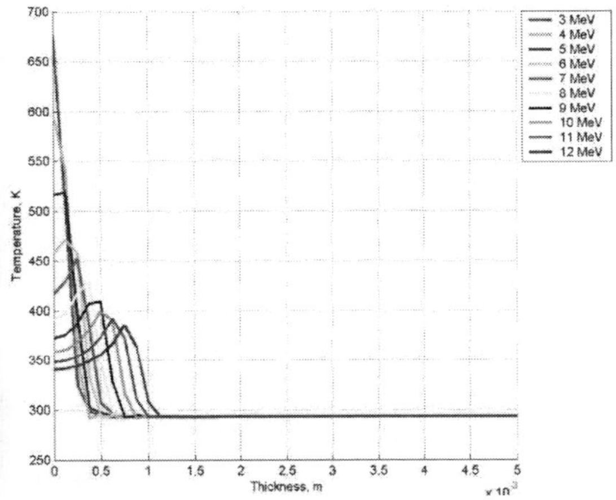

Figure 5. Vertical temperature distribution for 3 to 12 MeV protons stream

It is evident from the distributions that the temperature maximum is at the maximal penetration depth of the protons with the specified energy characteristics. This is due to the fact that the heat distribution is directly related to the protons stopping power. For clarity, we presentthe vertical temperature distributions for the combined energy values (Fig. 5).

We use the above resultsas initial conditions to determine the stress-strain behavior of the plate. The choice of the applied pressure is based on the extent of the incident floweffect on the aircraft.

The coordinate dependence of stress in the plate according to the Huber-Mises theory illustrates the magnitude of the emerging stresses and their distribution (see. Fig. 6). It is evident that the maximal stresses occur within the protons stream zone (that is, in the center) and over the edges of the plate. The reason for this is thatthe center of the sample,apart from themechanical stresses, is subjected to the thermal stresses from the heat of the protons beam. Stressesat the edges occur due to the rigid fixing conditions.

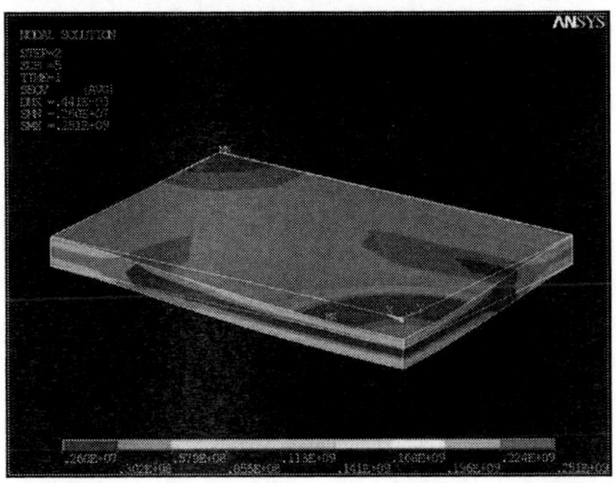

Figure 6. The stress distribution in accordance with the Huber-Mises theory of plastic yieldfor 3 MeV protons stream

The stress distributions do not clearly indicate whether the stress-strain parameters are close to critical. For this purpose, we use the maximum straincriterion, which helps identify the point where a sample is most likely to fail, with thethermal and mechanical effects parameters specified. The maximum strain criterion is defined as the ratio of the actual strain at agiven point to the failure strain for this type of load. If the maximum criterionis greater than unity in at least one point of the test sample,the critical stresses occurat this point, which could lead to the entire sample failing. This criterion serves as the basis for the further analysis. By way of example, Fig. 7 shows the criteria distribution for 3 MeV protons streams.

The calculation results are summarized in a graph showing the maximum strain criterion as a function of the proton stream energy at the point of the maximum heat dissipation (Fig. 8).

It is evident from the graphthat the value of the maximum strain criterion is greater than unity for the 3-6 MeV protons stream.

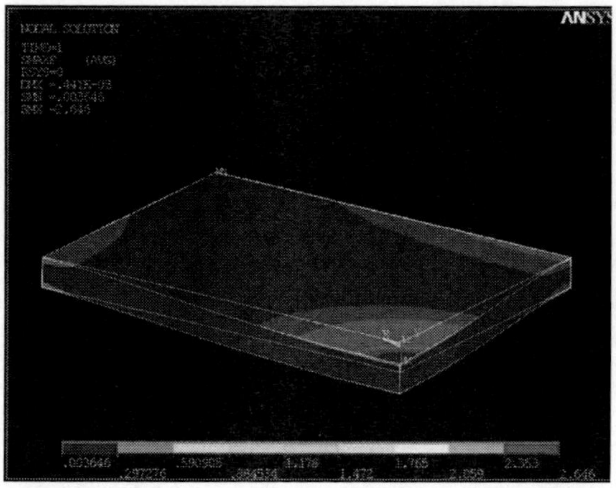

Figure 7. The maximum strain criterion for3 MeV protons stream

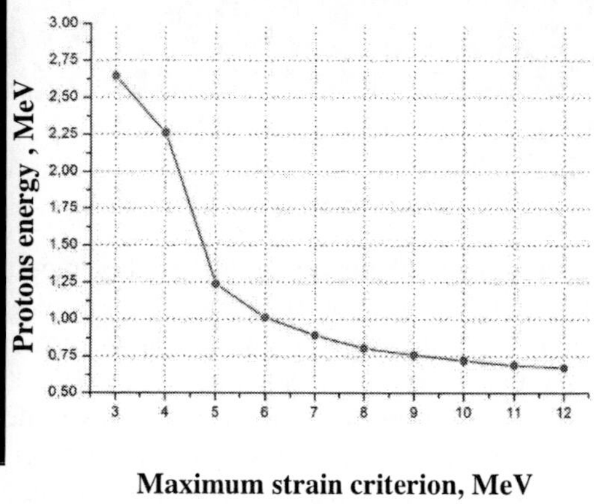

Figure 8. Maximum strain criterion as a function of the proton stream energy

4. Conclusion

In the course of research, the stress-strain behavior of a nanosatellite during launch was analyzed numerically andthe highest risk areas were identified. We also performed the analysis of a rigidly fixed element(a rectangular plate made of the OT 4alloy)that was simultaneously exposed to apulsed proton beam and mechanical loading. We estimated the possibility for the stress-strain parameters of the material to reach the critical failure values.

References

[1] Gerasimov Y V, Karetnikov G K, Selivanov A B andPhionov A S 2013 Assessment of the relative final weight of the nano-satellite, delivered in near-earth orbit with the help of pulse and the start of corrective devices*Vestnilk BMSTU "Mashinostroenie"***3**pp 69–76

[2] Gerasimov Y V, Karetnikov G K and Selivanov A B 2013 Stabilization of nano-satellites in the process of the electromagnetic launcher in the atmosphere and outer space*Engineering journal: science and innovations***8**

[3] Babichev A P, Babushkina N A, Bratovskiy A M et al 1991 *Physical magnitudes: Reference book*(Moscow)p 1232

[4] GOST 4401-81 Russian National Standard 2004 *Atmosphere Standard. Operation Factors* (Moscow)p 181

[5] Basov K A 2002 *ANSYS in examples and tasks* (Moscow)p 224

[6] Basov K A 2005 *ANSYS Reference book* (Moscow)p 640

[7] *ANSYS 5.7 2001Thermal Analysis Guide*(Moscow: Cadfem)p 110

The interference response of space microsatellitehaving the form of the Luneberg lens

V O Gladyshev[1], **A G Strunin**[2]
[1]Department of Physics, Bauman Moscow State Technical University, Moscow, Russia
E-mail: vgladyshev@mail.ru

Abstract. The calculations were performed for "Blitz" satellite parameters. Luneberg lens consisted of a central sphere of radius R_2=53,5 mm and an external meniscus of radius R_1= 85 mm, with half of the external meniscus having a mirror coating. Central sphere was made of heavy flint glass with a refractive index n_2=1,76470, and an external meniscus was made of light crown glass with a refractive index n_1=1,47290, the laser wavelength was λ=532 nm. The speed of the lens moving in its orbit was 7500 m/s, the orbit height was h = 835 km.

The task of increasing the accuracy of global satellite positioning systems can be solved with an optical segment which is, in general, a set of reference LEO satellites with a set of reflectors.

In [1] it is shown that spherical gradient-index Luneberg lens can be used as a microsatellite-reflector. The satellite optical response is comparable to the response of a point source and can be used to improve the accuracy of coordinatedetermining in the GLONASS system.

The theoretical model of the optical response formation by the moving Luneberg lens must include the equation of moving mediaoptics. This is because there is the additional ray deflecting and stirring in the rotating lens moving in the orbit [2, 3]. The result is a redistribution of the intensity of radiation reflected by the satellite "Blitz", in the recording plane on the Earth's surface.

The multipath interference response of Luneberg lens moving in the Earth's orbit is calculated in the work, with due account for the effects of the moving media electrodynamics.

The calculations were performed for "Blitz" satellite parameters [4]. Luneberg lens consisted of a central sphere of radius R_2 = 53,5 mm and an external meniscus of radius R_1 = 85 mm, with half of the external meniscus having a mirror coating. Central sphere was made of heavy flint glass with a refractive index n_2 = 1,76470, and an external meniscus was made of light crown glass with a refractive index n_1 = 1,47290, the laser wavelength was λ = 532 nm. The speed of the lens moving in its orbit was 7500 m/s, the orbit height was h = 835 km.

The intensity distribution of the reflected radiation on the Earth's surface was found depending on coordinates for the lens central aperture.

The calculation of parameters of the radiation passing through the moving medium is based on the dispersion equation

$$\vec{k}_e^2 - \frac{\omega_e^2}{c^2} - \left(\frac{\varepsilon\mu - 1}{c^2}\right) \cdot \frac{\left(\omega_e - \vec{k}_e \vec{u}\right)^2}{1 - u^2/c^2} = 0. \quad (1)$$

Solving the equationallows us to determine the wave vectorprojections and frequencies at all sections of the light raytrajectory from the radiation source to the receiver.

The theoretical model, following [2], uses the kinematic equations for the rays in geometric approximation.

To calculate the intensity distribution in the plane of the signal reception, angular deflection of the beam reflected by the satellite ,$\Delta\vartheta$, was introduced. Then, the beam deflection relative to the terrestrial detector, located near the source of radiation, after reflection by Luneburg lens is equal to

$$\Delta l_j = htg\Delta\vartheta_j . \tag{2}$$

For phase of $j-^{\text{th}}$ beam on the Earth's surface it can be written

$$\Phi_j(x) = \sum_{i=1}^{n} w_{ij} t_{ij} - w_8 \Delta t_j - \sum_{i=1}^{n} k_{ij} \Delta r_{ij} , \tag{3}$$

where $\omega_{ij}, t_{ij}, k_{ij}, \Delta r_{ij}$ are the frequency, the propagation time of $j-^{\text{th}}$ beam, the wave vector module and the length of geometric path at i section of the trajectory respectively, $i = 1..n$, where $n = 8$ is the number of sections in the optical path in the optical system under test between interface boundaries, $\Delta t_j = t_0 - \sum_{i=1}^{n} t_{ij}$ is the difference in transit time of the reference beam and calculated one.

The intensity distribution in the observation plane of the interference pattern depends on the number of rays (from the total number N), falling in each coordinate interval (for this task the interval was chosen to be $\Delta x = 1$ m).

$$I(x) = I_0 \left(\sum_{j=1}^{N} sin\Phi_j(x) \right)^2 , \tag{4}$$

where I_0 is single beam intensity of the order of 10^{-6} W/m^2.

Numerical calculations were performed with $N = 20000$, in the range of angles of beam incidence on the lens surface ϑ_0=0..7°. The graph of the reflected light intensity distribution $I(x)$ at the Earth's surface is shown in Figure 1.

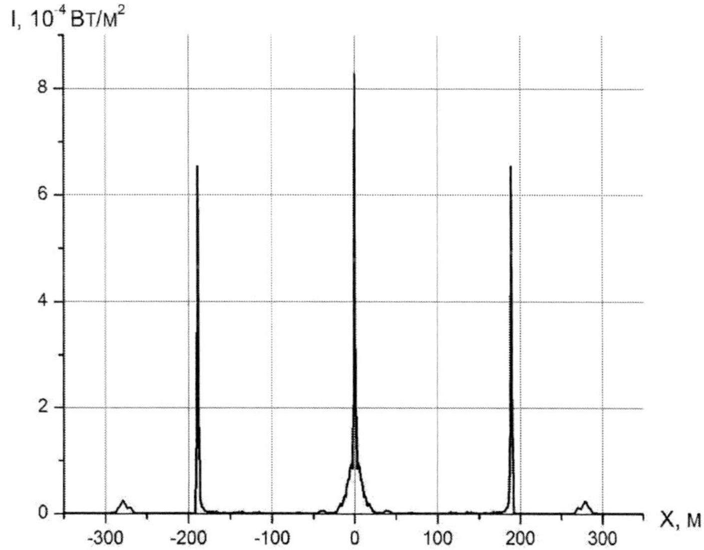

Figure 1. The dependence of the intensity $I(x)$ of the reflected radiation on the coordinates on the Earth's surface.

To analyze the dependence of the coordinates of the maximum intensity on the satellite velocity, the coordinate of the maximum intensity of the reflected radiation, x_m, is introduced. The graph of the coordinate x_m vs. the satellite velocity V is shown in Figure2.

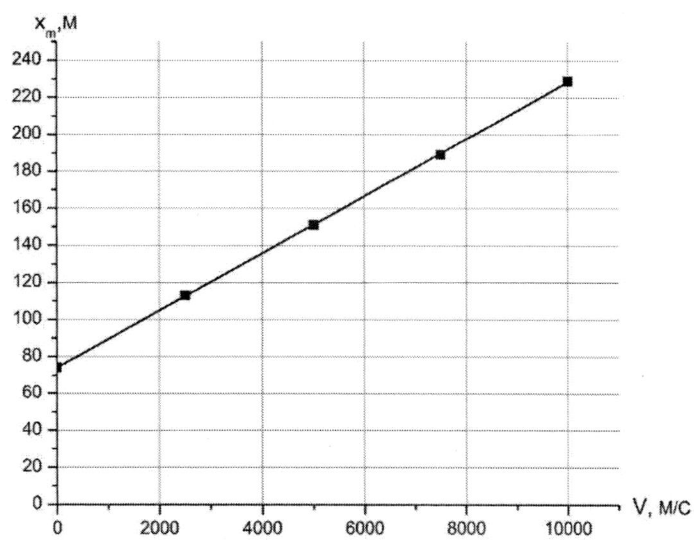

Figure 2. The graph of intensity peak coordinates x_m vs. the satellite speed V.

For correction of the form of the intensity distribution (intensity peak shift to the receiving area), it is possible to change the optical system geometrical characteristics. Below the results of calculations of x_m at different radii R of the external satellite meniscus are shown.

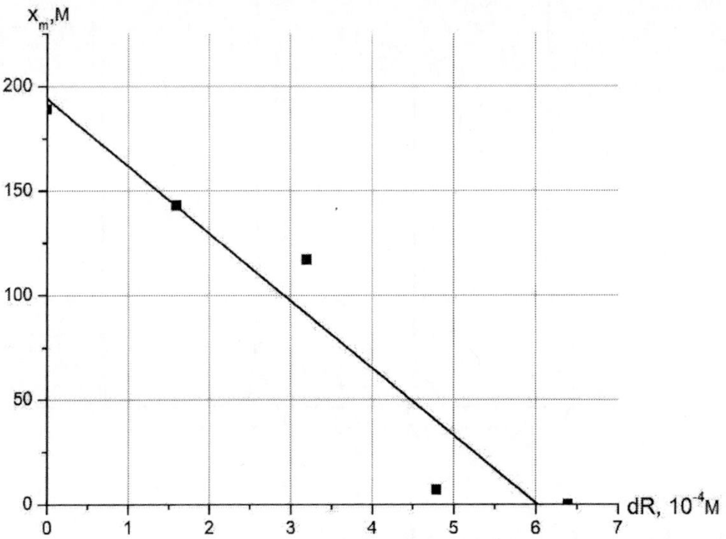

Figure 3. The dependence ofintensity peak coordinates x_m vs. changes in the radius of the meniscus dR.

As a result, for $x_m = 0$ $dR \approx 0.64$mm. The graph of the radiation intensity distribution at the external meniscus radius of 85.64 mm is shown in Figure 4.

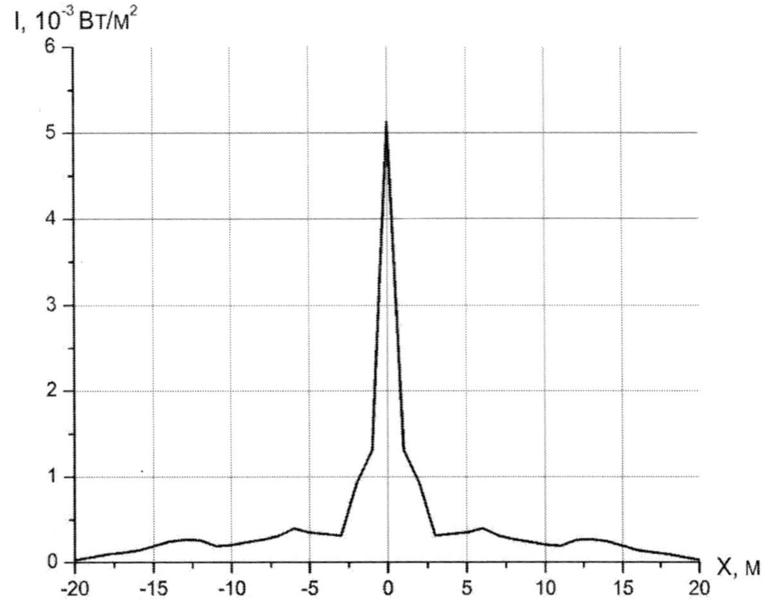

Figure 4. The light intensity distribution at the outer meniscus radius of the $R_1 = 85{,}64$ mm.

From the above graph it follows that the change in the outer radius of the meniscus can significantly affect the intensity distribution in the reception area and makes it possible to optimize the signal;so when the radius of meniscus is increased by $dR = 0.66$ mm, the signal strength in the receptionarea increases by more than 6 times .

As a result of the calculation, the dependence of the radiation intensity in the observation plane of the multipath interference pattern on the satellite velocity was found. Moreover, the inverse problem of finding the dependence of intensity peak coordinates on the lens outer meniscus radius has been solved.

Findings of the research show that a decrease in signal strength due to the presence of satellite motion velocitycan be compensated by changing the radius of the lens external meniscus, which will provide an effective signal/ noise ratio in the reflected radiation registration area.

The work was supported by Russian Foundation for Basic Research grants no. 16-02-00488 A and no. 16-08-00618 A.

References

[1] Kucharski D, Kirchner G, Hyung-ChulLim, KoidlF2011*Adv. Space Res.***48** pp 1335–1340

[2] Gladyshev V O, Tereshin A A 2016 A Luneburg Lens in moving Coordinates. *Optics and Spectroscopy* **120** no. 5 pp 773–780

[3] Gladyshev V O, Bazleva D D, Tereshin A A, Gladysheva T M 2016 Determination of Light Beam Curvature in a Rotating Luneburg Lens. *Technical Physics Letters* **42** no. 9 pp 948–950

IOP Conf. Series: Journal of Physics: Conf. Series **918** (2017) 012045 doi:10.1088/1742-6596/918/1/012045

[4] Shargorodsky V D, Vasiliev V P, Belov M S, Gashkin I S, Parkhomenko N N 2006 Spherical Glass Target Microsatellite. *Proceedings of 15th International Workshop on Laser Ranging* pp 566-570

IOP Conf. Series: Journal of Physics: Conf. Series **918** (2017) 012046 doi:10.1088/1742-6596/918/1/012046

The model of interaction between the elements of construction and the layered medium

V G Fomin[1]

[1]Associate Professor, Yuri Gagarin State Technical University of Saratov, Saratov, Russia

E-mail: foma777@rambler.ru

Abstract. The article considers the plane stress problem of the non-linear elasticity theory for a plate or a beam located in the aggressive medium which affects the physical properties of the material. In solving the problem we use the finite element method in combination with the successive approximations method and the variable elasticity parameters method.

The object of study is a subject to the bending deformation structural element such as a plate or a beam lying on the laminated base. The material of each k-th layer has different dependencies between the stress intensities and deformation $\sigma_i^{(k)} = f_k(\varepsilon_i^{(k)})$. Each of the base layers is assumed to be in a plane stress state, i.e. the law of plane cross sections remains valid. The last layer rests on the nondeformable base. The basic relations for stresses and deformations with due account for the formulated assumptions for an arbitrary k-th layer will have the following form:

$$\frac{\partial \sigma_x^{(k)}}{\partial x} + \frac{\partial \tau_{xy}^{(k)}}{\partial y} = 0,$$

$$\frac{\partial \tau_{xy}^{(k)}}{\partial x} + \frac{\partial \sigma_y^{(k)}}{\partial y} = 0.$$

$$\varepsilon_x^{(k)} = \frac{\partial u^{(k)}}{\partial x},$$

$$\varepsilon_y^{(k)} = \frac{\partial v^{(k)}}{\partial y},$$

$$\gamma_{xy}^{(k)} = \frac{\partial u^{(k)}}{\partial y} + \frac{\partial v^{(k)}}{\partial x}.$$

Here $\sigma_x^{(k)}, \sigma_y^{(k)}, \tau_{xy}^{(k)}$ - the components of the stress tensor, $\varepsilon_x^{(k)}, \varepsilon_y^{(k)}, \varepsilon_{xy}^{(k)}$ - the components of the strain tensor, $u^{(k)}, v^{(k)}$ - the components of the displacement vector of the k-th layer. It is assumed that for each layer material the ratios of the small elasto-plastic strains theory are valid, i.e. the relationship between the stresses and deformations has the following form:

$$\sigma_x - \sigma_0 = \frac{2\sigma_i}{3\varepsilon_i}(\varepsilon_x - \varepsilon_0), \quad \sigma_y - \sigma_0 = \frac{2\sigma_i}{3\varepsilon_i}(\varepsilon_y - \varepsilon_0), \quad \tau_{xy} = \frac{\sigma_i}{3\varepsilon_i}\gamma_{xy},$$

Here, σ_0 , ε_0 are the average stress and strain.

$$\sigma_0=\frac{1}{3}(\sigma_x+\sigma_y),\quad \varepsilon_0=\frac{1}{3}(\varepsilon_x+\varepsilon_y+\varepsilon_z),\quad \varepsilon_0=\frac{1-2\mu}{E}\sigma_0\ ,$$

where μ is Poisson's ratio, E is Young's modulus.

The further solution is performed using the variable elasticity parameter method and allows the original nonlinear problem of the elasticity theory to be converted to a sequential series of linear problems with the variable Young's modulus E and the Poisson coefficient μ.

In each approximation the finite element method is used in solving the linear problem, where the functional of the potential deformation energy is minimized

$$U=\sum_{t=1}^{N}U^{(t)}$$, where N is the number of elements.

The functional for each element has the form:

$$U^{(t)}=\frac{1}{2}\iint_{S^{(t)}}\Big[\frac{E}{1-\mu^2}\left(\frac{\partial u}{\partial x}+\mu\frac{\partial v}{\partial y}\right)\frac{\partial u}{\partial x}+\frac{E}{1-\mu^2}\left(\frac{\partial v}{\partial y}+\mu\frac{\partial u}{\partial x}\right)\frac{\partial v}{\partial y}+$$

$$+\frac{E}{2(1+\mu)}\left(\frac{\partial v}{\partial x}+\frac{\partial u}{\partial y}\right)^2\Big]dxdy$$

The solution on the basis of the iterative method of variable elastic parameters is completed when the maximum discrepancy between the two successive approximations does not exceed one percent [1]. Using this approach allows us to apply this model to calculate the stress-strain state in plates or beams interacting with the layered medium. In addition, it allows taking into account the effect of the temperature field on the change in the physical and mechanical properties of the material under the aggressive medium influence [2]. The whole variety of factors consideration makes it possible to bring the given model into close proximity with the real conditions of the structural elements behavior.

References

[1] Fomin V G 2016 Effect of variable thickness on the stress distribution in a doubly-connected plate under the influence of a temperature field // Mathematical modeling, computer and natural experiment in the natural sciences No 4; URL: mathmod.esrae.ru/4-24

[2] Fomin V G 2016 Simulation of a doubly-connected plate located in the temperature field under the influence of an aggressive medium. // Mathematical methods in engineering and technology-MMET-29. SciConf in 12 toms T 2 230 p, pp83-85

Combinational processes in uranil structures

V S Gorelik[1,2], V M Korshunov[1,2], A I Vodchits[3], Yu P Voinov[1], A A Loboyko[1,2]

[1] Lebedev Physical Institute of the Russian Academy of Sciences, Moscow, Russia
[2] Bauman Moscow State Technical University, Moscow, Russia
[3] Stepanov Institute of Physics of National Academy of Sciences of Belarus, Minsk, Belarus

E-mail: gorelik@sci.lebedev.ru

Abstract. It is established that the photoluminescence spectra of uranyl compounds have a shape of combination bands as equidistant Stokes satellites. These satellites correspond to the transitions from the excited electronic state to the vibrational states of fully-symmetric mode with the frequency of 854 cm^{-1}. The stimulated Raman scattering of light in uranyl compounds is observed when excited by the ultra-short powerful pulses of the laser YAG: Nd^{3+} (λ=532 nm). The detected combinational processes open the possibility for detecting the uranyl compounds in mixtures and in the environment. The article analyses the conditions for the realization of laser generation in the uranyl compounds.

1. Introduction

The registration of the uranyl compounds spectra is an important task for detecting these compounds in the environment and using them as the components for industrial purposes. In particular, it is of great interest to record the Raman-spectra (RS) of uranium structures containing the uranyl ion $(UO_2)^+$. In the vibrational spectra of such compounds there is a high-quality fully-symmetrical mode with the frequency of 854 cm^{-1}. The RS of analytes with the ion $(UO_2)^+$ relate to the corresponding combinational satellite. Accordingly, the RS can serve as a reliable indicator of the uranyl-ion. Previously the RS of compounds with the uranium dioxide powder were registered during the excitation by helium-neon laser with the power of about 10 mW and with the use of a double monochromator providing the 2 cm^{-1} resolution [1].

In [2] the RS were measured for the polycrystalline minerals with the uranyl-ion, and in [3] the RS of uranylacetates dissolved in water were recorded. The RS of sodium uranyl acetate $NaUO_2(CH_3COO)_3$ solid crystal were registered upon excitation with the argon laser (λ=514 nm), photomultiplier and spectrograph [4] as well as in the excitation radiation by the xenon lamp [5,6]. Theoretical calculations of the crystals $NaUO_2(CH_3COO)_3$ vibrational spectrum were provided in [7].

In this paper we posed the problem of express detection of compounds with uranium dioxide (UO_2) based on the modern methods of express registration of such spectra as RS and photoluminescence (PL). These methods involve the use of a micro probe and a small-size high-power spectrometer with a multiple element secondary radiation receiver.

IOP Conf. Series: Journal of Physics: Conf. Series **918** (2017) 012047 doi:10.1088/1742-6596/918/1/012047

2. Experimental part

The Raman-spectra of uranyl compounds were recorded using the micro probe and the continuous laser with the generation wavelength λ= 785 nm and the average power P=300 mW (see Figure 1). The samples for the study were the polycrystals of sodium uranyl acetate $NaUO_2(CH_3COO)_3$ and the uranium glass.

The laser radiation (6) passed by means of quartz fibers through the interference filter (5) and was focused with a collecting lens (4) on the surface of the investigated sample (1). The scattered radiation was incident on the photonic crystal (5). The filter passed the laser radiation with the wavelength λ= 785 nm and reflected the long-wave radiation. The scattered radiation entered the spectrometer (8) and the lens (4). Then the scattered radiation was analyzed by means of the computer (9). The resolution of the spectra was 2 cm^{-1}.

The luminescence spectra were recorded using the FSD-10 spectrometer with the sensitivity range of 200-1100 nm. The complexes were optically excited with the wavelength of 385 nm (3.24 eV). We used a quasi-monochromatic semiconductor LED (see Figure 2). The semiconductor LED consisting of a mirror (1), an electron and hole layers (2 and 4), p-n junction (3) and a collecting lens (5) generated the quasi-monochromatic radiation with the wavelength of λ= 385 nm. This radiation was directed to the surface of the sample (8) using a probe (6) and quartz fibers (7). The sample was fixed on the object table (9). The reflected or transmitted signal was directed to a digital minispectrometer FSD-10 (10) and then it was analyzed using a computer (11).

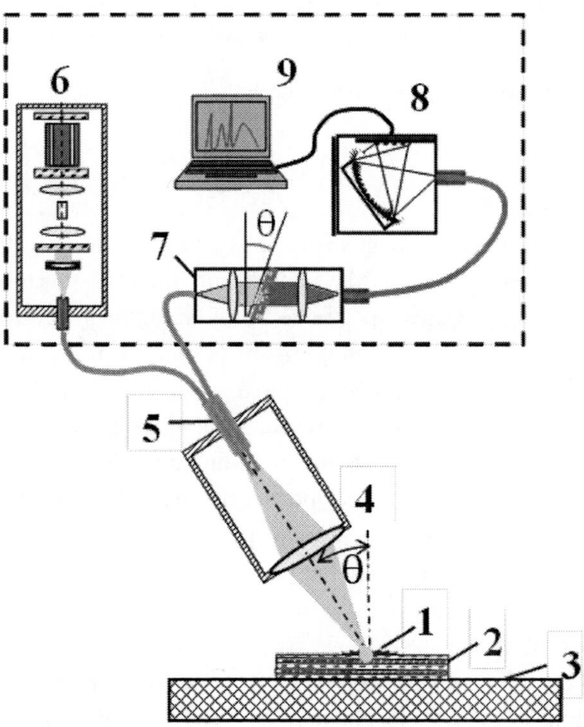

Figure 1. Experimental setup for the registration of Raman-spectra: 1 – sample, 2– substructure, 3 – object table, 4 – focusing lens, 5– probe, 6 – laser radiation, 7 – refractor, 8 – FSD-10 spectrometer, 9 – computer.

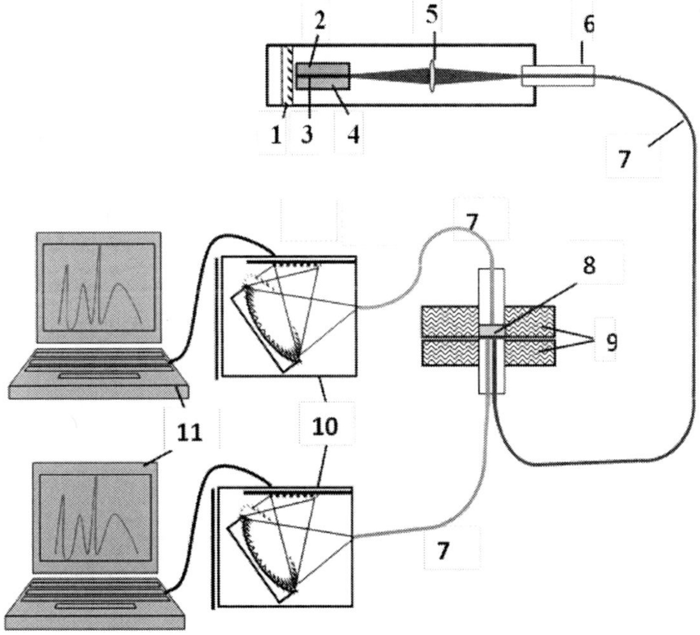

Figure 2. Experimental setup for the registration of uranyl-compounds PL: 1 – mirror, 2 и 4 – electron and hole layers, 3 – p-n junction, 5 – collecting lens, 6 – probe, 7 – quartz fibers, 8 – sample, 9 – object table, 10 – minispectrometer FSD-10, 11 – computer.

3. Results and discussions

The PL and Raman-spectra of sodium uranyl acetate $NaUO_2(CH_3COO)_3$ are shown in Figure 3. The PL spectrum of this crystal consisted of four equidistant (in frequency) lines with the 854 cm^{-1} shift (see Figure 3a). There were 14 lines in the spontaneous RS-spectrum - more than in the results presented in the paper [7]. The values of the corresponding frequencies of the Raman lines as well as their classification according to the vibration types are given in Table 1. In this table we compare our results with the literature data [7-10].

The frequencies 854 cm^{-1} (O-U-O), 956 cm^{-1} (C-C) and 1479 cm^{-1} (O-C-O) correspond to the valence type A_1 fully symmetric vibrations ν_s of the $NaUO_2(CH_3COO)_3$ crystal T point symmetry group. The frequency 931 см$^{-1}$ corresponds to the antisymmetric mode ν_{as} (type E). The frequencies 222 and 255 cm^{-1} (O-U-O), 676 and 689 cm^{-1} (O-C-O) correspond to the deformation mode (F).

The most intense line in the Raman spectrum of the uranyl acetate sodium crystal with the frequency of 854 cm^{-1} is also present in the Raman spectra obtained earlier for $Cu[(UO_2)_3(SeO_3)_2O_2]\cdot 8H_2O$ [7] and $CsUO_2(NO_3)_3$ [8,9]. This line can be attributed to the fully symmetrical UO_2 oscillation [11]. Thus, in the Raman and PL spectra of the sodium uranyl acetate crystal a valence fully symmetrical vibration is manifested. This oscillation can be used for the uranyl compounds detection.

IOP Conf. Series: Journal of Physics: Conf. Series **918** (2017) 012047 doi:10.1088/1742-6596/918/1/012047

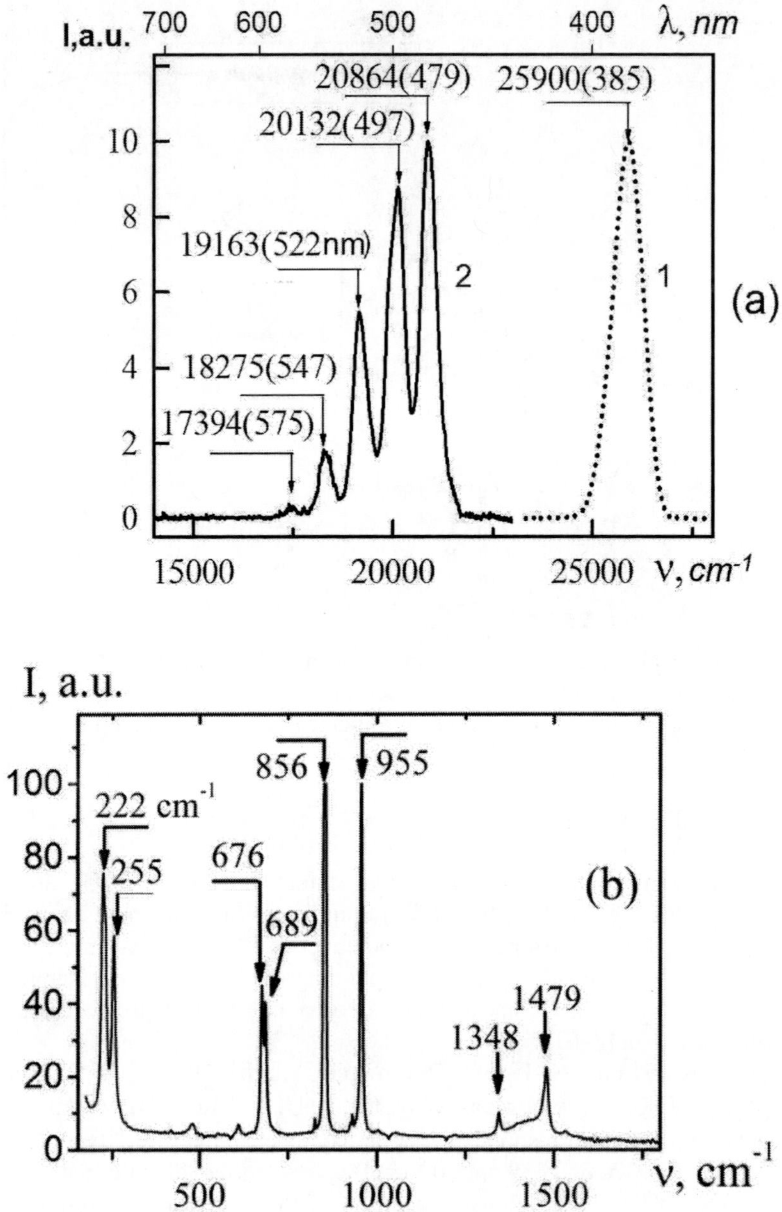

Figure 3. The PL spectra (a) and RS spectra (b) of polycrystalline sodium uranyl acetate $NaUO_2(CH_3COO)_3$. The semiconductor LED provided optical excitation of luminescence (curve 2) with the wavelength λ=385 nm (curve 1). The RS spectrum was carried out with the λ=785 nm CW laser excitation, and the acquisition time was 6 s.

The PL and Raman-spectra of the uranyl glass are shown in Figure 4. In the PL spectrum of the uranyl glass the phonon repeats with a spectral shift close to the frequency of the fully symmetric vibration of the uranyl group UO_2: 850 cm^{-1}. The photoluminescence spectrum shifts to the red region of the spectrum (see Figure 4 a) in comparison with $NaUO_2(CH_3COO)_3$.

IOP Conf. Series: Journal of Physics: Conf. Series **918** (2017) 012047 doi:10.1088/1742-6596/918/1/012047

Table 1: Crystal $NaUO_2(CH_3COO)_3$ frequencies in the infrared absorption spectrum (IR) [9] and Raman – spectra. The frequencies in the Raman – spectra (RS) were obtained theoretically [6]. RS* – the results obtained in this paper.

$NaUO_2(CH_3COO)_3$			Corresponding
IR[9]	RS[6]	RS*	
		222	Translational modes
		255	
		418	
470		480	
609		610	
675	677	676	O-C-O
	685	686	
		826	
	852	854	O-U-O
936		931	O-U-O
950	957	957	C – C
1005			
1053			CH_3
1350		1344	
1446			O-C-O
1472	1479	1479	
1518	1510		O-C-O
1537		1523	

In the uranyl glass Raman spectrum the band with the maximum intensity at the frequency of 1348 cm^{-1} is the most intense. This band corresponds to the excitation of the CH_3 group vibration degrees of freedom (see Table 1). The contribution of the UO_2 group vibration mode to the observed Raman spectrum is insignificant in comparison with the vibration mode of the CH_3 group. At the same time, the PL spectrum of the uranyl glass has phonon repeats with the equidistant phonon shift (850 cm^{-1}) corresponding to the fully symmetric mode of the group UO_2.

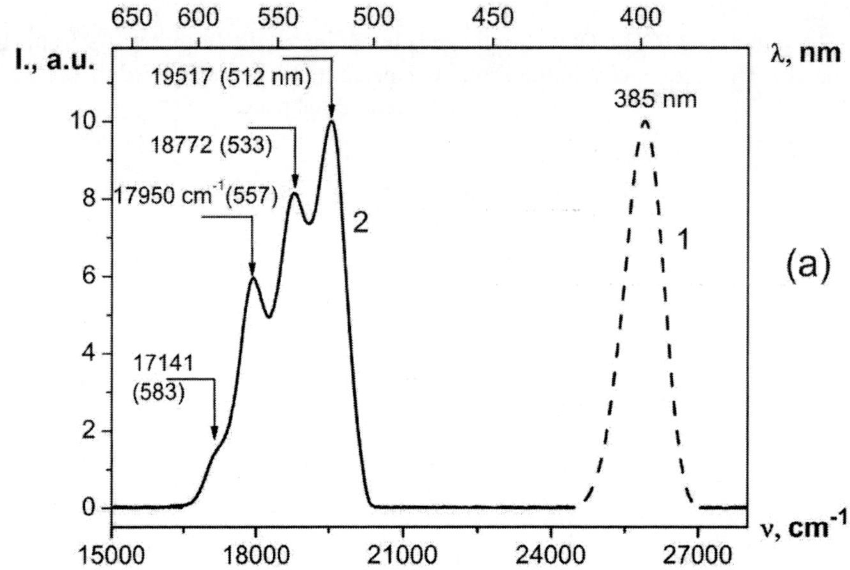

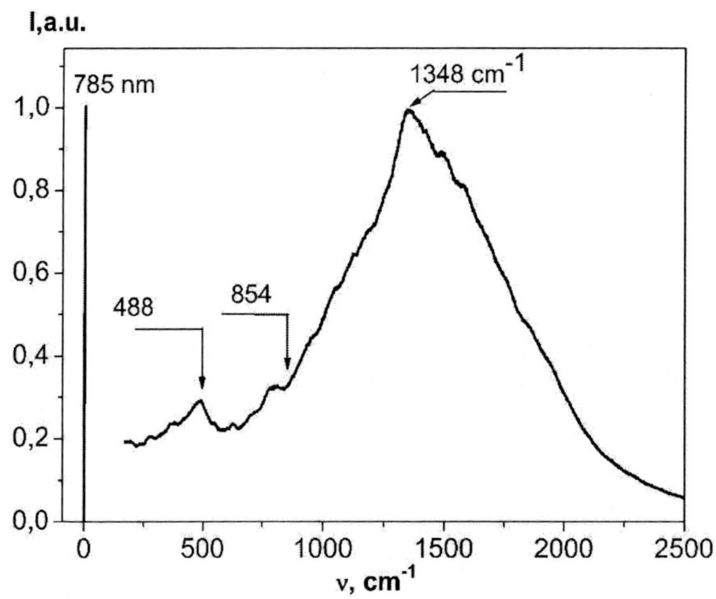

Figure 4. The PL spectra (a) and RS spectra (b) of the uranyl glass. The semiconductor LED provided optical excitation of luminescence (curve 2) with the wavelength of λ=385 nm (curve 1). The RS spectra were carried out with the λ=785 nm CW laser excitation, and the acquisition time was 3 ms.

When the sodium uranyl acetate crystal is excited by the intense laser (10^9 W/cm^2) with a pulse duration of 80 ps, a repetition rate of 10 Hz and a low pulse energy (1 mJ), intensive combined-frequency satellites with the frequency shift close to the uranyl group (854 cm^{-1}) were observed in the secondary emission spectrum. The intensity of the Raman satellites was comparable to the intensity of the laser radiation line. Thus, in this case the conditions for observing the stimulated Raman scattering (SRS) in the crystals of sodium uranyl acetate are realized. The stimulated Raman scattering was excited at the fully symmetric mode of the uranyl group (ν_s=854 cm^{-1}; see Figure 3 a, b). In the (SRS) spectrum the second ($2\nu_s$) and third ($3\nu_s$) Stokes satellites were observed. The first SRS Stokes satellite was strongly attenuated by the filter and was not detected (see Figure 5).

The excitation of sodium uranyl acetate crystal by small energy (0.1 mJ) laser pulses provides the nondestructive manner of SRS observation. The use of the laser sources with the short pulse time and high intensity opens the possibility for the creation of the uranium group UO2 coherent oscillations of large amplitude during SRS.

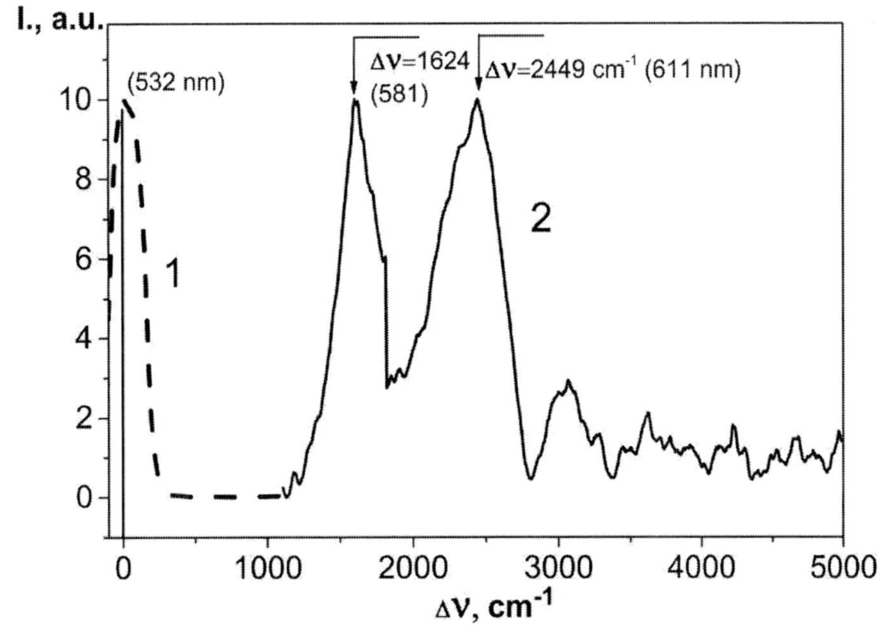

Figure 5: Normalized SRS spectrum of $NaUO_2(CH_3COO)_3$ crystal (curve 2) obtained under excitation by the giant ultrashort pulses of YAG;Nd^{3+} with the λ=532nm laser (curve 1).

4. Conclusion

Thus, in this paper it is established that in the Raman and photoluminescence spectra the high-quality fully-symmetric mode with the frequency 854 cm^{-1} is revealed. The developed technique opens the possibility of the uranium compounds express noncontacting detection. The recording of the Raman spectra is provided in a digital form for the small quantity (10^{-6}–10^{-8} g) substance during 1-10 s. The possibility of the stimulated Raman scattering excitation in the sodium uranyl acetate crystal by means of the YAG:Nd^{3+} (λ=532nm) laser with short (10-100 ps) and intensive (10^9 W/cm^2) pulses of laser radiation was estimated.

Acknowledgments

This work was supported by the Russian Foundation for Basic Research (grants 15-02-02882, 16-08-00618, 16-02-00488, 16-52-00026).

References

[1] Xiaoping Sun, Derrick R J Kolling, Hajer Mazagri 2015 Investigation of charge-transfer absorptions in the uranyl UO_2^{2+}(VI) ion and related chemical reduction of UO_2^{2+}(VI) to UO_2^{2+}(V) by UV–Vis and electron paramagnetic resonance spectroscopies. *Inorganica Chim.Acta* **435** 117–124

IOP Conf. Series: Journal of Physics: Conf. Series **918** (2017) 012047 doi:10.1088/1742-6596/918/1/012047

[2] Barraclough C G, Cockman R W, O'Donnell T A 1981 The vibronic and raman spectra of the uranyl ion in anhydrous hydrogen fluoride. *Inorganic and Nuclear Chemistry Letters* **17** 83-86

[3] Garg C L, Narasimham K V 1971 Energy levels and structure of ammonium uranyl fluoride. *Spectrochim. Acta* **27** 863-875.

[4] Volod'ko L V, Komyak A I, Sleptsov L E 1965 Polarization and the luminescence spectrum of crsystalline sodium uranyl acetate. *Zhurnal Prikladnoi Spectroskopii* **3** (1) 65-71

[5] Ray L Frost, JiříČejka, Godwin A. Ayoko 2007 Raman spectroscopic study of the multi-anion uranyl mineral schroeckingerite. *Journal of Raman Spectroscopy* **38** 1609-1614.

[6] Gorelik V S, Pyatyshev A Yu, Krylov A S 2016 Raman Scattering in Sodium Nitrite Crystals near the Phase Transition. *Physics of the Solid State* **58** (1) 170–176

[7] Boikov V N, Krasovskii A N, Umreiko D S 1982 Fine structure of the bands in the low-temperature spectra of sodium uranyl acetate. *J. Appl. Spectroscopy* **37** 1277

[8] Bastians S, Crump G, Griffith W P, Withnall R 2004 Raspite and Studtite: Raman spectra of two unique minerals. *J. Raman Spectroscopy* **35** 726-731

[9] Volod'ko L V, Sabilo K V 1966 Raman spectra of solutions of uranyl compounds. Journal of Applied Spectroscopy. *J Appl. Spectroscopy* **4** 39-42

[10] Quilès F, Burneau A 1998 Infrared and Raman spectroscopic study of uranyl complexes: hydroxide and acetate derivatives in aqueous solution. *Vibrational Spectroscopy* **18** 61-75

[11] Gorelik V S, Korshunov V M, Voinov Yu P 2016 Resonance excitation of photoluminescence in sodium uranyl acetate crystals. *Optics and Spectroscopy* **121** (6) 819-822

Institute of Physics
Dirac House, Temple Back
Bristol BS1 6BE UK

ISSN: 1742-6588
ISBN 978-1-5108-5361-4

CURRAN ASSOCIATES INC.
proceedings.com

8th International Conference on Applied Physics and Mathematics (ICAPM 2018)

Journal of Physics: Conference Series
Volume 1039

Phuket, Thailand
27 – 29 January 2018

IOP
Institute of Physics